Encyclopedia of Articles
derived from
Popular Mining Magazine

Volume I
(Originally written from 1984-1996)

First edition, 2003

Published by
Action Mining Services Inc.
Sandy OR 97055
Phone: 1 800 624-1511
Fax: 1 800 711-7807
Email: actionmining@att.net
Website: www.actionmining.com

Popular Mining

Volume 1

Over 100 DIY Articles for the Small Miner taken from the Popular Mining Magazine Series

Published by

Sylvanite Publishing
Florence, Colorado
719-466-4868
www.Sylvanitepublishing.com
info@sylvanitepublishing.com

Preface

Popular Mining was a bi-monthly publication dedicated to the miner who is independent..... willing to try new ideas and innovations presented by other miners in a do-it-yourself fashion....to those who want to learn.....to those who share ideas. It was circulated throughout the U.S., Canada, So. America, Australia, New Zealand, and Mexico. Our aim was to increase awareness and production in the precious metals field.

While we make every effort to ensure that the materials presented are factual and easy to follow, we do not try out each idea published, we take no responsibility in the making or use of any product or idea you may try to build or use. *Like all equipment, mining supplies hold certain hazards and dangers. It is the sole responsibility of the reader to take proper precautions and safety measures.*

The staff at Action Mining Services Inc.

Dedicated to:

Our mining friends and customers who lived, shaped, and contributed to the network of articles provided by the Popular Mining Magazine from 1984 to 1996

Introduction

For many years after we quit publishing Popular Mining magazine, we received calls requesting subscriptions and back issues. Many a time, I thought about doing this type of project, bringing the years of mining data together into a book format. I knew it would be a daunting project, 1000s of hours of work compiling the information, but once I started I realized I couldn't stop! *The information was so priceless, it just had to be preserved.*

For those of you new to Action Mining Services, you need to understand the background of Popular Mining magazine to appreciate fully what happened. *The articles you are about to read were written by our staff and miners from around the world.* **These are real people doing real mining!**

Action Mining was founded in 1979. We were mining a high sulfide, complex ore body at the time and started selling our newly developed, proprietary leaching compounds (CLS) and fire assay supplies. New products and services were added to our catalog as the years went by. In the early 80's we were holding seminars to teach people how to do different aspects of lab work and mining techniques. I could not imagine miners or prospectors wanting such a thing (our experience had been with 'loners', 'gold-fever independents', and 'secretive types'. But, guys (and a few gals) came from all over the world to participate. We were amazed at the number of people out there who wanted to learn and were willing to travel to Trona California (see if you can find that place on a map!) to get the data. That aspect of teaching sprouted the idea of publishing a how-to magazine for independent miners, who were becoming our main clientele. So in 1984, out came Popular Mining magazine! Miners from all over the world were sending in articles of things that they had made and were using in their projects! Granted, a lot of the 'articles' were nothing more than a note or rough drawing or maybe a picture or two, but I've always been good at deciphering cryptic notes! Exchange of information was happening in the mining world. - an amazing feat at the independent level.

We moved to Las Vegas, NV and business grew, while the price of gold went up and down through the 80's and the 90's. We were doing so much traveling around the world, consulting on mining projects and attending mining conventions, that something had to 'give'. The 'give' was the magazine, which absorbed hundreds of hours each issue.

There you have it....for 12 years Popular Mining magazine inspired miners worldwide in all aspects of their projects. Our attitude was then and still is 'anyone can do anything' as long as you educate yourself in an area and are willing to learn more and try new things. And once you have a successful method, just fine tune it.

Mining has been good to us. It's been a fun journey and it's not over yet! May mining be good to you too.

Enjoy these volumes and keep mining alive for the independent miner of future generations!

Sincerely,

Nancy Glenn
General Manager
Action Mining Services Inc.

Table of Contents

Table of Contents

     503 826-9330

HOW TO STAKE A CLAIM

by Jim Humble

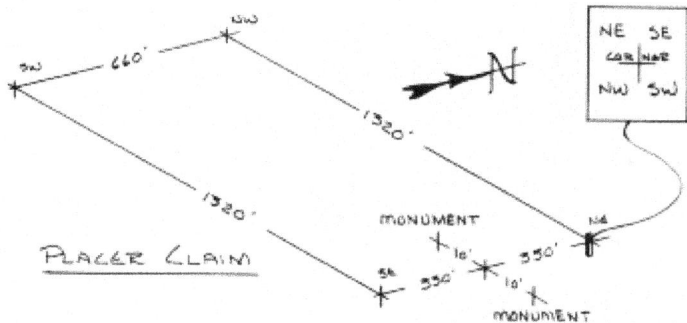

PLACER CLAIM

In this modern day and age staking a claim can be very complicated if one wishes to make it complex. It can become more complex as one begins to stake more than one claim. But the fact is that just the mere act of staking a claim and making it legal is very simple. We will try to stay out of the complexities of the subject at least for this article. We may touch on the more specific legalities in future issues if you request it.

There are four simple basic steps to staking a claim: (1) putting in the actual stake known as the Location Monument, (2) putting in the four corner boundary stakes* (3) recording the description and location at the County Recorder's Office* and (4) recording the description and location with the Bureau of Land Management (BLM). This must be done within 90 days of recording with the county. I will explain each step below, after which anyone should be able to stake his own claim.

(1) The Location Monument is usually a 4x4 post five foot tall or more stuck into the ground at least a foot. (A 4x4 post really is 3 1/2 inches by 3 1/2 inches.) Exactly what can be used for stakes varies from state to state. But the 4x4 usually is sufficient. In some states a pile of rocks 3 1/2 foot high is sufficient. Many of the larger companies are now using white 5 foot lengths plastic sewer pipes 3.5 inches in diameter. In any case you must post a location notice on this Location Monument. Miners use tin cans with lids or bottles with lids for this purpose to keep out the elements. See below for description of location notice.

The Location Monument is usually posted at one end of the claim near the center of the claim* and should be in an obvious position so that it can be easily seen.

(2) Corner stakes. The corner stakes have the same requirements as the Location Monument as far as size is concerned. They should be marked as to what part of the claim they represent. (For example: North West Boundary or NW boundary post for The Little Acres Lode Claim.)
LODE CLAIMS are claims where the precious metals are found in some kind of a vein. This vein is usually marked with the Location Monument right on the vein and the corner posts at each side along the vein. The posts may be 300 feet on either side of the vein. That makes the claim a maximum of 600 feet wide. The rules state that any one claim may run 1500 feet long. This means that you may have a long narrow claim 600 feet wide and 1500 feet long, running along the vein. Right? No definitely not!

Do not get caught on this one. Many claims now considered legal could be declared illegal. The rules state that a claim can be 1500 feet long and it can be 600 feet wide, but there is nothing in the rules that states that the claim can be 1500 feet long and 600 feet wide at the same time. If you want it 1500 feet long then you must have somewhat less that 600 feet wide. The reason for this is that no claim may be more than 20 acres. 1500 by 600 is 20.66 acres. You can use 290 feet on each side by 1500 feet, or you can use 300 feet on each side by 1450 and still be legal. But if you use or have used 600 by 1500 for your lode claims you can be had on a technicality. Better file an amended claim to be safe.

After you have posted your Location Monument you must then specify where it is. This is where you must find some natural or long standing monument and then describe where your claim is in reference to the natural monument. Generally you would give the direction to or from the natural monument to your monument in feet. You must also give the county and the section number (see below) in which your claim is located, as well as the Range and Township in which your section lies. These are found on the edges of a topographical map (topo map). (Example: R44E, T23S would mean Range 44 East and Township 23 South.) You should also describe the direction along which your claim lies and specify the Meridian. The Meridian is also on the topo map.

PLACER CLAIMS are claims where the precious metals are scattered throughout the dirt or rocks. There is no particular vein of rocks. Placer claims are claimed same as legal subdivisions. If the area has not been surveyed then you come as close to it as you can. Look on any topographical map and you will see the land divided into square miles. Each square mile is a section. The section numbers are listed in each section. You can then specify any 20 acre portion of any square mile as follows: The east 1/2 of the North West 1/4 of the North East 1/4 of section sixteen. This divides the section up into quarters and then each quarter is divided into quarters. Then the last quarter is divided into halves to make 20 acres. You can divide the last quarter four ways, using either the North 1/2 or the South 1/2 or the East 1/2 or the West 1/2. You must also give the Township and Range (see above) and County and Meridian.

In the case of placer claims or lode claims it is not necessary to survey the land. You do need to be as accurate as possible.

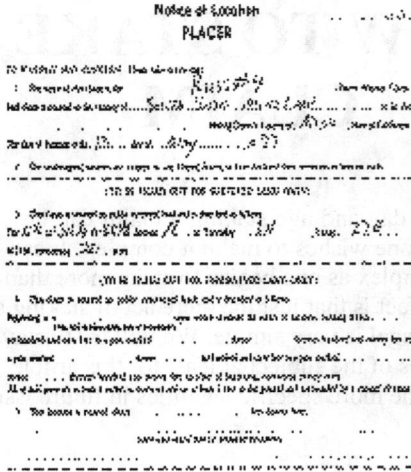

(3) Recording the description and location at the County Recorder's Office. Every county has a Office of Records where deeds and other transactions are recorded. They will record almost anything for a fee. You should first get a mining claim paper to fill out. You can usually buy such a paper from any large office supply store, or get the form from the Recorder's Office. This paper must provide the information given above and blanks are provided for that purpose. Bach state also has various other data that they require you to provide. If you buy a mining claims paper in your state it will usually state all of the data that is necessary. Just hand two copies of your mining claim paper to the clerk and the proper fee and he will stamp both copies and give you one copy back. Keep this copy. It must be sent to the BLM.

Be sure that you give all of the data asked for on the mining claim paper. It is not the Job of the Recorder's Office to see that these papers are accurate or even their job to see that you have not claimed someone's claim. They merely record anything you wish recorded. Some Recorder's offices have maps showing which areas are claimed, but most do not. So you must claim your area and hope that no one else has a claim there also. The Recorder will not tell you if someone already has your area.

Of course, if there were no monuments or other evidence of a claim you are fairly safe. But if someone does in fact have a claim on your area the fact that he was there first will usually give him the legal right to it. But you might be able to prove that he has not done his assessment work and that he had not marked his claim. In which case you would have the legal rights.

(4) Recording with the BLM. The paper that you receive back from the County Recorder must be sent to the BLM within 90 days along with $5.00 for each claim that you wish to register. The BLM will assign you claim numbers that you must keep. The rules state that you must mail or deliver your paper from the County Recorder to the local BLM office. However in some cases all claim papers are referred to the district office. Claims for Southern California are referred to the BLM office in Sacramento. The address is, Department of the Interior, Bureau of Land Management, 2800 Cottage Way, RM E-2820, Sacramento, California 95825.

Now that you know how to file a claim there are a few other important facts that you should Know. One new rule that some miners have missed is that you no longer have to do discovery work. In other words now you can file a claim without doing any kind of work on the claim. The only work that you would have to do is the assessment work the following year.

Check with your local BLM office to determine which lands are open for prospecting. (They call it open for location or locating). There must actually be valuable minerals on your claim. You will eventually have to prove it. Any US citizen or one who has declared his intentions to become a citizen can stake a claim.

You can stake a 5 acre claim for use as a mill site. It is done in the same way as outlined above. It must be in connection with a mine.

The exact method of staking claims vary from state to state. Write to your state geological survey or mineral division office for details. For topographic maps and geologic maps west of the Mississippi River write to, U.S. Geological Survey, Federal Center, Denver, Colorado 80225. For Maps east of the Mississippi, U.S. Geological Survey, 1200 South Bads Street, Arlington, Virginia 22202.

Later comments...........

IMPORTANT COURT RULING by Russ Letlow
The United States District Court in Reno, Nevada ruled on October 19, 1983 that abandonment mandated by the Federal Land Policy Management Act for failure to file an affidavit of assessment work is unconstitutional "denial of procedural due process".
It should be noted that this ruling presumedly doesn't declare the filing requirement unconstitutional; but rather the denial of constitutional rights of due process. Judge Bruce R. Thompson ruled that due process procedure included notice and hearing to determine abandonment. He further found that filing one day late was not a presumption of abandonment. — If you straggle in a couple of years late he may have a different opinion. -- This decision will be appealed to the U.S. Supreme Court.

I was wrong. Load claims do not have to have a minimum of 20 acres. That is true only of placer claims. In California it takes six stakes to complete a lode claim. The extra two stakes are required one at each end at the center. In Colorado 6 is also required but the two extra stakes are required at each side midways between the two corner stakes.

MEASURING DISTANCE
BETWEEN CLAIM POSTS
WITH RANGE FINDER..

STAKING CLAIMS WITH A RANGE FINDER

Calibrating the range finder on a football field

by Jim Humble

Using the range finder to mark out claims is much easier and simpler than most other methods of measuring distances. For the small miner or prospector surveying a claim can be quite an expense, and surveying is not a legal requirement and certainly not necessary at least at first. One may wish to survey his land after it has begun to make money for him. We have put in several hundred claims and with over 6,000 acres using the Rangematic MK5 1000. The claims were put in quite satisfactorily. Of course(accuracy is not as good as surveying but it is legal and if our land begins to produce we will then have it surveyed.

The Rangematic MK5 works like the focusing device on most 35 mm cameras. You see two images. You merely turn the knob until there is only one image. The yards are read out on the dial. Accuracy is plus or minus 1 yard at 100 yards, plus or minus 9 yards at 300 yards, plus or minus 25 yards at 500 yards, and plus or minus 100 yards at 1000 yards. The range is 50 yards to 1000 yards. That's plenty good for roughing in a claim. It can save you much money and many hours or even days or weeks depending upon how much land that you are claiming.

The oil companies and large mining companies are claiming hundreds of thousands of acres every week. Using a range finder such as this one is the cheapest way that you can get some claims before the oil companies have it all. Refer to other data in this magazine on how to stake a claim. The Rangematic MK5 comes with a 6 power prismatic scope that makes objects seem 6 times closer. The 6X18 monocular is removable for use as a high quality scope. The carrying case costs extra. Anyone would make a mistake not to buy the carrying case. It is a necessity out in the hills. The MK5 should be calibrated once when you first get it. Merely look at something which you know the exact distance such as the distance of your own lot or a foot ball field and then use a coin to set the distance. From that point on all other distances are correct. That's all there is to it.

Ranging Inc. manufacturers the MK5 as well as several other small instruments for shorter distances down to 6 feet. Their accuracies are similar (plus or minus 1% at the shorter distances to about 5% at the longer distances). Of course range finders are useful for many other things such as shooting, golfing, building, etc.

The body of the Rangematic (TM) is made of fiberglass reinforced plastic. It is shock resistant, weather proof, and thermally stable. It weighs approximately a pound and a half. It is black and carries a 2 year limited warranty. Cost is approximately $119.95 and the carrying case sells for $15.95 It can be purchased from many local surveying and hardware stores.

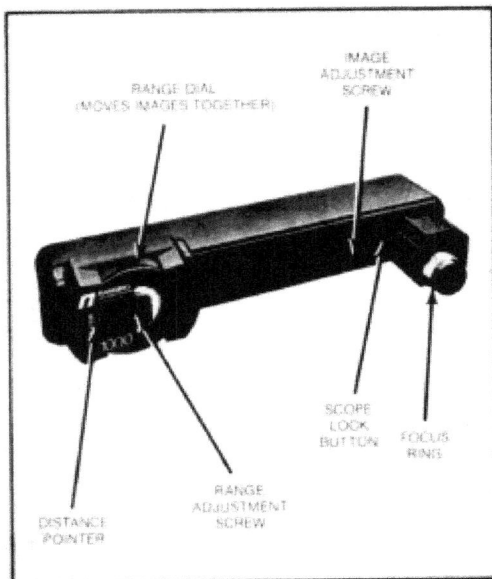

WHIRLPOOL CONCENTRATOR

WHIRLPOOL CONCENTRATOR
IN OPERATION...

by Ken Horn Written by Jim Humble

This cheapy concentrator can be built almost any size to handle 10's of pounds of ore a day or to handle hundreds of tons a day. It is meant to be used with very finely ground ore with a great deal of light material. 150 to 400 mesh is best for this concentrator. This concentrator will simply wash away the light parts of the ore and hold the heavy parts.

CONSTRUCTION — As you can see by the picture, the whole unit is made from a 5 gallon bucket and PVC pipe. The center pipe is 1.5 inch PVC. You can make the center piece nice and sturdy if you follow this carefully. The 1.5 inch elbow fits below the bottom of the bucket. Cut a piece of 1.5 inch pipe 1 1/4 inch long. Glue this piece into the elbow. This will leave about 5/8" inch pipe sticking out of the elbow.

Next, use a piece of the 1.5 inch PVC pipe to mark a circle in exactly the center of the bottom of the bucket. Now drill a hole and then use a knife to cut the hole to the exact size as the 1.5 inch pipe. Next, add PVC glue to the pipe sticking out of the elbow (mentioned in the last sentence of the paragraph above). Add PVC glue to a 1.5 inch PVC sleeve. Take the PVC elbow with the pipe with the fresh glue and push it through the hole from the bottom side of the bucket. Quickly push the PVC sleeve onto the pipe from the inside of the bucket. The elbow should now be clamped to the bucket from the bottom side and the sleeve holding from the top side.

On the inside, add a five inch piece of 1.5 inch PVC pipe to the 1.5 inch sleeve. This makes the water about 6.5 inches deep before it begins to drain down the center. To the elbow under the bucket add a 1.5 inch PVC pipe as long as you wish to drain off the tailings.

Now you are ready to add the swirling mechanism, which as you can see, is simply the water added at a tangent on each side about 1.5 inches from the bottom of the bucket. You should be able to see from the pictures exactly how we did it. There are no doubt better ways of doing it, but we just used silicon rubber glue to hold the PVC pipes in place in the bucket. Most buckets are made of polyethylene and silicon rubber is the only thing that will stick to it. If you could find a PVC bucket it would be much better. The only thing that you can't see is that we cut a small piece of pipe and glued it in the end of each of the water jets into the bucket to leave only about 1/3rd opening to get more water pressure and thus better water swirl. The concentrator will have to be set or bricks or blocks to allow for the elbow and escape pipe on the bottom for the tailings. Parts for the complete concentrator are:
1-5 gallon bucket
1-1.5" elbow PVC
1-1.5" sleeve PVC
1-1.5" pipe PVC 3'

INSIDE OF WHIRLPOOL
CONCENTRATOR...

4-3/4" elbows PVC
1-3/4" Tee PVC
1-3/4" plastic hose valve with PVC fitting one end

One thing that we did not do was put an outlet at the bottom of the bucket for the concentrates. It would thus be possible to have a continuous flow concentrator. One could also use a 55 gallon barrel to do the same thing and have a very large many tons per day concentrator.

To operate the concentrator, turn on the water and wait for the whirlpool to form. The speed of your whirlpool will determine the amount of concentration. Remember, this concentrator is for finely ground ores but it will work to some extent on ores only ground to 60 mesh. The greater the speed of the whirlpool the less heavies will be retained. If you want only a few heavies then adjust the water for maximum spin, but if you intend to leach your heavies then use a slower spin to make sure you don't lose anything. You can add ore to this concentrator anywhere along the side of the inside of the bucket. You can use wet or dry ore. As you increase the ore flow into the bucket you will need to increase the water flow to keep the whirlpool action going.

This is a great concentrator for small batches of ore. Let us know how it works. Send some pictures. We'll try to publish them, and any modifications you have made.

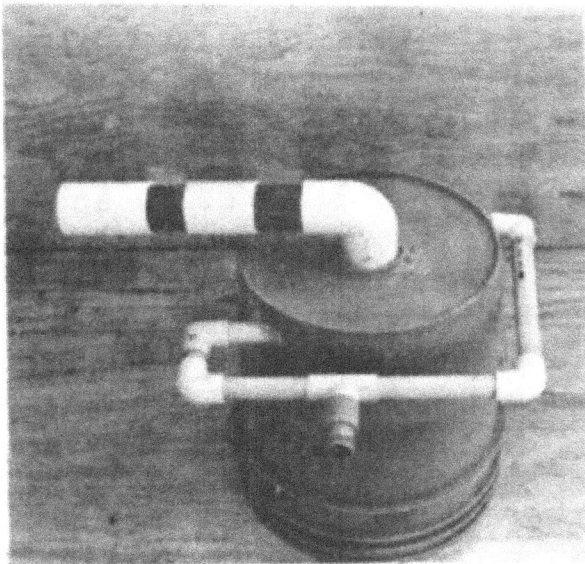

> "this concentrator will simply wash away the light parts of the ore. . . "

BOTTOM VIEW OF WHIRLPOOL
CONCENTRATOR

Whirlpool Concentrator
Dear Ken and Jim:
I built your "whirlpool concentrator" the way you did. Then I revised it by putting a 90 inside the pan at the end of the water pipes. Then I plugged them and drilled out a 3/8" hole, causing the water to move faster.
I also did not glue the 5" center pipe in place. What I did do is cut one at 5" and one at 3". I start with the 5" and wait for the water to clear then I replace it with the 3".
In sifting the sand, I used a screen from a window since nobody here in Las Vegas could tell me how to find mesh. Window screen is 14 strands x 16 strands or 200+ per sq inch. Right?
I am trying to build a separator with 3 rings - one inside the other and using 1/4 mesh, window screen, and a nylon stocking - all junk you can get anywhere.
If you can help with this, I would enjoy it. Thank you, Paul Osborn - Las Vegas, Nevada

CENTRIFUGAL CONCENTRATORS

by Jim Humble

Homemade version of a centrifugal concentrator

A "Nelson-type" concentrator is a centrifugal concentrator similar to that made by Knelson. (Knelson is a trade make of a Canadian company) We have come to the conclusion that these concentrators do indeed work. However nothing is perfect and thus there can be problems connected with the Nelson concentrator. After visiting several sites where "Nelson" type concentrators are in use we have come to the conclusion that the major problem with such concentrators is simply not using them properly or not using them for the purpose for which they were designed.

The only other problem that we can see with the Knelson concentrator is of price. Knelson charges over twice the price that other Nelson type concentrators are sold for. From all the evidence available and in observation under actual working conditions we could see no better results from the Knelson concentrator vs. the "Nelson-type" concentrators. In fact there was some evidence that there was slightly better results from one of the others.

The two other units that we observed were "Hy-G" concentrators and as yet an unnamed concentrator designed directly from the Knelson. After panning the concentrates from the Hy-G unit, using a variety of ores, we found microfine gold from each ore we tried. Other concentrators did not fare so well in other tests. There is not doubt that the increase in gravity (60 G's or more) does indeed increase the recovery.

THEORY OF OPERATION — The ore is injected into the center of the spinning bowl and drops to the bottom of the bowl and then is washed out by the water that is forced into the bowl. As this ore is washed out it becomes 60 times heavier than it is normally under standard gravity. Of course, the water is also 60 times heavier than normal.

The bowl has rings in it which essentially form riffles for the ore to pass over. Each one of these riffles fills up with rocks and ore. Since the gravitational force within the bowl is actually centrifugal force and that force is outward 60 times more than the pull of gravity is downward. The actual bottom of these riffles is the inside of the sides of the bowl. In other words the bottom of the riffles is actually outward. In the bottom (the sides of the bowl) of each of these riffles is a number of holes through which water is forced. This water forces its way out through the rocks and ore and causes the ore itself to become fluid, essentially a quick sand bed. Some engineers would call it a dilated bed. This quick sand bed in each riffle is what catches and holds the ore.

Each riffle bed will contain the larger size rocks at the bottom of the bed with progressively smaller and smaller pieces of ore or sand outward towards the top portion of the riffle bed. (The top portion of the riffle bed is actually the very inside of the bowl). The outer layer of ore will be the micro-fine pieces of ore. With the larger pieces of rocks laying somewhat over the holes which are emitting water, the water will be evenly dispersed throughout the riffle bed. It then moves outward keeping all of the pieces of ore just slightly separated. This is what causes the quick sand bed. The microfine particles of gold are from 5 to 15 times heavier than the ore particles and thus the gold particles sink into the quick sand and are trapped.

In addition to the quick sand each gold particle has a difference in weight from the ore particles around it that is much greater than when it is under just plain gravity. In fact the difference in weight is 60 times greater. This makes it much more likely to get trapped in the lighter quick sand.

The principle of operation is a replacement process. The riffles will hold about 12 pounds of sand or ore. So, after the first 12 pounds of ore has been fed into the machine the riffles will be full. From that point onward the operation is a replacement process. Heavier particles move in and displace the lighter particles.

USING THE "NELSON-TYPE" CONCENTRATOR FOR PRODUCTION -- Of course, the "Nelson-type" concentrator must be used correctly or it is of very little use. The biggest mistake that we have observed is in not cleaning out the black sand that has been caught in the bowl often enough. In some areas there is no gold in the black sand and thus one would be justified in not cleaning it out. But in most areas one should be interested in saving his black sand because it can have 1 to 50 ounces of gold

Running the Hy-G concentrator on various ores - Hy-G unit comes with vibrating screen & centrifugal cone with easy clean out facility

per ton trapped within the particles. In some cases the black sand can be so heavy that it will pack down and prevent the quick sand from forming. This will then let much of the fine gold escape.

The Hy-G concentrator 8" bowl will hold about 12 pounds of concentrates in the riffles. If one wants to save his black sand he should know about how often he must clean his bowl. If he does not clean it often enough he will lose black sand. If he cleans it too often he will simply be working harder than he should and he will find that his concentrates are not very concentrated. There will always be some lighter colored sand with the black sands. So one does not wait until the quick sand is completely black sand. This would cause him to lose black sand. It is best at first to pan the output to see when black sand is beginning to show up. Shift would indicate that black sand is being lost.

If one does not care about the black sand but wants only the gold, he can then run much longer before he must clean out the riffles. The fine gold will displace the fine black sand. The only thing to look for is that the black sand is not packing. If it packs one will begin to lose gold.

There are several signs of packing. Turn off the machine And wait until the spinning stops. If the sand falls out of the riffles it is not packing. But if the sand (ore) remains upright in the riffles it means that packing is taking place And one is losing gold. Another indication is if there is a greater concentration of black sand at the surface of each riffle rather than at the bottom of the riffle. This would mean that the heavier black sand has not sunk towards the bottom of the riffle. Of course, the larger pieces of ore will be at the bottom of the riffle because of their size.

The actual water pressure Applied to the bowl itself is 25 to 30 pounds of pressure. The water flow is about 25 gallons per minute for the 8" unit. This is about 60% of the water required for sluicing. The 8" unit will handle from 2 to 5 tons of ore per hour. This depends upon the size of rocks in the ore And the consistency of the ore and other factors.

EFFICIENCY -- The manufacturer claims 95% recovery. Most manufacturers of recovery equipment claim At least this good and usually better. But each ore is different. This machine like all the rest first depends upon how much gold in the ore is liberated. Large chunks of ore with small particles of gold in them will usually be lost. If most of the gold is still contained within ore particles then most of it will be lost. But if most of the gold is liberated then there is a good chance of this machine getting 95% of the liberated gold. No machine made will remove all of the gold.

In some cases one can liberate more gold by grinding the ore to a finer mesh size. But in some cases that will produce gold that is too fine for any concentrating machine and thus gold will be lost. In such a case it would be best to set the machine to concentrate lightly and then regrind and leach the concentrates.

We expect to have further articles giving details on building your own centrifugal concentrator. But for those in a rush here is the formula for determining the RPM of the cone. This formula is based on *60* gravities (60 G's). RPM = the square root of 32X60/R then divided by 6.28 and then multiplied by 60. The way you would work this formula is to multiply 32 times 60 (60 G's), divide that by the radius of the cone in feet (6 inches diameter = .5 foot and the radius would be 1/2 of that or .25) Now get the square root of this answer and divide that by 6.28 and multiply that answer by 60 for RPM.

In any case, as far as concentrators are concerned, we believe "Nelson-type" concentrators to be the very best there are at this time.

Open Your Own Assay Office

by Jim Humble

The gold and silver mining field has been increasing in size for many years and it appears that it will continue to do so for some time. The need for fire assays is greater now than it has ever been. There are very few communities in the western part of the US that could not support an assay office. There are areas even in the eastern part of the States that could support an assay office. Many assay offices work by mail and thus it matters little where they are located.

There are reasons for doing fire assays that have little to do with mining and there is always someone who needs a fire assay done. In large cities there are jewelry sweepings that need assaying. Every time a plating company sells or buys gold or silver plating liquids, the liquids need assaying. (This is a simple procedure and we will tell how to do it next issue). But, of course, the main reason for fire assays is mining, or the locating of gold and silver minerals in the ground.

There are many fancy new devices that detect gold and silver. Yet in the final analysis the fire assay is still used to prove the point. And the fire assay is a simple procedure that anyone can learn. It takes time to know all of the ins and outs about the subject. But one can learn to do a very good job at fire assaying in several days. The amount of equipment necessary to do a very accurate assay is minimal. In fact for less than $1500 one can buy everything necessary. And if one is willing to build some of his own equipment, less than that.

At this time starting an assay lab can be a very good opportunity. The time required to learn the subject is minimal and the cost to get into business is very low. There are a number of people doing assay businesses now who simply bought an assay kit from GAME Company and hung up their shingle.

One couple bought an assay kit one week and had a full time business the next week with all the business they could handle for weeks to come. But this is not the usual case, people do not simply start businesses and be an overnight success. Usually it takes a lot of work and time, and if you have money, a lot of money. One should expect to get only a little work at first, and then depending upon the service and professionalism, more as he goes along. Advertising is a must. One should put his card out in mining areas and where miners are likely to look. One should also advertise in the local paper and phone book as well as mining magazines.

LEGAL REQUIREMENTS? There are surprisingly few legal requirements for assayers. Very few states require that one do much more than buy a local business license and if one is not doing it as a full time profession, he can often get away without doing that. In California and most states a certified assayer is an assayer who has a certificate from somewhere. Since few schools teach assaying, a certificate from most any school is okay.

When you do a fire assay for anyone you are taking on a legal contract. You are saying, in essence, that you are an assayer and that you can do accurate assays. If you should goof up and that someone can prove that you did not know what you were doing then you can be sued. You can even be found guilty of fraud. So when a person does go into assaying for the public he should know what he is doing, which is not hard to accomplish at this time. In years past assaying was held as a deep dark secret. But now assaying is well understood and the basics can be learned in several days time.

Assaying is an opportunity to make mining pay for itself. It is an opportunity to be in the know about mining properties and mining strikes. Almost all mining data finds its way into the assay office first. The guy that knows more about mining in any area or the guy that learns more about mining in any area fastest is the assayer.

All in all if one expects to go into mining he should learn assaying and his own assay office will pay off in many ways.

Pete McLaughlin's
Johannesburg CA Assay Office

Estimating Gold Values
In Placers

by Dave Parkhurst

ESTIMATING GOLD VALUES IN PLACER GRAVELS

Gold per Cu. Ft.		penny	Value per Cu. Yd. in placer fineness						Gold Troy Oz
Grns	grams	weight	700	750	800	850	900	950	Cu. Yd.
.1	.006	.004	$1.89	$2.03	$2.16	$2.30	$2.43	$2.56	.0067
.2	.012	.008	3.78	4.05	4.32	4.59	4.86	5.13	.0135
.3	.019	.012	5.67	6.08	6.48	6.89	7.29	7.69	.0202
.4	.025	.016	7.56	8.10	8.64	9.18	9.72	10.26	.0270
.5	.032	.020	9.45	10.15	10.80	11.47	12.15	12.83	.0336
.6	.038	.025	11.34	12.15	12.96	13.77	14.58	15.39	.0404
.7	.045	.029	13.23	14.17	15.12	16.06	17.01	17.95	.0472
.8	.051	.033	15.12	16.20	17.28	18.36	19.44	20.52	.0540
.9	.058	.037	17.01	18.23	19.44	20.65	21.87	23.08	.0607
1.0	.064	.041	18.90	20.25	21.60	22.95	24.30	25.65	.0675
2.0	.129	.083	37.80	40.50	43.20	45.90	48.60	51.30	.135
3.0	.194	.125	56.70	60.75	64.80	68.85	72.90	76.95	.202
4.0	.259	.166	75.60	81.00	86.40	91.80	97.20	102.60	.270
5.0	.324	.208	94.50	101.25	108.00	114.75	121.50	128.25	.336
6.0	.388	.25	113.40	121.50	129.60	137.70	145.80	153.90	.404
7.0	.453	.291	132.30	141.75	151.20	160.65	170.10	179.55	.472
8.0	.518	.333	151.20	162.00	172.80	183.60	194.40	205.20	.540
9.0	.583	.375	170.10	182.25	194.40	206.55	218.70	230.85	.607

The single, most important question to be answered when evaluating a placer gold deposit is; "How much is it worth?"

Accurate sampling of placer gravels is essential to obtain a reasonable estimate of the total gold values contained in the deposit. Because it is difficult to estimate a cubic yard of gravel, samples can be taken in units of one cubic foot. Due to the size of the samples, it is advisable to take a large number of samples over the entire placer area. A bulk test run of between 500 and 1,000 cubic yards is the best method to prove a placer deposit, but this is not always practical. Therefore, to obtain a reasonable estimate of the deposit's value by hand methods, a minimum of 10 to 20 samples should be taken, panned carefully, and the recovered gold values weighed separately.

Placers should be sampled from the top to the bottom of the gravels at several locations in the deposit. A box measuring 12x12x12 inches inside measurement can be constructed from wood to standardize the one cu. ft. samples, or a line can be drawn inside a bucket to represent this volume of gravel. All materials from the sampling area (including large rocks) should be taken in the sample. As the gold from each sample is weighed, its approximate value per cubic yard can be found in the accompanying Gold Chart.

Once the sample values have been obtained, they should all be totaled together and averaged. If one sample is exceptionally rich or poor, it can be discounted. If 2 or more are exceptionally rich or poor, they should be included in the average unless they all came from the same location. If one area shows rich or poor, it should be calculated separately.

Values in the table are calculated according to relative "fineness". Pure gold is 1000 fine, or 100%, and 850 fine gold is 85% gold. If you don't have an assay value for the placer gold, fineness can be approximated by the average fineness of the gold found in the same area. If not that use 850 fine until you know exactly.

To calculate values not shown: if sample weight is 2.6 grains add amount shown in 2.0 row to amount shown in the .6 row. If weight is 20.7 grains, take 2 times the amount shown in the 9.0 row, and add this to the amounts shown in the 2.0 row and the .7 row.

CONVERSION FACTORS:
1 troy oz = 31.103 grams = 480
grains = 20 pennyweight = 1.097 oz. avoir
1 grain = .0648 grams = .04167 pwt. = .00208 troy oz.
1 penny weight = 1.55515 grams = 24 grains = .05 troy oz.
1 gram = 15.433 grains = .643 pwt. = .032 troy oz.

The Gold Chart is based on the amount of gold in 1 cubic foot of placer gravel. There are 27 cu. ft. in 1 cu. yd. of in place gravel. When gravel is loosened, however, it expands or "swells" by about 20%. This means there are approximately
32.4 cu. ft. of loose gravel in 1 cu. yd. of in place gravel. The chart takes this "swell" into account, and gives the dollar value for each cu. yd. of in place gravel. Remember you art taking the gold from a cubic foot to find the amount of gold in a cubic yard.

Because gold prices vary, the chart is based on a price of $400.00 per troy ounce. The dealer price is always below the market price for gold (which is sold at 995 fine or higher), so whatever this price happens to be, it can be converted to a percentage of the amount shown in the chart. For example, if gold is $420 per oz, add 5% of the amount shown in the table. If gold is $380 per oz, subtract 5% of the amount shown in the table. Never base placer gold value at the market price for gold, as this will artificially inflate the value of the placer deposit.

Keep in mind that the value obtained by use of this method is an estimate of the recoverable gold values contained in the placer deposit. If the estimate comes out too low, too high or marginal, then repeat the sampling procedure. If 2 or more sampling runs product nearly the same results, it is a good indication that your estimates art reasonably accurate.

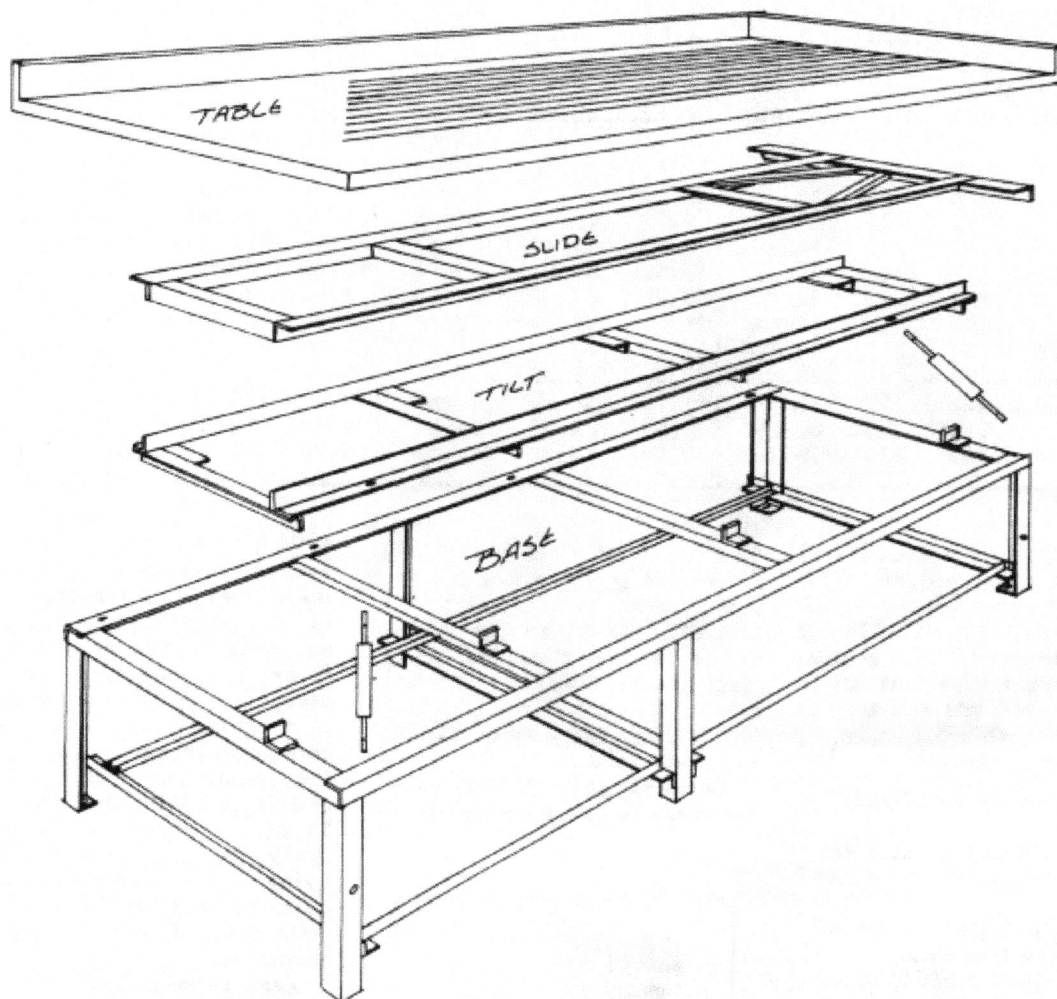

A Quality Concentrating Table
You Can Build

Designed and Built by Mike Glenn
written by Jim Humble

Note: This is the first of a series of six issues on the construction of a quality concentration table. Anyone wishing to see the actual table is invited to drop by Popular Mining. We believe that this table is well worth your time to build and it will be, when complete, equal to a table costing thousands of dollars.

Concentrating tables are perhaps the most complex of all milling equipment for the recovery of gold, silver, and other heavy minerals. There are a number of cheap, medium, and highly expensive tables on the market to choose from. Action Mining Company had the experience to build a number of tables over a period of several years using each table extensively. Thus Mike Glenn and others at Action Mining are responsible for the design which is the accumulation of Knowledge over that period of time.

One very important consideration when building a table is stability. A light cheap table will vibrate in an up and down motion as it moves back and forth. This vibration will continuously cause the fine gold to break loose from it's connection to the table and then wash away with the water. This is also true of the very fine heavy particles in the ore. Thus much gold is lost with most cheap tables.

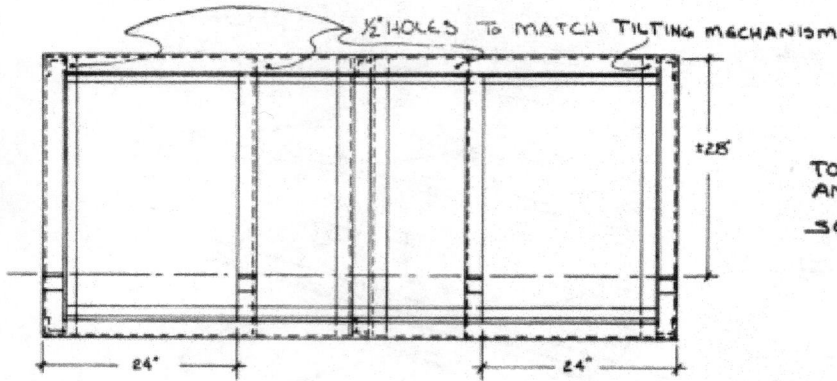

½" HOLES TO MATCH TILTING MECHANISM

TOP TO BE WELDED FLUSH
AND GROUND FLUSH

SQUARE ALL CORNERS

TOP VIEW

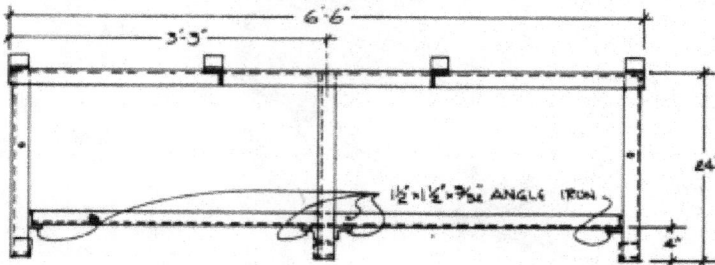

FRONT VIEW

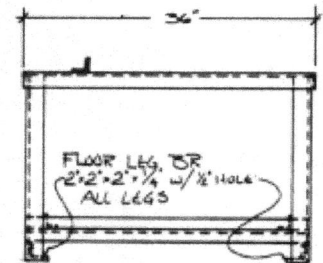

END VIEW

BASE

¾" IN   RIFFLE   ¾" OUT

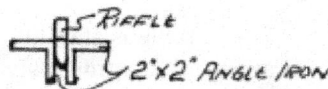

RIFFLE

2"x2" ANGLE IRON

The shaker mechanism on a cheap table is also a source of vibration. This mechanism causes the table to bump, bump, bump, against the tailings end of the table which moves the ore across the table. Bach bump causes a vibration in the table and breaks away the fine gold. When the fine gold is vibrated up from its contact with the table the water then carries it away.

A perfect table would move back and forth carrying the ore and gold along but there would be no vibration at all. The gold would slide slowly towards the opposite end and into the wash out area. There all of the sand and light materials would be washed away leaving pure gold to drop into the gold chute. But the perfect table does not exist. We can only try to approach that perfection.

To eliminate the vibrations the more expensive tables are built sturdy and heavy. The mechanism is much more expensive and it does not go bump, bump. Instead it moves the table back and forth as smoothly as possible but pulling the table back faster than it moves it forward. You can try this kind of a motion yourself. Take a board and lay a pencil on it. Now to simulate a bump, bump, mechanism try bumping the end of the board against the wall. You will note that the pencil moves towards the wall as you bump the board. Next you can make the pencil move by Just moving the board forward and then jerking it back. If you are careful you can make the pencil move with a great deal less noise and vibration by Jerking the board than by hitting it against the wall. So, the more expensive table has a heavier base, a heavier table and a more expensive mechanism. This results in a great deal less vibrations and better gold recovery. Of course there are other important factors, some of these will be covered as we go along.

Action Mining has built the base of this table to hold 1200 pounds of solid cement blocks. Yet when the blocks are removed the table can be moved to other locations easily. The base is made of welded steel for sturdy construction. When complete, it will have • incorporated the important features that we have found necessary to recover the finest of gold even from sulfide ores. Cost of the complete table will be between $300 and $500 depending upon your source of supply and the amount of scrap that you already have.

CONSTRUCTION
THE RIBS - It does seem funny to make the RIBS first, but the reason for this is that angle iron is used in sawing out the RIBS. Once the RIBS are sawed then the angle iron can be used in the construction of the BASE.

THE RIBS are made from 18 pieces of bay shoe. See the list of materials. Bay shoe is bought at the lumber yard. (It is similar to baseboard but has no grooves along the back side). Be sure to buy the hardwood bay shoe (oak). The shortest RIB is 42 inches. Cut out 18 RIBS with each RIB being cut one inch longer than the one before it. This will make the longest RIB 60 inches long.
Make the RIBS as follows:
(1) Lay two 2 inch angle iron by 78 inches in your vice so they can be clamped at one end.
(2) Cut a piece of bay shoe 42 inches long.
(3) Using the vice and 4 or 5 "C" clamps, clamp the bay shoe between the two angle iron so that the bay shoe sticks completely out at one end and is completely in at the other. This leaves the bay shoe clamped in the angle iron on a diagonal line. Remember when you are doing the clamping that the rounded part of the bay shoe must be on the upside and towards the open side of the table. When the ore drops onto the table it must not see a rounded edge at the top of the RIB.
(4) Using a handsaw laid on its side cut the bay shoe so that the saw runs along the angle iron. This will produce a RIB that begins 3/4 inch high and runs down to zero. Carefully sand these RIBS, especially the first several. It is important that they taper to zero smoothly and that a sharp top edge is always towards the ore or the boxed side of the table.
(5) Cut all 18 RIBS in the same way but be sure to cut each RIB one inch longer than the one before. When all 18 have been cut put them aside and you are ready to start on the BASE.

THE COMPLETE TABLE - The completed table consists of five basic assemblies. (1) The BASE, (2) the TABLE TILT MECHANISM, (3) the TABLE SLIDE MECHANISM, (4) the TABLE TOP, and (5) the WATER FLOW MECHANISM AND, (6) the MOTION GENERATOR.

THE BASE - the list of materials for the BASE is as follows:
1) 2 - rails 78"   2x2"x 1/4" angle
2) 4 - cross 36"   2x2"x1/4" angle
3) 6 - legs 24"   2x2"x1/4" angle

4) 4 - stops   2"   2x2"xl/4" angle
5) 6 - floor bracket 3" 2x2"xi/4"angle
6) 2 - braces 75.5" 1 1/2"xl 1/2"x5/32" angle
7) 4 - braces 36"   1 l/2"xi 1/2"x5/32" angle

The easiest way to get the above pieces is to just order them already cut from the metal supply company. On the other hand you may have to cut them yourself. You can do this with a metal cutoff saw, or you can buy a composition metal cutting blade for your Skill saw. But if you do that you should realize that you will probably ruin your Skill saw unless you put some kind of filter medium over the lubers on the saw motor. The metal particles and sand from the blade will ruin a motor after a few hours if the motor is not protected with filters. Just tape a filter cloth across the luber openings. Do not use plain cloth* or if you do» do not let the motor get hot.

STEP BY STEP INSTRUCTIONS:

(i) Now that all the pieces are cut start with the top of the BASE. Use the two rails (78") and the four cross pieces (36"). Notch all pieces so that the top surfaces of the angles will be even. The drawing showing the BASE gives enough details for you to see where the notches must be made in the ends of the cross pieces. Be sure to get the two inside cross pieces in

pl

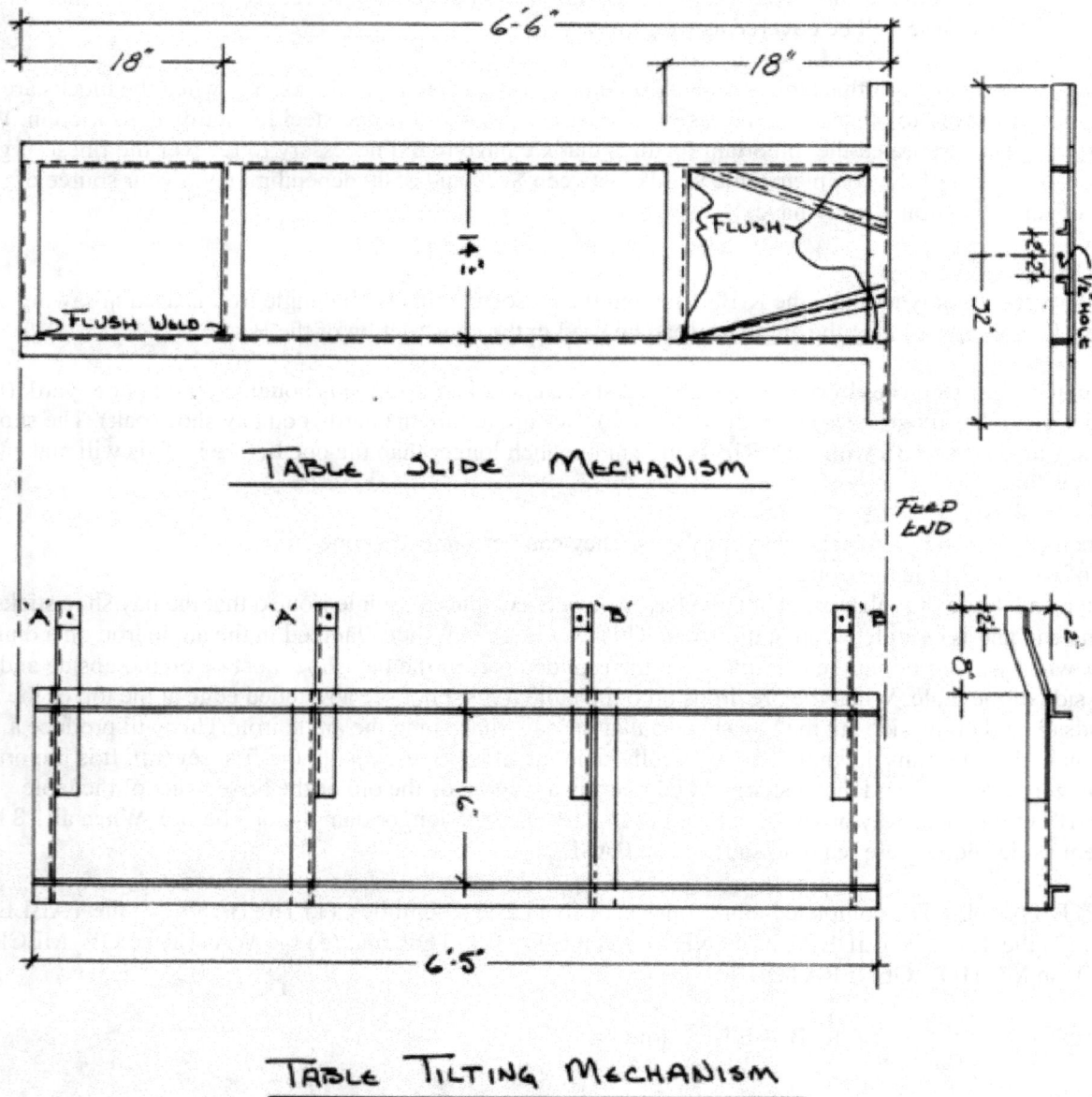

TABLE SLIDE MECHANISM

TABLE TILTING MECHANISM

ace properly. Note that they lay opposite one another. This will be utilized to keep the table solid.

(2) Lay all of the rails and cross pieces in place and make sure that they are sitting square and level. Use a piece of plywood if you don't have a level floor. Use a square to square the corners before welding. Just before welding each corner be sure to again check the corner for squareness. Weld all joints.

(3) Weld on the six legs. First clamp each leg in place with a "C" clamp. Use your square to check that the leg is straight up and down. Do this by checking squareness from several different directions. Then tack the leg in place and again check for squareness. Then tack in a second place, check for squareness and then do the complete weld.

(4) Clamp each bottom brace in place with "C" clamps and then tack. When all of the braces have been tacked in place properly then weld each joint.

(5) Weld the floor brackets as shown on the print.

(6) Weld the stops in place as shown on the print.

(7) Clean each join and sand for a finished product.

(8) Painting the BASE at this point is important. Use a spray can to spray on a non-rust type of undercoat first and then finish off with the color of your choice. Spray cans are the easiest.

THE TABLE TILTING MECHANISM -

This is the frame that sits directly on top of the BASE. It holds the TABLE SLIDES and is mounted so that the table can be tilted under actual operating conditions.

The list of materials for the TABLE TILTING MECHANISM is as follows:
1) 2 - rails   78" 2x2"x1/4" angle
2) 4 - cross   20" 2x2"x1/4" angle
3) 4 - tilt bracket 17" 2x2"x1/4" angle
4) 4 - slide mounts 24" 1 1/2"x1 1/2"x1/8" angle
5) 4 - slides heavy duty ball bearing slides (for computer drawers). Buy slides from Cal Aero Supply Co., 13840 Paramount BLVD, Paramount, Calif. 90723 or from any other electronic surplus store. Surplus drawer slides should be made from .062 sheet steel and should have 30 to 40 ball bearings in each slide. Almost any length slides will do but they should not be wider than 1 1/2". The actual motion of the slide on this table will not be more than one inch.

TILT BRACKET

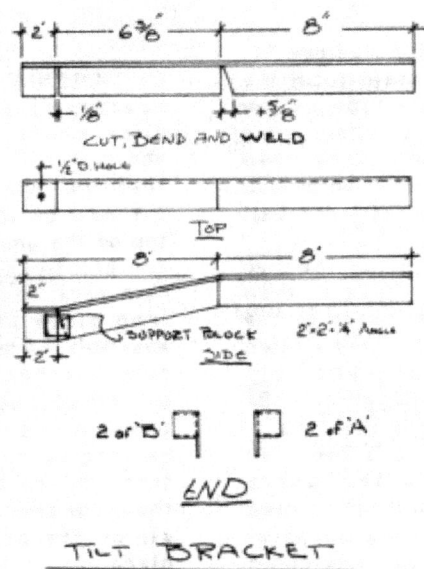

CUT, BEND AND WELD

TOP

SIDE

END

TILT BRACKET

Build the TABLE TILT MECHANISM as follows:

(1) The cross pieces (4) are laid in place up against the stops on the BASE. Be sure to lay these cross pieces in place properly as shown on the drawings. The two inside cross pieces in the TABLE TILT MECHANISM lay directly on top of the two cross pieces in the BASE. They are opposite one another as are those in the BASE. The two end cross pieces lay on top of the BASE cross pieces, but are opposite to the BASE cross pieces.

This is a very important point, the cross pieces must be laying in place properly when they are welded. The MOTION GENERATOR will try to move this TABLE TILT MECHANISM and that will be prevented only if these cross pieces are layed in tightly before welding.

(2) Lay the two rails in place as shown in the drawing. One rail must be laid up against the stops on the BASE and the other rail is laid opposite at the other end of the cross pieces. When everything is laid in place and carefully squared and checked to see that there is no slop between the BASE cross pieces and the TABLE TILT MECHANISM cross pieces, then carefully tack everything in place.

(3)   Weld all Joints.

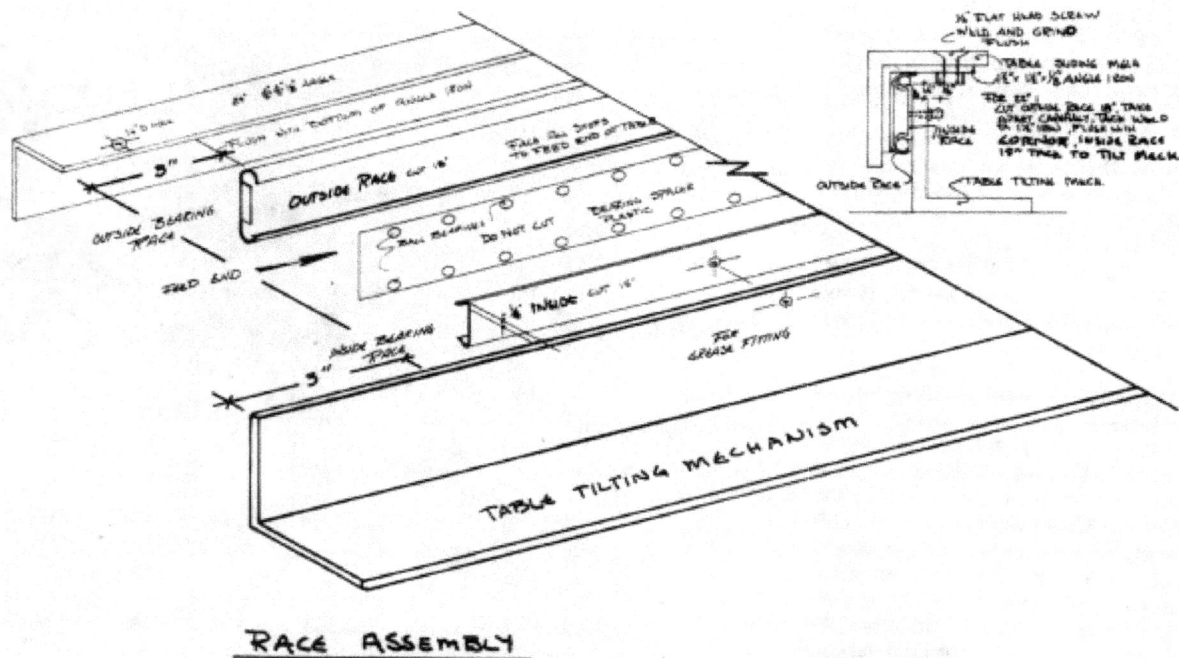

RACE ASSEMBLY

(4) Construct the tilt brackets. as given on the prints. There are four tilt brackets but two are mirror images of the other two, because they are mounted opposite on the TABLE TILT MECHANISM.

(5) Weld the tilt brackets to the TABLE TILT MECHANISM as shown on the prints.

(6) Installing the ball bearing slides can be a bit tricky so be sure to study the prints. The end result is a TABLE TILT MECHANISM with the ball bearing slides mounted on it and with the 24" slide mounts mounted on the ball bearing slides. This presents four 24" slide mount surfaces on which the TABLE SLIDE MECHANISM is mounted. Do this as follows:

(a) Use a hacksaw to cut the end off of one side of the ball bearing slides. This allows the slides to be taken apart. Do this over a wide plastic pan to catch the ball bearings.

(b) The smaller of the two slide bearing races is then welded to the inside of the rails at four places (two slides at each end, six inches from each end). Be sure to have the stop end of the slide towards the head end of the table. Every effort must be made to keep the bearing race parallel with the rail but just about 1/4 inch above the rail. Use 1/16" welding rods and be quick. Practice with the parts that you cut off until you can weld this thin metal properly then weld the bearing races in place.

(c) Weld the opposite slide bearing race into the 24" slide mount angle iron as shown in the prints. Be sure that the race sets against the wall of the angle or the top of the angle as shown on the print. This makes the race parallel with the angle iron piece and makes the assembly strong. Weld this race 3 inches from the end.

(d) Finally, put the bearings one at a time into the plastic bearing spacer at the same time you are pushing it into the outer bearing race. When all of the bearings are in place push the outer race over the inner race thus putting the slide back together and completing the slide mount. Do this to all four slides. This leaves the TABLE TILT MECHANISM complete.

The TABLE TILT MECHANISM mounts on the BASE and is held in place mostly by gravity. Each of the cross angles of the TABLE TILT MECHANISM are between the cross angles of the BASE. After the TABLE TILT MECHANISM has been properly adjusted these cross angles will be clamped together to form a sturdy solid unit.

THE TABLE SLIDE MECHANISM mounts to the table and attaches the table to the TABLE TILT MECHANISM.

The list of materials for the TABLE SLIDE MECHANISM is as follows:
1) 2 - rails 78" 2x2"x1/4" angle
2) 3 - cross 14.5" approx. cut to fit from 2x2"x1/4" angle
3) 1 - cross end brace 32" 2x2"x1/4" angle
4) 2 - end angle brace 18" approx. cut to fit from 2x2"x1/4" angle

Before starting construction of the TABLE SLIDE MECHANISM grind the sharp corners of the 24" slide mounts to somewhat of a rounding edge. This is to allow the rails of the TABLE SLIDE MECHANISM to fit more closely. This is because the inside corner of the rail is not as sharp as the outside corner of the slide mount.

TOP VIEW
TABLE SLIDE MECHANISM

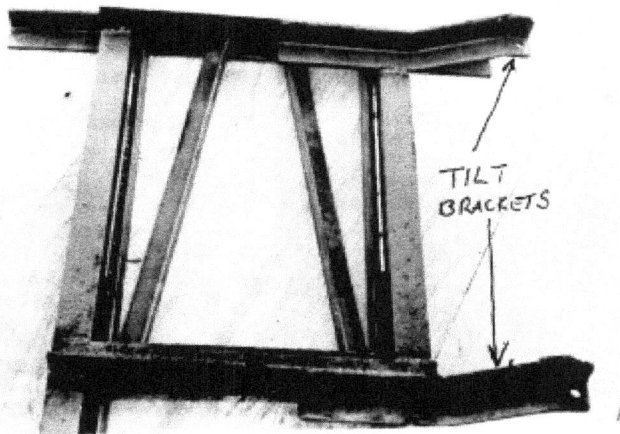

TILT BRACKETS

Construct the TABLE SLIDE MECHANISM as follows:

(1) Lay the rails in place on top of the slide mounts. Line up the end of the slide mounts with the ends of the rails. Using a 1/4" drill, drill three holes through the rails and through each of the slide mounts. This makes three holes at each end of the rails or a total of 12 holes. Counter sink each of these holes for a 1/4" 20 flat head bolt and install the bolt from the top with a nut and lock washer on the bottom.

(2) Spot weld each flat head of each of the bolts installed in the above paragraph. Grind off the spot weld for a smooth rail. (This prevents the bolt from working loose later and making a hole in the wood table). To remove the TABLE TOP at any time merely remove the 12 nuts on these bolts and the entire TABLE TOP will lift off.

(3) Cut to fit three cross angles as shown on the drawings. Weld these cross angles in place and grind the welds smooth.

(4) Notch the 36" cross end brace to fit at the head of the TABLE SLIDE MECHANISM so that it fits evenly with the rails. Weld in place and grind the welds smooth.

(5) Cut the end angle braces to fit and weld in place. Grind the welds smooth.

(6) Remove the 12 nuts from the mounting bolts to the TABLE SLIDE MECHANISM and lift it off. Paint as per above instructions and re-install when dry.

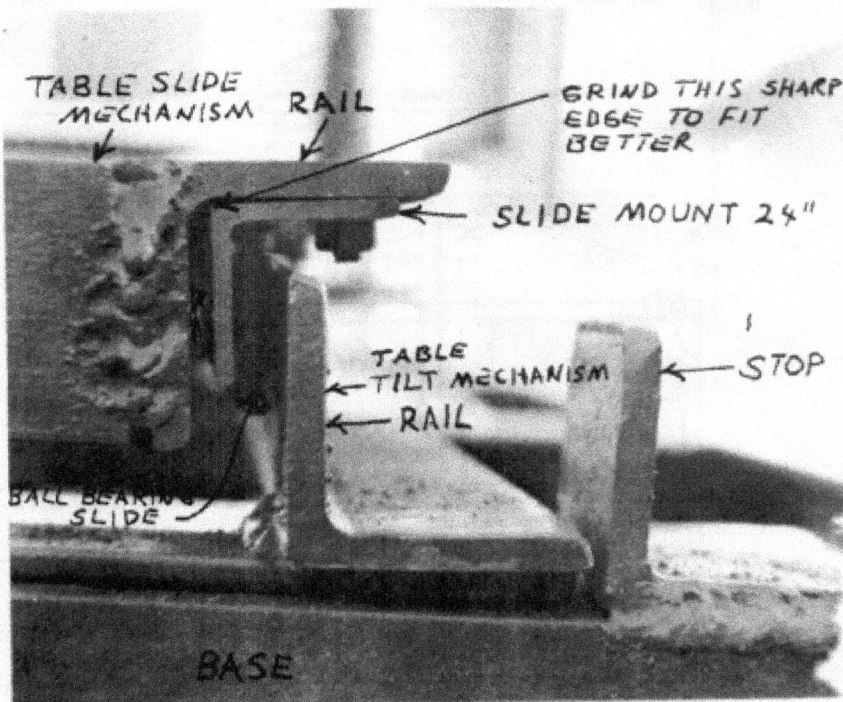

TABLE SLIDE MECHANISM   RAIL

GRIND THIS SHARP EDGE TO FIT BETTER

SLIDE MOUNT 24"

TABLE TILT MECHANISM

RAIL

STOP

BALL BEARING SLIDE

BASE

This completes the first section of our concentration table. Issues to come will contain (2) table top details, (3) motion generator, (4) water flow mechanism details, (5) ore feeding mechanism, (6) secrets of an expert table operator. Continued next issue.

PART II

WE HAVE ADDED two turn buckles to help hold the TABLE TILT MECHANISM down in the back. These turn buckles are at each end of the BASE and attach from each BASE back leg to the TILT MECHANISM. They clamp the TILT MECHANISM securely up against the STOPS.

WE HAVE ADDED four grease fittings for the ball bearing slides. Drill one 11/32" hole at about the middle of the each slide in the TILT MECHANISM rail. Be careful not to ruin the slide bearings (the hole must go through the rail and through one side of the ball bearing slide). Screw in on 1/8" straight grease fitting in each of these holes. Buy the grease fittings from any automobile parts store. Ask for self-tapping grease fittings. Merely drill the hole and screw in the fitting.

Part 2

As you know last issue we covered the BASE, the SLIDE MECHANISM, and the TABLE TILT MECHANISM. This issue we will cover just the construction of the table top. It is important that the table top be constructed properly and that it be built waterproof. A properly constructed table should last for years.

No doubt you have visited one milling site or another where you saw an old table warped beyond use. This is because, first the water got to it and then the dry elements got to it to cause dry rot. There are very few table tops that will not remain in good condition during the first use even if it is used for several months the first time. The major problem usually occurs when the table is turned off. If the water got to it, it will warp as it dries. And if it is not properly finished, the dry rot will set in. Then, of course, the sun will destroy almost any table if the table is allowed to sit unattended in the sun.

A table is usually a reasonably expensive investment. It must be treated properly. However, if the table is built properly to begin with, it will last a great many more years even when mistreated. So do not try to shortcut the finishing and preparation stages of the table top.

We used T&G subflooring plywood for the table top. Buy enough 1 1/8" T&G subflooring and other wood from any good building supply store to make the following cuts:

1 piece T&G subflooring 1 1/8" thick by 35"x83"
1 piece 1x2 lumber cut to 90"
1 piece 1x2 lumber cut to 42"
1 piece 2x6 lumber cut to 90"
1 piece 2x6 lumber cut to 42"
20 pieces of Bay Shoe 72"x1/2"x3/4" if you have not already bought them. Ask for them at any good building supply store. Plus the following parts:
10 TEE nuts 3/8"
10 bolts 3/8"x 1 1/4"
10 lock washers 3/8"

Notch the two pieces of 2x6 along the entire length with a 3/4" by 1 1/2" notch as shown.
Before going any further, coat all wood surfaces with water sealer. Be sure that the water sealer soaks into the wood as well as possible. See instructions below. Use Varathane diluted 50% with denatured alcohol or Thompson's Water Seal. We used Varathane 93 Satin, but from past experience Thompson's Water Seal works as well. The reason that the Varathane is diluted with alcohol is to cause it to soak into the wood making a much better seal than it would be Just on the surface. Be sure to coat the table ribs (made in the last issue) with Water Seal. The best way to do the water sealing is as follows:
(a) Buy a piece of 12'x12' polyethelene plastic film such as a drop cloth for painting.
(b) Lay this film out on a flat surface and put a 2x4 under each side. This then forms a shallow pool.
(b) Pour your Varathane mixture or Water Seal into this pool until the liquid is 1/8" to 1/4" deep. Lay in the TABLE TOP and the RIBS and all the other pieces of wood required for the TABLE TOP (see List of Materials). The TABLE TOP piece itself should be placed into the liquid face down. Then finish up by pouring the rest of the gallon of mixture over the pieces of wood.

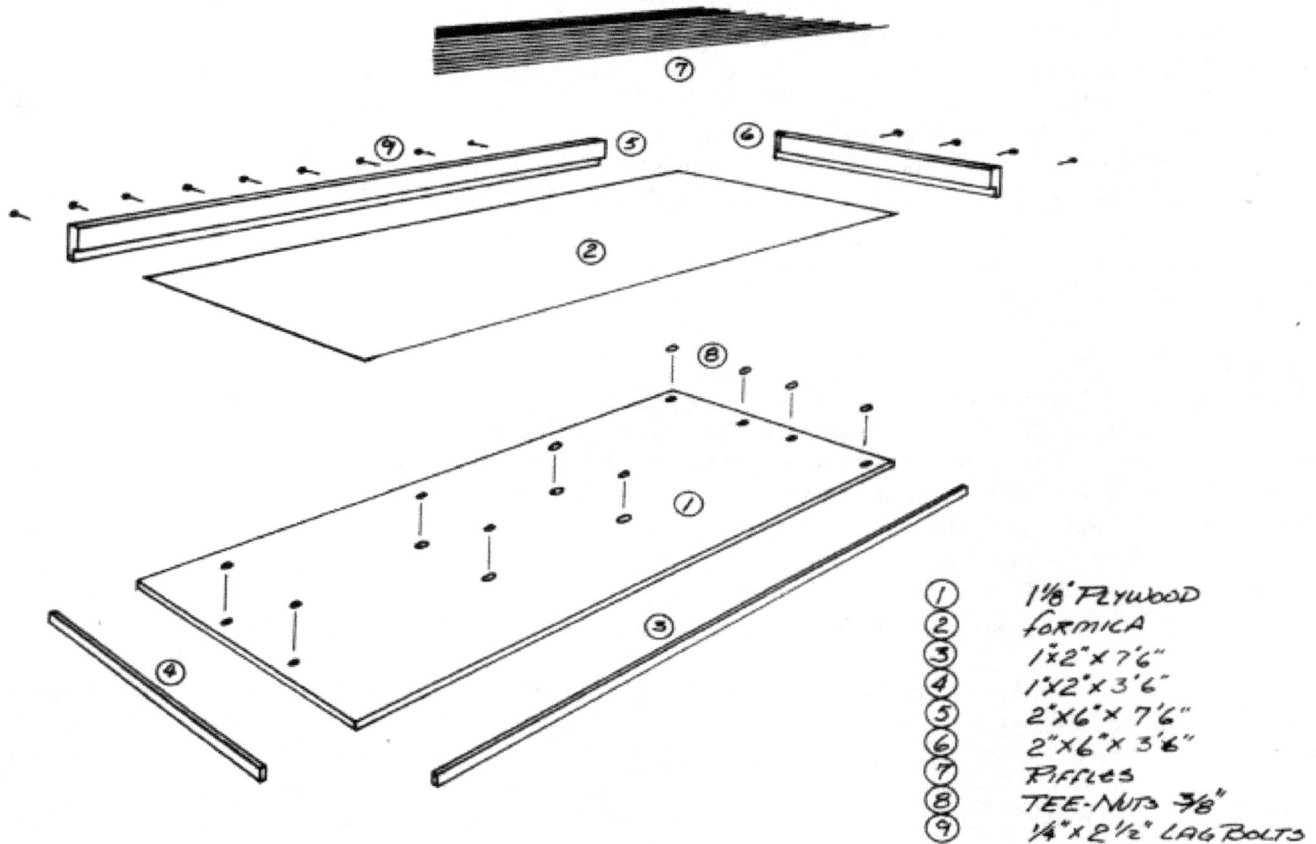

Legend:
1. 1 1/8" PLYWOOD
2. FORMICA
3. 1"x2"x 7'6"
4. 1"x2"x 3'6"
5. 2"x6"x 7'6"
6. 2"x6"x 3'6"
7. RIFFLES
8. TEE-NUTS 3/8"
9. 1/4"x 2 1/2" LAG BOLTS

(c) Allow the wood pieces to soak for a period of 12 hours coming back every few hours to turn most of the pieces over to lay them back into the mixture. Be sure all the pieces are well soaked. The TABLE TOP piece should be turned over at least twice.

d) Remove all pieces of wood from the mixture and stand them on end for drying.

CUTTING THE RIFFLES
Please refer to prior pages for instructions on cutting the TABLE TOP RIBS. Cut the ribs before waterproofing.

MOUNTING THE TABLE TOP
(1) Place the i 1/8" plywood piece on the TABLE SLIDE MECHANISM. Pull the plywood forward towards the end where the MOTION GENERATOR connects until the plywood extends about 1/2" beyond the TABLE SLIDE MECHANISM. Make sure that the plywood piece is centered above the table BASE. It should extend on each side equally.

(2) Now that you are sure that the plywood piece is placed exactly as given above in step (1), take a can of spray paint and spray the underneath of the plywood piece at each cross bar on the TABLE SLIDE MECHANISM. This marks exactly where these cross bars will mount to the table.

(3) Turn the plywood piece over and drill two 1/8" holes in the plywood at each cross bar. These holes should be 2 1/2" from the outside edge of the plywood.

(4) Place the plywood piece back on the SLIDE MECHANISM and set it in exact position by use of the spray paint shadow. Then using the holes drilled in the step above, drill 1/8" holes in the cross bars of the SLIDE MECHANISM. *Note: The plywood piece which is to be the TABLE TOP must be bolted to the SLIDE MECHANISM securely. 3/8" TEE nuts (which have a 1 1/8" flat surface) are used for this purpose.*

(5) On the top side of the TABLE TOP use a 1 1/8" spade wood bit in your electric drill to drill a hole 3/8" deep at each point where you have drilled the 1/8"

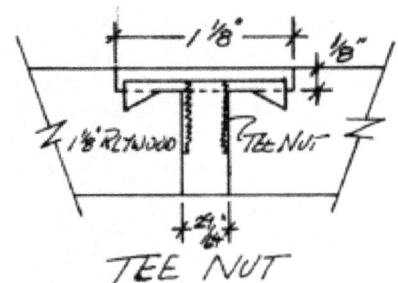

TEE NUT

holes. This is the hole that the TEE nut will set in. Be sure that you complete this step before going to step (6); since, once you have drilled larger holes the spade bit can no longer be used.

(6) Drill each of the 1/8" holes in the TABLE TOP out to 29/64" or 1/2" if you don't have the 29/64" bit. Now the TEE nuts will fit into the TABLE TOP and set below the surface of the table.

(7) Drill a 3/8" hole in each of the 1/8" holes drilled above in the cross bars on the SLIDE MECHANISM.

(8) Put the TABLE TOP back in place on the SLIDE MECHANISM and bolt in place using a 3/8" by 1 1/4" bolt with washer into the TEE nuts. This is two bolts in each cross bar or a
total of 8 bolts. Be sure that the bolt does not extend through the TEE nuts past the table top.

(9) Epoxy glue each of the TEE nuts in place. Pour the 1 1/8" hole drilled for the TEE nut full of epoxy after first soaping the end of the 3/8" bolt to prevent the epoxy from sticking to the bolt. Sand the epoxy flush with the table top.

(10) Refer to drawings. Epoxy glue and tack with nails the two 1x2 lumber pieces on one side and one end. This is the open side and end of the TABLE TOP.

FORMICA TABLE TOP FINISH

There are many types of table top finishes to choose from. Some choose rubber which is quite expensive and some prefer wood and others prefer plastic. We have found that formica is a good plastic and it seems to give the gold a very good surface to ride against. We have no idea if formica is the best or if it is better than rubber, but the fact is that we have used it successfully. We have also found that gold is easy to see against a white background.

Buy a 4' x 8' piece of non-glossy white formica and a gallon of epoxy. The epoxy should be standard epoxy used for fiberglass work. It should not be the finish coat but the epoxy that is used with the fiber glass. When mixing the hardener with the epoxy be extremely careful to mix well. A little extra mixing could save you a great deal of trouble. If the epoxy is not mixed properly it will not harden. Use an epoxy that sets up in two toeight hours, but nothing that sets up quicker than that.

(11) Cut a piece of white formica about 1/8" larger than the table top. (If you have a router a formica cutting blade is best; otherwise, you can buy a formica trimmer for several dollars. To use a formica trimmer you must score along the line you wish to cut and then break the formica. Be sure to score very deeply running the trimmer along the line 10 or 12 times or more.)

(12) Mix about a quart of epoxy and cover the TABLE TOP with it. It should be about 1/16" thick on the table. Next mix a somewhat less amount and paint the bottom of the formica piece.

(13) Place the formica on the TABLE TOP and carefully position in place. There is plenty of time to get it right so be sure it is positioned correctly. If everything is not exactly square do not try to trim at this time. The formica should extend over the edge on the open side and end of the table at least 1/8". Now, weigh the formica down with plenty of weights for a period of 24 hours. Be very careful to use enough weights so that the formica is exactly level especially at the edges of the table.

(14)   After 24 hours cut two pieces of formica, one piece 2"x42" and one piece 2"x90". Using epoxy, cement these pieces in place over the 1x2 lumber pieces at the open side and end of the TABLE TOP. It will be necessary to clamp the formica in place while the epoxy is hardening.

(15) Go over the whole job to this point looking for cracks and holes that water could possibly seep into. Fill these cracks and holes with silicon rubber from the hardware store. Buy a large tube of silicon rubber to use with a caulking gun. You will need it. Next go up underneath the table and caulk along the lower edge of the wood at the point where the formica extends beyond the wood with the silicon rubber. The water will run down the formica and off into a plastic gutter. If you have caulked the back side of the formica strips the water will not be able to splash, or drain, or seep uphill and get under the formica.

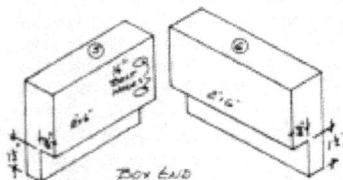

Box End

Sand the formica top flush with the edges of the table. Keep a careful lookout for holes and cracks that water could get into. Force silicon rubber into any and all such holes and then wipe the excess silicon away with a paper towel before it dries.

THE BOX END PIECES are the 2x6 lumber. One piece is 3'6" and the other is 7 '6".(17) Set the two box end pieces in place and clamp them or tack them so they will not move when drilled.

(18) Drill one 3/16" hole 2 1/2" deep every 12 inches through the 2x6 into the TABLE TOP edge. Mark the BOX END PIECES to be cut so that they fit the table top exactly.

(19) Remove the two BOX END PIECES and drill out each 3/16" hole to 1/4" in the BOX END PIECES only, do not touch the holes in the TABLE TOP edge. Cut the BOX END PIECES where marked so that they will fit the TABLE TOP.

(20) Cover the inside of the BOX END PIECES with formica. This is the same side that has the notch lengthwise. Use epoxy to glue the formica in place. When installed the long piece will overlap the shorter piece so leave 1 3/4" wood bare (not covered with formica) at the end where it will overlap the shorter piece.

(21) Put a liberal bead of silicon rubber along the edge of the TABLE TOP where the BOX END PIECES will be installed and at the one point where the long BOX END PIECE overlaps the short BOX END PIECE. Use enough silicon to completely fill all cracks.

(22) Using flat washers install the BOX END PIECES mounting bolts in the holes drilled for same. The bolts are 1/4" by 2 1/2" lag bolts. Quickly wipe away with paper towels all excess silicon that has squeezed out from under the BOX END PIECES. Do not let the silicon dry before attempting to remove the excess. It will be practically impossible to remove once dry and can possibly ruin the water flow characteristics of the table if not removed.

(23) Drill one hole 3/16" by 2 1/2" deep through the long BOX END PIECE at the point where it overlaps the small BOX END PIECE and into the small piece. Use one of the 1/4" by 2 1/2" lag bolts in this hole.

(24) Go over the entire operation looking for holes and cracks in which you cannot see silicon rubber. Force silicon rubber in every such hole or crack wiping away any excess with paper towels. Run a small bead of silicon rubber in all corners, then wipe away with your finger to make the bead smaller. Leave a small 1/16" deep bead showing in the corners and edges of the BOX END PIECES. Leave no openings for water to get into.

INSTALLING THE TABLE TOP RIBS

(25) The first rib is the shortest rib. All ribs should be the same height at the head of the table. The first rib slopes down to zero height in approximately four feet. We have suggested 3/4" height on the ribs, but some people who wish to run high amounts of ore use ribs as high as 1 1/2". Take your choice; it will still work. Install the first rib 7" out from the side BOX END BOARD (which is at the back of the table). See picture.

(26) Lay the rib in place and mark it with a pencil. Drill four or five holes for 1 1/2" finishing nails while the rib is in place. Remove the rib and smear the underside of the rib with silicon rubber. Put the rib back in place and nail down with finishing nails. *DO NOT TRY TO PUT A FINISHING NAIL ANYWHERE THAT YOU HAVE NOT DRILLED A HOLE THROUGH THE FORMICA. THAT WOULD CRACK THE FORMICA.* Wipe away all excess silicon rubber very carefully making sure that there is no silicon showing anywhere.

Note, the flat side of the rib is towards the backside of the table. Do not leave any silicon rubber sticking out from under the rib as this would spoil the flow characteristics of the ore, gold and water flow.

(27) Install the remaining ribs in the same manner. Each rib is progressively longer than the one before. The distance between ribs is 1 1/2". When everything is dry the TABLE TOP is ready to be used.

This completes the TABLE TOP installment. Next issue we will cover the MOTION GENERATOR. For those who can't find ball bearing surplus slides for the table, Accu-Ride in Santa Fe Springs, California (213) 944-0921 manufactures such slides.

Part III

THE MOTION GENERATOR

The motion generator that we have designed it patterned after the Wilfley motion generating unit. The design is fairly simple. Describing exactly how to make it may not be at simple. If you have any problems in assembly just give us a call or if you wish drop by and see our table.

Before going on with building THE MOTION GENERATOR you must at this point install your table in operating position if possible. If you cannot, that is you are a distance from the mill, keep in mind that the table is going to be a great deal heavier when you have completed this section. 80 if possible go to the end of this section first and install the table.

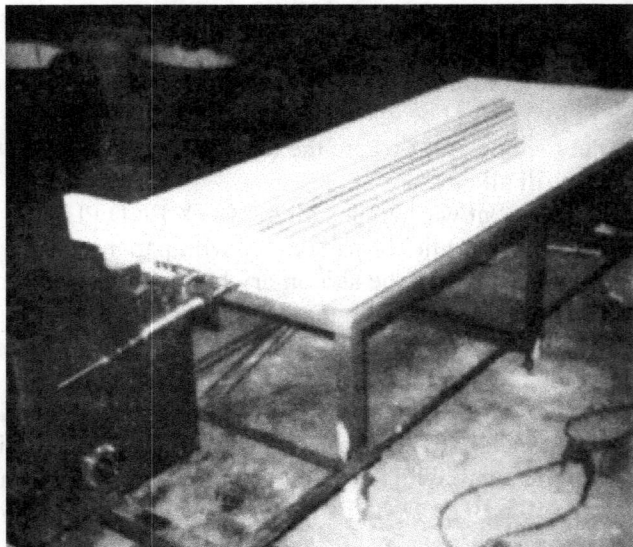

We welded most of the steel but a great deal of the MOTION GENERATOR could be bolted together if you have the time and patience. If that were the case, the one or two welds that are necessary could be done at a welding shop. But do not try bolting this unit together unless you are well experienced at such operations. Too small or too few bolts would prevent the unit from working at all. Better to fumble through welding it, providing you can make substantial welds.

The PARTS LIST is shown on the drawings. Drill the 20"x20"x1/4" plate before welding. The two channel irons #2 and #5 space the plates the correct distance apart. Be sure to drill your holes to fit the 3/4 bore flange bearings. So buy your bearings before drilling the holes.

The gear motor that we have bought to drive the MOTION GENERATOR can be varied in speed from 150 to 800. This has contributed greatly to its high price ($250) used. It would be much more new. We have done this to get a precise speed adjustment in the beginning. It has been our experience that once the speed is adjusted, it seldom, if ever needs adjusting. Thus if you should wish to wait until next issue, we will tell you the exact speed that this table should run. Then you can buy a much cheaper gear motor and just vary the pulley size if you need any real adjustment. But in any case it will be close to 220 and you could still vary the speed with pulleys. So you don't really need to wait.

Once the two 20"x20"x1/4" plates are drilled and welded, then do all the welding as shown in the pictures and drawings. Weld the frame on which everything is mounted to the table legs at the head end.

The motion generating parts themselves are shown on drawing #3. The 1-7/16" cam shaft part #21 impart motion to part #22 pillow block. The pillow block is bolted to angle #23 which is in turn connected through bearings to parts #24 and #25. These bearings are porous bronze bearings and they have zert grease fittings for each bearing. When you install these bearings be sure to install the grease fittings. They can be bought from any auto parts store.

You should have a machinist drill the 2-7/16" cam for the 3/4" shaft and while you are there, if he has welding facilities have him go ahead and weld the 3/4" shaft in place and grind the weld smooth.

The one piece #24 which is part of the motion generating parts is not secured at the rear end. It pushes against the rear end of the MOTION GENERATOR box and is held in place because of the tension of the two RETURN SPRINGS. Piece #17 (on drawing 2) holds this piece (24) in a position that is adjustable and this is what adjust the travel of the Table. The travel is adjustable from 0 to 1-3/4".

The TABLE RETURN SPRINGS are springs from a motor cycle front fork. We used the springs from a 150 dirt bike. Any such springs about 8" to 10" long and about 1-1/4" in diameter should do the trick. They have to be powerful enough to return the table about 1" approximately 220 times a minute.

When everything is assembled you should adjust the tension on the springs so that it easily returns when it is moved by hand, or so that there is a reasonable tension on the springs. Turn the table on and see if it moves smoothly without chatter-

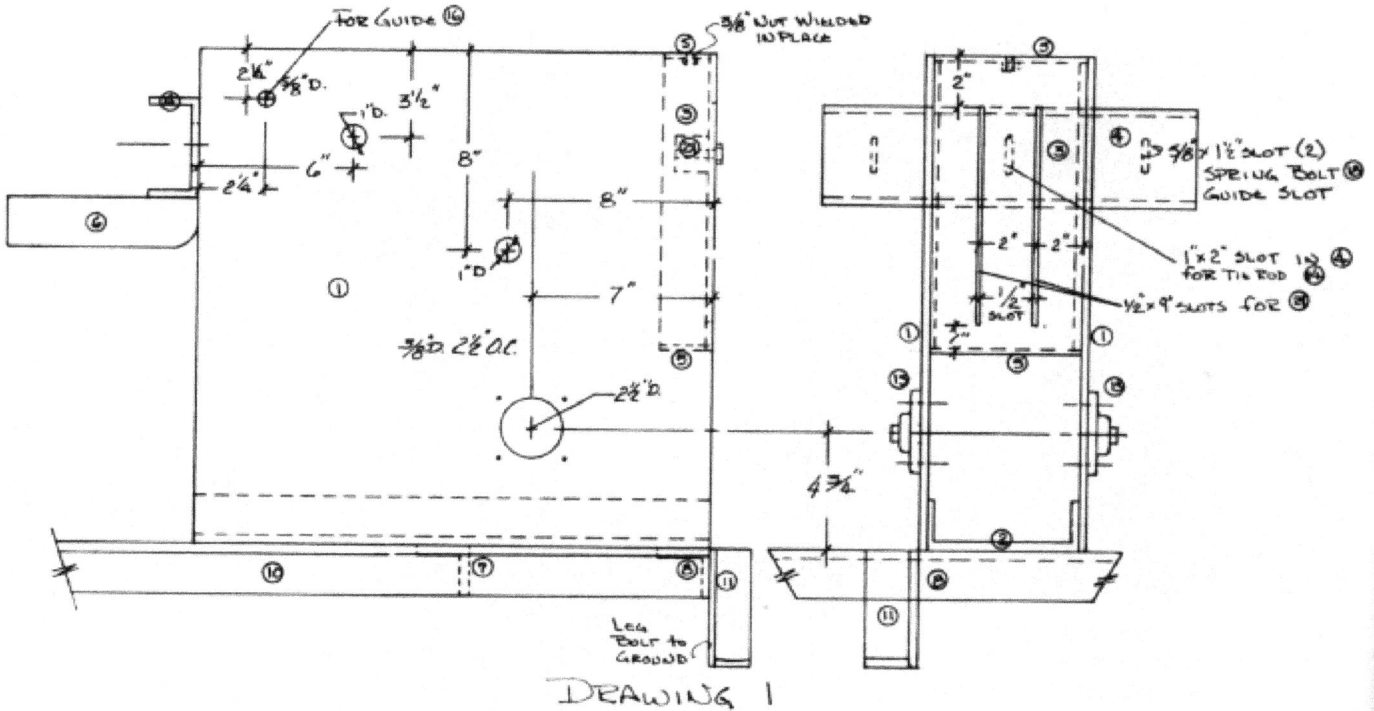

DRAWING 1

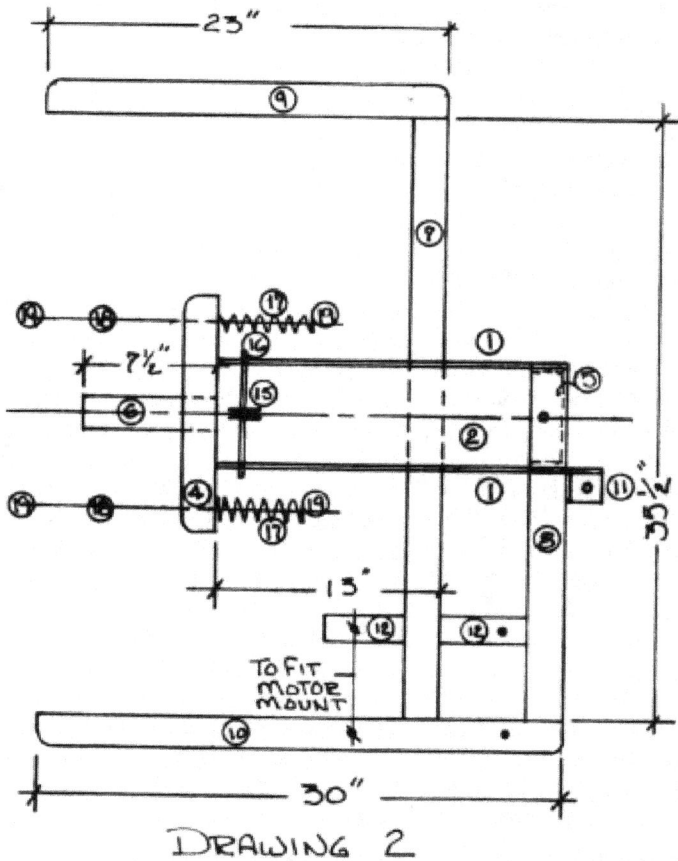

DRAWING 2

① 20" × 20" × ¼"
② 6" × 2" × ¼" × 20" CHANNEL
③ 6" × 2" × ¼" × 12" CHANNEL
④ 3 × 2" × ¼ × 14" CHANNEL
⑤ 2 × ¼ × 6" FLAT
⑥ 2" × 2" × ¼" × 7½ ANGLE
⑦ 2" × 2" × ¼" × 35½" ANGLE
⑧ 2" × 2" × ¼" × 21" ANGLE
⑨ 2" × 2" × ¼" × 23" ANGLE
⑩ 2" × 2" × ¼" × 30" ANGLE
⑪ 2" × 2" × ¼" × 8" ANGLE
⑫ 2" × 2" × ¼" × 6¾" ANGLE
⑬ ¾" BORE FLANGE BEARING
⑭ ¾" TIE ROD
⑮ 2" PULLEY WITH BEARING
⑯ ⅝" SHAFT 8" LONG
⑰ SPRING
⑱ ½" ALL THREAD 18" LONG
⑲ ½" NUTS AND WASHERS

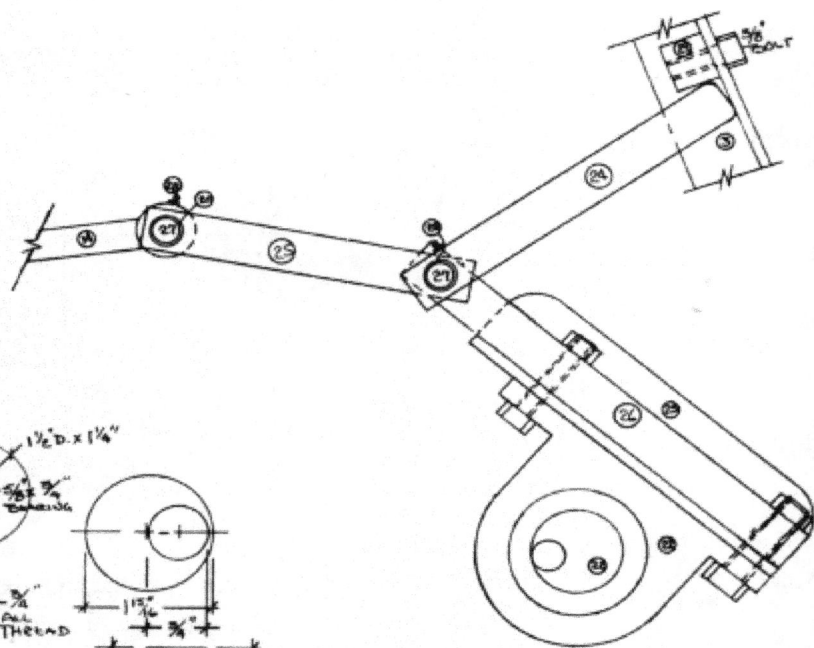

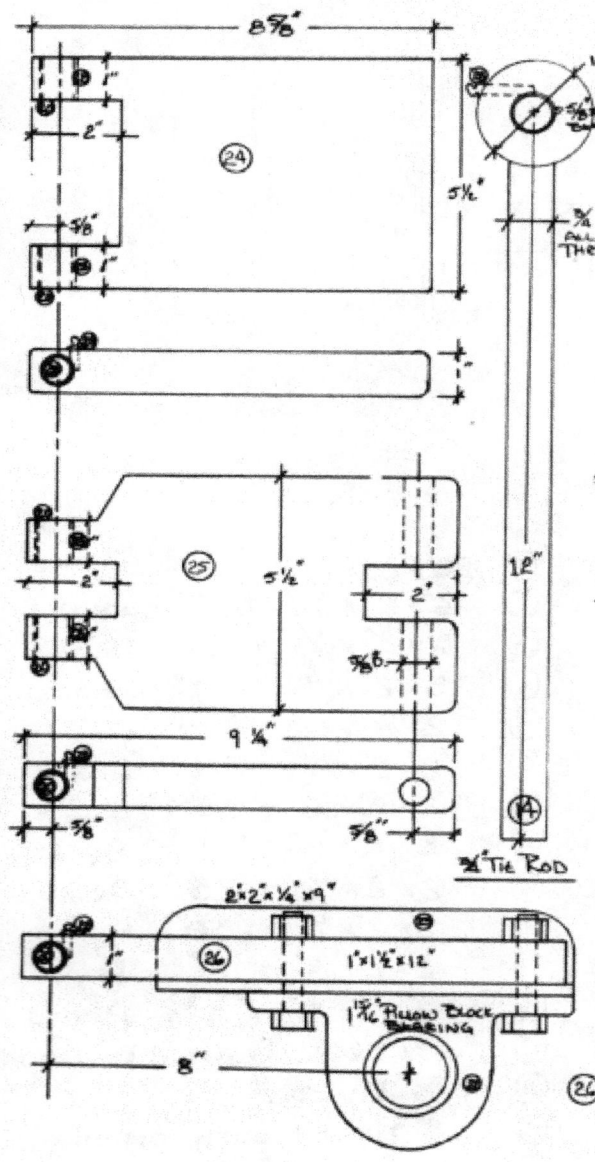

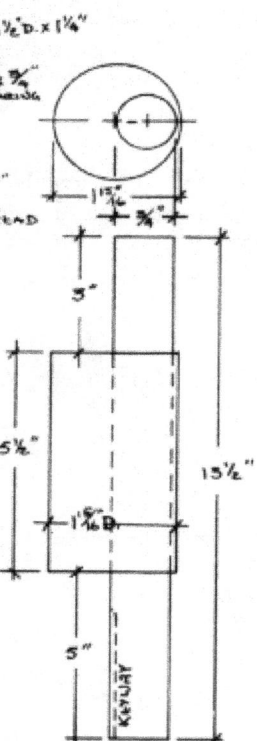

23 ⟨5/8" I.D x 3/4" O.D OIL LITE BEARING
20 ⟨1"⁵⁄₁₆" CAM ¾" SHAFT
22 ⟨1"⁵⁄₁₆" PILLOW BLOCK BEARING
23 ⟨2"×2"×¼"×9" ANGLE
24 ⟨1"×5½"×8⁵⁄₈" ALUMINUM
25 ⟨1"×5½"×9¼" ALUMINUM
26 ⟨1"×1½"×12" ALUMINUM
27 ⟨5/8"D. SHAFT 5½" LONG
28 ⟨GREASE FITTING
31 ⟨1"×1½"×5½" ALUMINUM

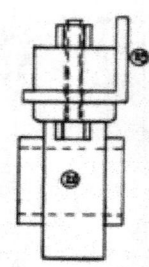

ing of piece #xx mentioned above. The point where this piece pushes against the back plate should not chatter. If it does increase spring tension.

Next place a coin, a quarter or nickel, on the table. It should under dry conditions move right down the table from one end to the other in about 30 seconds. If this does happen you are ready for the next installation. That will be next issue covering the connection and distribution of water to the table.

INSTALLING YOUR TABLE where you plan to use it. The installation should be well done thus eliminating many problems that can crop up from poor installations. We have provided no adjustments for end to end tilting. We do have the side to side adjustment as part of the table but you must provide end to end tilt by your installation. This is done as follows:

You will need six 18"x1/2" threaded rods, twelve 1/2" nuts and 12 large flat washers. Set the table in the exact location that you plan to install it. Mark the position of each leg of the table. Remove the table and dig a hole at each leg approximately 12" deep and 12" wide. Put a small piece of canvas or cloth at the bottom of each hole.

Tip the table on one side and install three threaded rods in the feet in the hole provided (from issue #1). Use the nuts and large flat washers for this purpose. Let each 18"xi/2" threaded rod stick 12" below the foot of the table. This will leave six inches of rod above the foot for adjustment of table tilt as necessary. Of course, you will never need six inches, but it seems like a good figure). Tilt the table (on the three rods) in the opposite direction and install the other three threaded rods. Pick the table up and set it in the six holes you have dug. Adjust the table somewhat level and fill each hole with cement. Let the cement harden. Now you are ready to adjust the table for tilt.

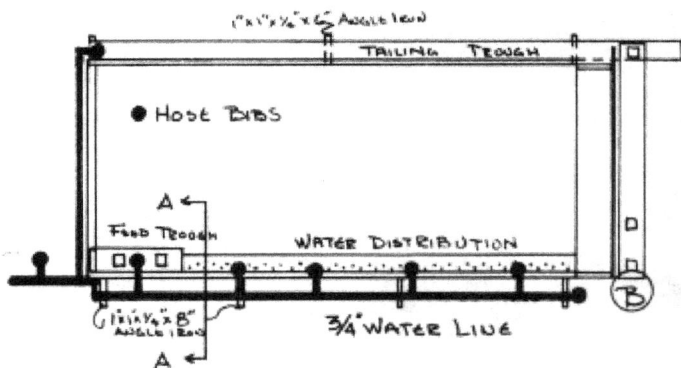

Finishing The Table - Part IV

This is the final article on building the Concentrating Table. It looks very good now. We are pleased with our table. It is being set up at this time to be fed directly by the Ball Mill also shown in this issue. We will have an article on operating the concentrating table next month.

As you can see from the pictures, we have used PVC gutter troughs that can be bought at any large hardware store or building supply store. There are a number of different fittings for the PVC gutter troughs. Do not try to buy and use the metal troughs. We tried that a couple of years ago. All of our waterworks were leaking within weeks. We spent most of our time fixing the water ways.

If you have followed our articles, you will see that everything on this table that touches water is plastic with a few wooden pieces that can be replaced. Most of the drawings in this article are self explanatory.

When you are buying the gutter trough be sure to get some corrugated plastic (as shown in the pictures). Put corrugated plastic pieces at the ends and sides as illustrated. This technique has a number of advantages. First it prevents most of the water splatter from going over the sides of the gutters. Secondly, it allows you to put small containers under the various troughs in the corrugations to determine exactly where your precious metals are coming off of the table. You can

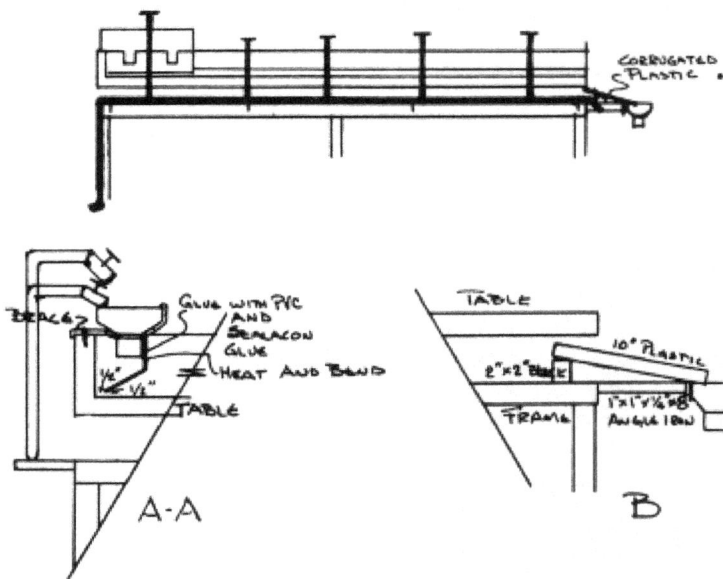

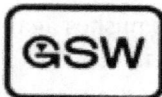

Position the enclosed compression clips in the gutter directly above the fitting seal. Snap clips into position as illustrated below.

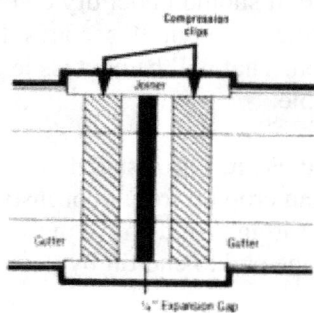

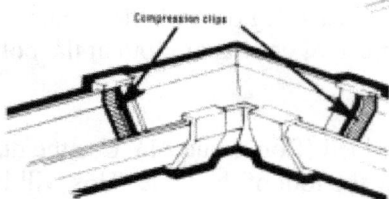
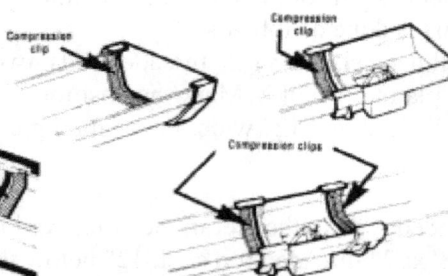

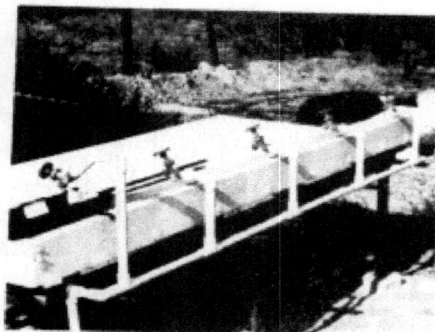

then set your trough down-spouts at the points that you wish to pick off the concentrates. These downspouts can be routed to any lower location that you wish. We think that you will find this an excellent method for handling the water flow from your table.

When installing the water distribution trough that lies along the top edge of the table, we used a little heat on the PVC and bent the trough edge up to give us a V shaped bottom. A number of holes were then drilled in the bottom of this trough. About one 3/16" hole every 1/2". Heat the trough carefully with any heat gun. A good hair dryer would probably work. You can use a torch, but you have to be very careful not to burn the PVC. Use a board to push under the trough to keep it bent up while cooling.

You should have no trouble installing the PVC water pipes and valves as shown. We used 3/4" white PVC pipe and fittings. Metal brackets were welded to the non-moving parts of the frame to hold both the pipe and PVC gutter troughs. Follow the pictures and if you have any questions give us a call. Good luck. Jim

Pump Applications

Typical uses for pumps

1. - Water supply for summer camps
2. - Hi-pressure water for car wash
3. - Circulating water in milk coolers
4. - Pumping septic tanks, cesspools
5. - Making home swimming pools
6. - Draining cow lots, barns, etc.
7. - Emptying washing machines
8. - Irrigating gardens—yards
9. - Weed and tree spraying
10. - Fire fighting protection

Industrial Uses

1. - Circulating machine shop coolant
2. - Circulating air conditioner water
3. - Pumping dyes, soaps, ink, etc.
4. - Pumping brine, syrups, paint
5. - For use as a booster pump

Design Of Pumping System

The selection of pump and pipe for a given pumping application embraces the following steps:

Step # 1 - Determine the amount of water in gallons per minute desired; also the pressure at the outlet.

Step # 2 - Determine the elevation in feet the outlet will be above the water level. Multiply by 0.43 to get pressure required to lift water to the desired elevation.

Step # 3 - Determine (from pipe friction table) the pressure loss in pounds (multiply head loss in feet by 0.43) for the volume to be carried and for the size pipe to be used. For example: Assume 300 feet run, 10 g.p.m. and ¾" pipe. From table, $27.10 \times 3 \times .43 = 34.9$ lbs. pressure required to force 10 g.p.m. through 300 feet of ¾" pipe.

Step # 4 - Add up total pressure requirements to determine pump to use and driving speed. For example:

Pressure desired at outlet	20 PSI
Pressure to lift water 100'	43 PSI
Pressure to overcome friction	34.9 PSI
Lbs. Per Square Inch	97.9

Hence, for the example given, a pump capable of producing 10 g.p.m. at 100 pounds pressure would be needed.

Conversion Table (Head/Ft. to PSI)

Water develops a head pressure at 0.43 lbs. for each foot of elevation. Water in a vertical pipe 100' high would develop a pressure at the base of the pipe of 43.31 lbs. This table can be used to convert from feet of head to lbs. of pressure.

Head/Ft.	P.S.I.	Head/Ft.	P.S.I.	Head/Ft.	P.S.I.	Head/Ft.	P.S.I.
1	0.43	6	2.60	20	8.66	70	30.32
2	0.87	7	3.03	30	12.99	80	34.65
3	1.30	8	3.48	40	17.32	90	38.98
4	1.73	9	3.90	50	21.65	100	43.31
5	2.17	10	4.33	60	25.99		

Suction Line Problems

• the suction pipe line should be made up tight so that there is no leak, as a very small leak will impair the efficiency of the pump or might prevent the pump from lifting water. Pipe lines should be kept as straight as possible, as sharp bends cut down the flow of water.

• In the case of shallow well water system pumps, we recommend that the suction line be the same size as the threaded opening in the pump. Where an excessively long suction line can not be avoided, use next larger size pipe.

• It is desireable to keep suction piping as short and direct as condition permit, and that it slope upward to the pump. Overhead loops create an air trap resulting in priming difficulties.

Centrifugal Pump Performance

• Engine operated self-priming pumps have a maximum output when working against minimum head pressure. For example: a typical 2" pump will put out 6540 gallons per minute at 20 foot head (zero suction), or 2290 gallons per minute at 80 ft. head. The table below shows typical performance characteristics of a 2" self-priming pump.

CENTRIFUGAL PUMP – CAPACITY IN G.P.H.

TOTAL HEAD	SUCTION LIFT IN FEET					
	0	5 Ft.	10 Ft.	15 Ft.	20 Ft.	25 Ft.
20 Ft.	6540	6210	5990'			
30 Ft.	5975	5610	5380			
40 Ft.	5360	5040	4640	4340		
50 Ft.	4700	4410	3980	3600	3190	2763
60 Ft.	3995	3660	3210	2760	2300	1780
70 Ft.	3215	2785	2290	1770	1115	
80 Ft.	2390	1770	1110			

Discharge Of Nozzles (GPM)

The amount of water that will flow out of a hose or pipe nozzle in terms of gallons per minute will depend on the pressure at the nozzle end and its diameter. This is shown for various size nozzles in the table below:

NOZZLE (P.S.I.)	NOZZLE DIAMETER				
	1/16"	1/8"	1/4"	1/2"	3/4"
10	0.37	1.48	13.6	5.91	53.1
20	0.53	2.09	33.4	8.35	72.1
30	0.64	2.56	40.9	10.2	92.0
40	0.74	2.96	47.3	11.8	106.0
50	0.83	3.30	52.8	13.2	114.0
60	0.90	3.62	57.8	14.5	130.0

Common Pump Troubles

Common causes of trouble in water pumps, particularly the centrifugal type, are leaks in the suction piping and humps in the suction line causing air pockets. Even a small air leak will prevent the pump from priming or, if primed, it will soon lose its prime and stop pumping. Air leaks in suction piping can be prevented by the generous use of "Permatex" gasket compound on all pipe and pump joints and by screwing the joints up tight.

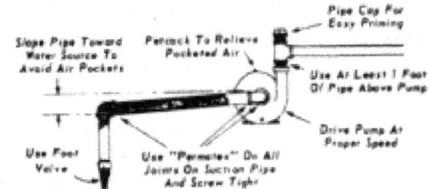

Suction piping should be installed with a slight slope toward the intake as shown to prevent air pockets.

Pumps will give better service if used with a foot valve. When priming a centrifugal or other type pump, fill the piping system and pump completely full of water. Be sure air is not trapped in top of pump.

Packing glands around shafts must be tight enough not to leak, yet sufficiently loose that they do not bind on shaft and cause excessive wear.

PIPE FRICTION LOSS OF HEAD (given in feet) PER 100 FEET OF GALVANIZED PIPE

Gals. per Minute	¼ inch Pipe	⅜ inch Pipe	½ inch Pipe	¾ inch Pipe	1 inch Pipe	1¼ inch Pipe	1½ inch Pipe	2 inch Pipe	2½ inch Pipe	3 inch Pipe	4 inch Pipe	5 inch Pipe	6 inch Pipe
1	28.0	6.4	2.1										
2	103.0	23.3	7.4	1.9									
3		49.0	15.8	4.1	1.26								
4		89.0	27.0	7.0	2.14	0.57	0.26						
5		126.0	41.0	10.5	3.25	0.84	0.40						
10			147.0	38.0	11.7	3.05	1.43	0.50	0.17	0.07			
15				80.0	25.0	6.50	3.0	1.08	0.36	0.15			
20				136.0	42.0	11.1	5.2	1.82	0.61	0.25			
25					64.0	16.6	7.8	2.73	0.92	0.38			
30					89.0	23.5	11.0	3.84	1.29	0.54			
35					119.0	31.2	14.7	5.1	1.72	0.71			
40					152.0	40.0	18.8	6.6	2.20	0.91	0.22		
45						50.0	23.2	8.2	2.80	1.15	0.28		
50						60.0	28.4	9.9	3.32	1.38	0.34	0.11	
70						113.0	53.0	18.4	6.21	2.57	0.63	0.21	
75							60.0	20.9	7.1	3.05	0.73	0.24	0.13
100							102.0	35.8	12.0	4.96	1.22	0.41	0.16
120							143.0	50.0	16.8	7.0	1.71	0.58	0.23
125								54.0	18.2	7.6	1.86	0.64	0.30
150								76.0	25.5	10.5	2.55	0.88	0.39
175								102.0	33.8	14.0	3.44	1.18	0.48
200								129.0	43.1	17.8	4.40	1.48	0.62

NOTE: VALUES ABOVE HEAVY LINE ARE RECOMMENDED FOR NORMAL OPERATION

FOR PLASTIC PIPE MULTIPLY ABOVE VALUES BY .54

FITTING FRICTION LOSS (figures give equivalent in feet of pipe) ADD to PIPE LENGTH

	⅜"	½"	¾"	1"	1¼"	1½"	2"
90° Wrought Elbow	.8	1.0	1.5	1.8	2.4	2.9	3.6
45° Elbow	.4	.5	.8	1.0	1.3	1.6	2.0
Straight Tee	2.5	3.0	3.6	4.6	6.4	7.2	9.6
Gate Valve	.25	.30	.36	.46	.64	.72	.96
Angle Valve	4.0	6.5	9.5	12.0	16.0	19.5	22.5
Globe Valve	7.0	12.0	17.2	22.5	32.0	36.0	48.0

(Tables Courtesy of WAYNE PUMP CO.)

MERCURY POISONING - WHAT YOU CAN DO

by Jim Humble

A great deal more miners have mercury poisoning than they realize. Mercury poisoning can come from working with mercury and it can come merely from siting a non-mercury mine as well as a mercury mine. Mercury poisoning can come from being a standard fire assayer. Anytime a furnace does not have a well ventilated hood that can happen; because, almost all gold ores have some mercury with them. And there are at least 50 non-mining occupa-ons in which people are susceptible to getting mercury poisoning.

Symptoms of mercury poisoning are: insomnia, shyness, nervousness, dizzi-ness, loss of memory, loss of appetite, depression, irritability, anxiety, fatigue, headaches, numbness, muscle weakness, vision difficulties, speech disorders, and then hallucinations, paralysis, coma, and death. These symptoms all apply to other less severe disorders so do not get excited at first if some of them ap-pear. Of course, all of these symptoms will not be present in any one case of mercury poisoning. One symptom is usually nothing to worry about, but sev-eral symptoms can really indicate mercury poisoning in some cases.

There are some medical doctors that have as part of their specialty, the treat-ment of heavy metal poi-soning. Most medical doctors will say that very little can be done about it. But that really is not the case. There are nutrients that can be used to wash out the poison and for a number of years now doctors have used a chemical called EDTA. This chemical is injected di-rectly into the vein in the arm. About one quart of this chemical is drained through a person's veins for a period of several hours. This is done several times a week for a number of weeks depending upon the severity of the problem.

The EDTA is a chemical that chelates the poisons. That is to say that the EDTA surrounds the poison molecules rendering them harm-less. The poison can then be easily washed from the body. The EDTA treatments have been used with hundreds of thousands of pa-tients with a great deal of success. EDTA can also be taken orally. EDTA is a chemical that is inert in the body. It is a substance that has no nutritional value nor any harmful effects. It merely washes through the body and out again. But it does chelate poisons as it goes. If bought from a drug store one must have a prescription. If bought from a chemical supply house one does not need a prescrip-tion. The druggist usually buys his at the chemical supply.

What can you do for yourself? There are a number of nutrients that are known to remove poisonous heavy metals from the body. Sele-nium tablets contain Selenium compounds that complex with mercury making a neutral complex that is easily washed out of the body. Sulfated Amino Acids will make mixtures with heavy metals that help wash out poisons. Zinc tablets contain Zinc compounds that complex with heavy metals to make neutral complexes.

Vitamin C will in some cases neutralize poisons, but its main function is to wash poisons out of the body. Vitamin A will repair dam-age done by poisons. Vitamin E will see that vitamin A gets to where it must be to do the repairs.

Apple Pectin and Sea Kelp Alginate are preventatives. When one is taking these supplements the poisons have a much greater tendency to wash out of the body. These substances also help remove the poisons that are already in the body, but probably their main function is to prevent the poisons from hanging up in the body in the first place.

All of the nutrients mentioned above can usually be bought from any health food store. They can also be bought from GAME Company (advertisers in this Magazine). You must make your own decision on quantities one should take. It is generally accepted in the health food field that 5 times the recommended daily allowance is what is necessary for robust health. When one is trying to handle poisons probably 2, 3 or 4 times above the 5 times would be better.

MERCURY POISON PREVENTION

Mercury is a very poisonous substance if handled incorrectly. Mercury can also be a fairly safe substance to use if handled correctly. I have known a number of persons who have suffered mercury poisoning over the past few years. Questioning of these people and un-derstanding the characteristics of mercury has given us a set of safety rules. All of the people we know that have been convinced that they can circumvent following safety rules have become poisoned with mercury. Only people and miners who have been around mercury who are know-it-alls will try to get by these rules. One of the nicest guys we know lies right now in a hospital at death's door because he broke these rules. That's the way he always handled mercury. It didn't bother him at first, but then all of a sudden it got him.

RULE 1. Do not let mercury touch your skin. The slight acid or alkaline condition of your skin will leach tiny amounts of mercury away and the mercury will absorb into your skin. If you touched mercury when you were a child you will still have that mercury that was leached into your body now 50 years or more later. It accumulates until it destroys the nerves of the body.

RULE 2. Do not leave mercury sitting in open pots or containers. If you do not have a lid for the container cover the mercury with water. Mercury will out gas mercury fumes when it is not covered. Water will catch and prevent this. Mercury fumes are much more poisonous than mercury itself. In fact many people have swallowed mercury without serious effects. The undissolved mercury goes right through the body and out the other end. But that mercury that is dissolved by stomach acids will remain in the body and accumulate. Several times would probably do the job.

RULE 3. Do not under any circumstances allow mercury salts in liquids to touch your skin. A major example is mercury dissolved in nitric acid. This should never touch your skin. There are some mercuric salts that could Kill from a single contact with the skin.

RULE 4. Do most work with mercury out of doors. Never work with mercury in a closed room unless under a ventilation hood. It is likely that you will breathe fumes in a closed room.

RULE 5. When mercury is spilled cover the area with powdered sulfur and then vacuum the entire area (and any small mercury beads). Sulfur coats mercury and prevents out gassing of the mercury fumes.

RULE 6. Never, never squeeze mercury through a chamois using your hands. Wear rubber gloves. And better still use an air press to force the mercury through the chamois.

RULE 7. When retorting mercury do it outside. Do not stand downwind from the retort. Do not work near the retort. Any leak in the retort could cause excess mercury fumes. The fumes are the deadly part of mercury.

RULE 8. Follow the above rules, but do not be afraid of mercury. Respect mercury. If you follow the above rules there is very little danger. Thousands of people have worked with mercury with no trouble. However do not break these rules even once.

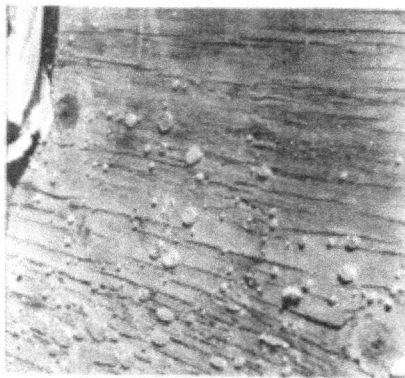

MICRON GOLD AMALGAMATOR

By Jim Humble

We feel that we have come upon a real break-through in amalgamation for the small and medium size miner. Now for the first time a small miner can make a profit on ore containing .18 ounces of gold per ton. And what is more than that, chances are very high that if the ore is assaying at .18 ounces that this amalgamator will get twice that or up to 10 times that. A single amalgamator might not make you rich, but if you own one you should never be broke. We call it the Micron Gold Amalgamator Model 2000 or just MGA 2000.

The Spaniards took shipload after shipload of gold out of the United States. There are old Spanish mines throughout the West and even in the East. There are old Spanish mines in places where no one nowadays can even find gold. Yet there are records of several hundred Spanish Galleons sinking with cargoes mostly of gold. If that many sank, think of how many didn't sink. And remember all the gold that the Spanish took was taken before the California Mother Lode was found. How did they do it? Well, most of that technology has been lost. But we do know that they used mercury. Could they have been getting the elusive micron gold? We think that they probably were. And you can get it too.

What is more than that, we guarantee it. Bring your ore by. We will test your ore for $400 for a 1000 pound test. If you buy a Micron Gold Amalgamator Model 2000 and it does not work on your ore we will refund your purchase price (if you return the machine). We want you to win, not lose. If it doesn't work on your ore it costs you nothing more than the cost of a test. Now, where can you find a guarantee like that in mining? You gamble the cost of a test and your time. When have you not been willing to gamble that? You're in mining, ain't you?

HOW IT WORKS

Ore and water are delivered in the MGA 2000 at the center of a large steel plate. In order to get out of the machine, the ore and water must pass under the plate and through many inches of mercury. This mercury is contacting the steel plate with many square inches of contact. The ore is rolled and pushed and moved through the mercury until it comes out from under the plate. In order to do this it must travel through at least 10 inches of mercury and in many cases as much as 20 to 30 inches of mercury.

We don't know the full theory of why amalgamators work and why this one works. But we feel that in addition to just plain amalgamation there may be a certain amount of electrical amalgamation. As any high school science book will tell you when two dissimilar metals contact one another in an electrolyte a voltage is developed. And this is true of this or any other iron amalgamator. But in this amalgamator (MGA 2000) the area where the mercury and iron and ore all contact at the same time is many times greater than any other amalgamator. Does the small voltage developed at the exact spot where the three come together help the amalgamation of the gold? We can't guarantee it, but we think it does.

In a standard amalgamator the ore is ground into the mercury. In the MGA 2000 this is not the case. The ore is forced through the mercury but no grinding takes place. In a standard amalgamator the water sits on the mercury and there is nothing to force the water through the mercury, but in the MGA 2000 the water must go through the mercury to get out. Because no grinding of the ore takes place there is much less flouring of the mercury.

William McDanel the inventor of this MGA 2000 spent 1000's of hours working with this machine. He wanted it easy to use and the MGA 2000 is built so that it is easy to use. As you can see in the pictures, the drive mechanism tilts up out of the way when you are ready to remove the mercury. The machine can be cleaned and mercury removed in less than ten minutes. Merely tilt up the drive mechanism, lock in place, slide the tub forward against the stops, tilt the tub and remove the mercury plug. Out comes the mercury into your bucket in front of the machine. You don't even have to climb under the machine.

Another unusual feature of this Micron Gold Amalgamator is the amount of mercury that it uses. A full charge of mercury is about 12 pounds. That's about l/5th as much as the standard amalgamator that handles the same amount of ore. This makes cleanup easier and recovery of the gold from the mercury easier.

Now with Action Mining's new mercury technology many ores can be amalgamated that could not be amalgamated in the past. This includes sulfide ores as well as other complex ores. Acid ores require one kind of treatment and alkaline ores require another kind of treatment, and with the proper handling, many ores can be amalgamated. Of course, there will always be that terrible ore that cannot be handled, but that one, we expect, can be concentrated and leached.

The price of the Micron Gold Amalgamator (MGA 2000) is $9,995. Sounds kind of high at first glance. But how many pieces of mining machinery have you seen sitting around not being used? What are you willing to pay for a machine with an unconditional guarantee that it will work on your ore oryour money back. Test it first here at our place. You will Know that it is going to work when you hook it up. If not bring it back for a refund. $9,995 for a machine that works and does the job is cheap. The cost of the machine includes two days training on how to handle it and how to handle, retort and clean mercury.

The capacity of this Micron Gold Amalgamator varies with different ores. We specify this machine at one ton per 8 hour shift. Now that's a lot lower than the specifications of most amalgamators of equal size and most of them grind the ore as well. We bought an amalgamator several years ago that was supposed to do a ton per hour, it wound up doing 80 pounds per hour. We guarantee ours to do what we say, but we say something reasonable. And you are going to get more gold per ton than any other amalgamator that we know of. But the thing to do is test your ore and find out. We have used this Amalgamator at 4 tons per 8 hour shift and removed all of the gold that we expected to get, but all ores will not be this simple. But we only guarantee one ton per 8 hours. You probably will be able to do better. Yes, you can put one ton per hour through it, but you won't get complete amalgamation either,

If you have ore with gold in it there is no reason for you to not have that gold. All you need for full production is a ball mill and our Micron Gold Amalgamator (MGA 2000). Well almost, you might need a jaw crusher. Gold recovery can be simple. Let us show you how simple. Call Action Mining Company (619) 372-5850 P.O. Box 533, Trona, Calif. 93562

INSIDE OF AMALGAMATOR...

CLEANING THE MICRON GOLD AMALGAMATOR TAKES 10 MINUTES...

SUPER HOT TORCH FROM PIPE

by Jim Humble and Mike Glenn

We have tried a number of torches using propane, butane and diesel. Most of these torches require special nozzles for different gases. Most of them are somewhat touchy in their adjustments. But this torch we are showing you here is no problem at all and it will work with propane, butane, acetylene or any other gas that you can come up with. It will also work with diesel if you allow it to get hot with a gas first.

This torch requires a source of air. We have used vacuum cleaners connected backwards to produce a good flow of air. One can usually buy an old vacuum cleaner that runs but is otherwise broken for $5 to $15 dollars from any vacuum cleaner repair shop. We have never paid more than $5 for such a unit. Remember the torch needs air not a vacuum. So the vacuum must be hooked backwards to produce the air. If you explain this to the man at the repair shop he will usually find you one in his junk pile. The tank type vacuum seems to be one that does the job well, but any vacuum that blows will work.

We have used both 1" and i 1/2" pipe for the throat of the torch. But 1 1/2" seems to be best. Probably you could use 2" or 3" pipe depending upon how much ore that you intend to smelt. But this 1 1/2" torch will smelt over 50 pounds of ore and 100 pounds of flux at one time.

This torch can be used to fire any size furnace up to a 55 gallon barrel furnace. How to make the furnace, if you do not have one, will be covered in the next issue of this magazine.

BUILDING THE TORCH

The picture is mostly self explanatory except for a few details. Use pipe threading compound to put all of the joints together. Of course, the gas goes into the small pipe in the rear and the air goes into the big pipe that goes to the tee. The air valve is a gate valve to allow maximum air to flow when necessary. The small valve is a needle valve. Be sure to buy a needle valve that will take up to 80 pounds of pressure if you plan to work directly from a butane tank with no regulator. It is best, of course, to use a regulator in which case you could buy a cheaper needle valve.

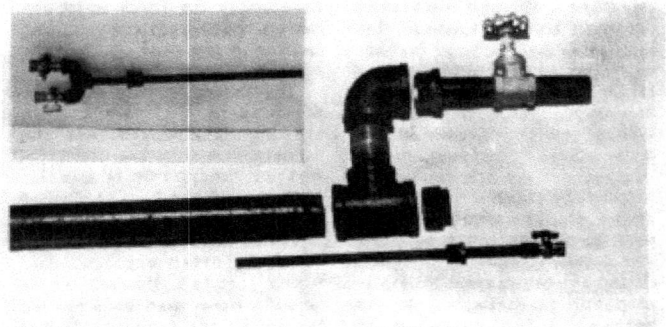

SHOWS BOTH DOUBLE INPUT DIESEL FEED & SINGLE INPUT PROPANE FEED...

**Now this is the important part.** Be sure to allow the inside l/8th" pipe to extend at least 6" to 8" beyond the tee where the air enters the torch throat. This is what gives the proper air gas mixture. The 1/8" pipe should end about 1/2 the way down the torch throat for the torch shown in the picture. You may wish to make the torch throat longer to reduce the temperature back in the valve area. In this case leave the l/8th" pipe the same length and just extend the 1 1/2" pipe throat 5" or 10".

There are several ways to get the inside pipe to fit in place properly. The important pipe piece that does this job is the 1/2" bushing that fits into the i 1/2" bushing that fits into the tee. (This is the 1/2" to 1/8" bushing) This bushing is not being used as pipe bushings were made to be used. There is a 1/8" pipe fitted into each end of this bushing. One pipe goes to the gas needle valve and the other goes inside the torch throat. Some bushings are threaded so that a pipe can be screwed in both ends. (These bushings were not meant to be made this way - it just happens because the guy that cut the threads cut them a little deeper than necessary.) Most big hardware stores will be glad to cut the threads in the opposite side of the bushing. If not then just try a number of bushings until you find one that a 1/8th" pipe can be screwed into from both directions.

After the pipe torch has been all screwed together tightly using pipe thread compound or teflon tape it is ready to use. The torch must fire into some Kind of a chamber. Round smelting furnaces are often used. One can use a small furnace the size of a 5 gallon bucket or one as large as a 55 gallon barrel. The chamber should be made of fire brick or of refractory materials.

LIGHTING THE TORCH

The above picture shows a good method of lighting the torch. The pipe with the bottom end closed off has several inches of Kerosene or diesel oil in the bottom. A 1/4" threaded rod with several rags tied to the bottom is kept in the diesel oil. A large jar cap is punctured and mounted at the top of this rod to provide for a hand shield and

TORCH LIGHTING DEVICE...

to act as a lid for the pipe when the rod is in place. When one is ready to light the torch one removes the rod with rags, lights the rags with match, then used the lighted rags to light the torch. This leaves one far enough back from the torch that flare ups will not singe him. As soon as the torch is lit one merely replaces the burning rags into the pipe where they are immediately extinguished.

The second way to light the torch is to just throw some burning newspapers into the furnace and turn the torch on. Not as fancy, but it usually gets the job done. The lighted rod is the way to go for anyone interested in production. Furnaces often explode when being lighted. Usually it is nothing more than blowing the lid off of the furnace. But it can be worse or someone can be hit by a flying lid (UFO). The way to prevent this is to make sure that a flame exists in the furnace before and during the time that the gas is turned on. If you forget and allow the furnace to fill with gas before throwing in the flame it is sure to go bang. *SO DON'T FORGET, THE FIRE MUST BE THERE FIRST.*

When using diesel to fire the torch, the torch must be fully hot. This usually takes about 1/2 hour of operation and in some cases it might be as long as an hour. The diesel must enter the torch in the same area that the gas enters the torch. The way we did this was to put a 1/8th" tee at that area. Then put the gas in from the top of the tee and the diesel from the bottom. If the diesel should come in from the top it would drain down into the gas tube. Use a needle valve for both the gas and the diesel. Always turn off the gas before turning on the needle valve.

This torch must always fire into a chamber of some kind. It is not efficient in the open air. The torch requires a very slight back pressure. This back pressure is not enough to detect but it is required for proper operation.

Keep in mind that all torches are dangerous to operate by those people who are not familiar in their operation and by those people who are not careful. If you have not operated a torch before then work with someone who has. There are thousands of things that can go wrong with any kind of equipment new, used, or home built. Anyone building this torch or any other equipment in this magazine must do so totally on his own responsibility. END

Building a Furnace Part II

by Staff

This is the second of a series of articles on furnaces and smelting. Fire assaying will be covered in other articles. This article will cover a very inexpensive furnace that you can build for less than $20. And including the torch given in the last issue and even a propane tank you should be up and going for less than $100 dollars.

Next issue we will cover a somewhat larger furnace that you can build for smelting concentrates. And in the issue following that we will cover a very large but still cheap furnace for smelting over one hundred pounds of concentrates plus necessary fluxes for a total of 300 pounds. This furnace will provide a method of lifting, tilting, and pouring.

Keep in mind that smelting is very dangerous to the inexperienced or the uncareful. Popular Mining takes no responsibility for accidents, inexperience of the operator, or for bad design. If you elect to build any project by Popular Mining it is assumed by Popular Mining that you have the sense to evaluate your own expertise, evaluate the dangers of the particular design, and handle all with the necessary precautions.

THE 5 GALLON POT FURNACE

The POT FURNACE is designed to work with the pipe torch given in the last issue. It will hold a crucible big enough to handle about a quart of smelted materials. It can attain temperatures as high as 3000 F.

STEP BY STEP INSTRUCTIONS

(1) Obtain a 5 gallon steel bucket. Almost any 5 gallon can will do. A junk yard would be the most likely place, but many service stations would have an empty oil or grease bucket lying around. Get an extra can or two if you can for mixing the cement in. Clean the cans of oil or grease.

(2) Drill a 1" hole in the middle of the bottom and a 2" hole 2" up from the bottom in the side of the 5 gallon bucket. Use a hole saw for metal for this purpose.

(3) Mixing the fire clay - Fire clay can be purchased from almost any building supply store. It is used for fire places. The mixture will be one part fire clay to two parts 60 mesh silica sand also bought at building supply stores. If you have a cement mixer you will need 5 gallons of silica sand and 2 1/2 gallons of fire clay. If not, you will have to mix smaller quantities several different times to complete the job.

You will need about 6 quarts of water. Add water until the mixture begins to ball up in the mixer then add just enough more water to stop the balling up. If you are mixing by hand add enough water to allow the mixture to be packed easily by hand but not enough water that will cause the mixture to sag when stacked against a wall.

(4) Buy a piece of 8" and 12" flue pipe (stove pipe) from a hardware store. Cut a notch 2" deep by 3" wide in one end of the 8" flue pipe. This is easily done while the flue is still a flat piece of steel before it is pushed together to form the pipe. Cut a 3" strip off of the 12" flue pipe to make a 3" by 12" circle. (See drawing)

(5) Pack the bottom of the 5 gallon bucket up to the bottom of the hole in the side with the fire clay mixture. This gives you a 2" deep bottom fire clay mixture. Push a hole through the middle with your finger at the point where the 1" hole has been drilled in the 5 gallon bucket.

(6) Put a 1 1/2" pipe (just about 2" outside diameter) into the 2" hole and then move it so that it bends the hole slightly. If the pipe were a torch it must point in on an angle so that the flame will swirl in a circle. (See drawing)

Furnace parts

Packing with fire clay

(7) Put the 8" flue pipe into the 5 gallon bucket so that the notch made in step (4) fits over the 1 1/2" pipe. Position it so that it is centered in the bucket.

(8) Pack the area between the flue pipe and the bucket sides with fire clay mixture. Pack it so that the mixture comes up and is even with the top of the bucket. Let it dry for a few minutes and then remove the flue pipe and remove the 1 1/2" pipe in the torch hole.

(9) MAKING THE LID - Lay the 3" high by 12" diameter circle made in step (4) on a flat surface and fill it with fire clay mixture. Imbed a 4"x1/2" bolt in the top of this lid. This bolt with be used to pick up the lid when it is hot. Make a hole 2 1/2" in diameter in the center of the lid. This is a vent hole and is also required to observe the smelt.

Refer to the picture. The flat rocks setting on top of the furnace lid are molded from clay mixture. They are used for setting crucibles on in the furnace during smelting or assaying. Make up a number of them for your own use in the furnace.

FIRING THE FURNACE

The furnace must dry at least 24 hours before firing. The safest way is to let it dry for a week if possible. When ready to fire» take the pot furnace outside and set it on bricks so that it is 2" or 3" off the ground. Prop the torch on several bricks so that it will fire into the furnace level but on an angle to produce a swirling action of the flames. (Same as the pipe in STEP (6)). The torch should not extend into the furnace beyond the thickness of the walls of the furnace. If this happens the end of the torch can melt.

When the torch is safely propped so that it will not accidentally fall out of the furnace and the air supply and propane supply is connected, you are ready to light the furnace.

LIGHTING THE FURNACE -

Be sure to light this furnace outside the first time. Wear some eye protection. Take the lid off the furnace.

(A) Either use your torch lighter made from instructions in last issue or light a crumpled newspaper and throw it in the furnace.

(B) Quickly turn on the propane valve. You should get an immediate fire in the furnace.

(C) Next turn on the air valve. (Your blower should already be on). If the air valve causes the fire to go out you just didn't have the propane valve open enough. So start over and turn more propane on this time. As you open the air valve flame will shoot out the top of the furnace in a swirling motion and it will begin to roar.

(D) At this point put the lid on the furnace. Use gloves and a pliers. Don't drop the lid and be careful. The lid should quiet the roar down appreciably.

(E) It is not likely that the furnace is fully turned on at this time so Just let it run at a low flame for an hour or so. You have to wait an hour the first time only. (After the first use you can turn it on full blast to start with).

(F) After an hour (first time only) turn up the propane and air for maximum roar. If you turn up the air too high the flame may go out. Just turn the air down and it will catch again. But the furnace is working at maximum efficiency at the loudest roar and it will get the hottest in that condition. It takes a bit of practice, but you will soon learn to turn the maximum roar down to the point that will produce the heat that you wish to have.

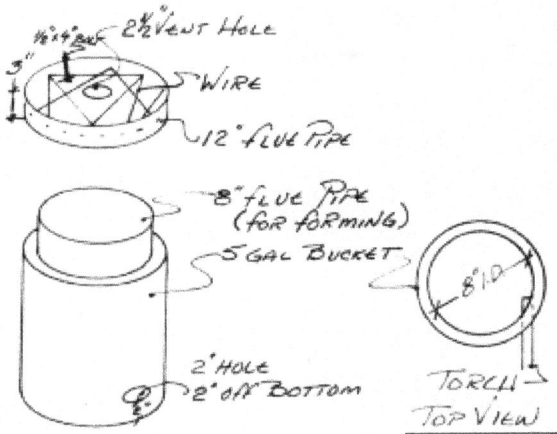

CREDITS: designed by Jim Humble, built by Ken Horn and Tracy Sarrett

AN ASSAY BALANCE

by Paul Wallace

This assay balance can be built for less than $20 in materials. The one shown and described is my first attempt on this type of device. I was not ready to go out and pay $2000 to $2500 for one* so here is the results of one week of work. This balance is sensitive to 1/2 milligram.

Purchase approximately six feet of 1/4th inch hexagon brass, and a six inch square of beryllium copper from a surplus store. Cost should be about $5.00. If you can't find beryllium copper you can use brass shim stock from an auto supply store. You will also need a foot or two of 5/8 inch steel tubing, three panes of window glass, scraps of hard wood and silicone rubber glue.

Note: Three knife edges are required on any assay balance. There is one main balance for the entire mechanism in the middle of the balance bar and a Knife edge at either end of the balance bar for the weighing pans to balance on. This eliminates almost all friction in the moving parts of the balance.

BEAM ASSEMBLY (-2) - Use two pieces of brass rod for the top of the beam to give room to mount a clamp for 1/2 of a Gillette razor blade to serve as the main knife edge of the balance. Only one piece of brass rod is used below to take the flexibility out of the beam. Silver solder all brass rods as shown on the drawing to complete the beam assembly.

KNIFE EDGES - Cut three pieces of brass rod 1/2 inch longer than a Gillette razor blade. Slit each piece of brass longwise through the center. Drill each end out far enough to allow the razor blade to fit between the screws. Use counter-sunk screws. Use a #36 drill and a 5-40 tap. This should be done so that the razor blade can fit between the screws and be clamped in place with the screws. Use 1/2 blade on each knife edge. Brake the razor blade by clamping into a vice and hitting on the side with a piece of wood. When clamping in place leave only a small part of the blade sticking out.

MOUNT THE MAIN KNIFE EDGE - On the underside of the two top brass rods locate the exact center. Do this by measuring carefully or by balancing the beam on a knife edge. (Use your own knife or a kitchen knife). Use four pieces of brass rod 1/4 inch long to set on each side of the piece with the knife edge. Carefully silver solder in place.

THE BEAM POINTER - Use a 3/32nd brass welding rod as the pointer. It should be sharpened to a long point before soldering in place. Be sure to allow an extra 1/4 inch in length for the end to stick through the brass hexagon rod. Silver solder it in place after drilling a hole at the point of the hexagon rods directly under the main knife edge.

THE BALANCING WEIGHT -Drill a 1/8th inch hole in the hexagon brass rod 1 3/16th inch to the left of the pointer. Drill this hole so that it angles slightly backward. Thread and bend a 1/8 inch brass welding rod as shown on the drawing. Solder this rod in place with soft solder.

BEAM END KNIFE EDGES -Solder a 7/16th inch hexagon piece to the center of each remaining Knife edge assembly. Use this piece to solder the knife edge in place as shown on the drawing.   The higher up towards the double beams that you solder the knife assemblies, the more sensitive the balance will be and the harder to balance (it will be). The cutting edge of these two outer knife edges is upward. (The opposite of the main knife edge).

KNIFE EDGE RESTS - Visit a rock shop and order three pieces of agate cut to 1/2 inch by 3/16 inch by 15/16 inch. Have a slot cut lengthwise .015 inch wide and .015 inch deep on one side of the agate. This slot is where the knife blades will rest.

KNIFE EDGE REST MOUNT -Cut a piece of copper 1/16th inch thick and just slightly bigger than the KNIFE EDGE REST. This is the KNIFE EDGE REST MOUNT. Solder this mount in place (3 places). (See view -8 and view -10 on the print.) After the mount has been soldered and cooled, carefully glue the agate KNIFE EDGE RESTS in place. Be sure that all surfaces have been wiped clean with acetone before gluing. The best glue to use is silicone rubber glue bought at any hardware store.

Popular Mining Magazine   February 1984   Page 36

An Assay Balance

Popular Mining Magazine   February 1984   Page 37

THE PAN HOLDER (-6) - The PAN HOLDER, in actual operation, connects to the KNIFE EDGE REST HOLDER by the hook at the top of the PAN HOLDER. THE PAN HOLDER is made from brass welding rod as shown on the drawing.

THE PAN (-4) - Cut two 2 inch discs of copper or brass shim stock. Use .060 thickness or greater. Heat to a dull red and drop in water. This will anneal the metal so it will form easily. If you have trouble do it again. Form over a 1.82 metal washer.

Construct all other items on the print as shown.

THE CASE. - Use a birch BOTTOM 1/4 inch by 5 1/2 inch by 14 15/32 inch. Across the front of the birch bottom make a cut 3/16 inch deep and about 1/64 wider than the front glass that you intend to use. Do the same to each end of the birch bottom. If you don't have any

THE CORNER POSTS - Cut four pieces of birch 3/4 by 3/4 inch by 8 3/4 inches long. Use your saw blade to cut grooves in these posts for the glass you have. The two rear CORNER POSTS will have only one grove in them. This groove must be cut l/8th inch from the edge as shown on the drawing. This groove will then fit directly above the cut made in the birch BOTTOM end (described in above paragraph). The two front CORNER POSTS must have two grooves cut in them. One groove will be just the opposite as the back grooves for the side glass. The other two grooves are for the front glass. Be sure to do them as shown. Leave l/8th inch wood area to hold glass.

Glue and screw the four CORNER POSTS in place. Be sure that the glass will fit properly before marking exactly where the posts must go with a pencil. Then hold a single post in place and drill a hole from the BOTTOM into the post and put a screw into the hole. Do this to all four corner posts.

Install the BEAM BALANCE BAR using three bolts as shown on the CASE print. Drill holes and install the three 10-32 leveling screws as shown.

THE BACK - Make the BACK from hardwood plywood l/8th inch thick. (Buy from a hardwood supply store or a hobby shop). Cut this plywood 9 1/2 inches by 15 3/16 inches long. Glue in place on the CASE being careful to keep everything squared. Use several brads to hold the BACK in place while drying. Glue a piece of quarter rounding to the BACK and flush with the top edge as shown. (Quarter rounding can be bought at any lumber store.)

THE TOP - Cut a piece of the 1/8th inch plywood 15 3/16 inch by 5 5/16 inch wide. Glue a piece of quarter founding to the front of the TOP so that it will fit between the two front CORNER POSTS. Next install the top using brads only but be sure to slip the glass sides in first.

THE GLASS FRONT - Cut a l/8th inch window glass to fit. Drill glass with a l/8th inch diamond drill. You can also use a cement drill if you are very careful. Most glass companies will drill the hole for you. Install a small brass handle 3/4 inch from the bottom of the glass at the center.

Now you are ready to install the BEAM ASSEMBLY and associated parts. With the BEAM ASSEMBLY in place next install the KNIFE EDGE REST HOLDERS and then the PAN HOLDERS and finally the PANS. With all of the knife edges working properly adjust the BALANCE WEIGHT adjustment for zero. In order to operate this balance you must have a set of milligram weights. These weights are used to balance out any thing that you may wish to weigh. Good Luck.

THE FINISHED PRODUCT

TIPS ON STAYING ALIVE UNDERGROUND

by Don Boston

Underground is a different environment and has certain rules for survival peculiar to itself. It is essential to maintain a healthy amount of respect. Humility is a good word here.

A person with real knowledge of his environment is usually the most respectful. A seasoned Arctic man is not foolhardy in the Arctic, which is how to become seasoned. Ever talk to a man who has been blasted? I have. These individuals are rare because blast victims usually tell no tales.

What does it feel like to get slabbed (slabbed means to have a slab fall on you)? I can answer that one. It's different each time but never boring. I'll describe a couple of extremely different reactions. First - Wham! No noise, no, nothing. You're down and down hard. You're down so hard and so fast you don't know how you got there - an instant effect. Atom bombs, earthquakes, what have you, there is a maximum force and that is what nailed you. If you survive this one you know who is boss - forever. Then there is the gradual awakening. You seem to have time to think. You have an awareness. You know that it's got you. You know that it is big, too big.

Crunch! She's down. The world, that is, and square on my shoulders. The pressure was unbelievable. The first conscious thought was surprise at still being conscious. This was too big for that and the feeling of incredibility was very distinct. Next was the awareness of my partner who had been passing wood to me from below. I could hear him clearly. He was cursing and I could follow his activity from his voice. There was tension in his voice and he was moving fast. I could follow his voice right up under me and on up above me to where the action was.

He told me later that all he could see was one leg protruding and that he thought it was severed. This explains his reaction later. He began moving boulders up above me. The ground was broken up pretty well. Each boulder he moved felt to me like 100 tons leaving me. This was taking place right above the sill and he was sort of rolling them off and they dropped on down the hole.

The last big boulder came off and I straightened up - some. My partner's reaction to this was the next big awareness in this sequence. He sort of blanched back - the best word I can find to describe it. I both saw and sensed it. He just wasn't prepared to see me move. My next concern was to let my partner know I was ok because I could see that he didn't quite think I was real.

Problem was, I couldn't talk. Lungs were deflated and I couldn't let out a peep, though I was trying. We eased on down to the sill over the muck pile and I hobbled around a bit and finally got things working. Croaked out that I was ok. I wasn't ok in every sense of the word. I never have been but I was sure ok relative to what I should have been.

He mucked up 9 tons from that spill the next shift and the timber was knocked out of whack.

I've fallen considerable distance, been seriously gassed and came close to drowning in New Mexico. Just happened to get slabbed on a high spot this time. The water in those headings was usually 3 ft. deep. Get slabbed in one of those and you drown before anyone can move a rock!

I relate these events to hopefully give some substance and credence, perhaps indicate enough experience to give a few tips....

1) Ventilation - smoke from a blast is one of the subtle ways of taking you down. (The lungs don't have nerve endings like the other organs and can take all kinds of abuse without any big complaint). Rock that has been blasted is in very small particles and you can breathe this type of dust deeper than regular dust. The body does have some defenses and can

catch some of the heavier stuff and cough it back out. Powder creates all kinds of gases, some of which can kill. Certain combinations of moisture and powder gas create * lethal gas, sometimes observable as a yellow haze.

The dust takes its time but is just as thorough. It just slowly cuts up the lungs. You can even get consumption driving a tractor on a farm. Farmers used to call it dust pneumonia.

Water is the best defense. Use lots of it. Wet down the muck pile and repeat as often as necessary during the muck cycle. This settles the gas and holds the dust to a minimum. Wet down the timber - soak it good. Wash down the face and walls but not the back or a hanging wall. This heavy wetting down tends to collect dust from the air and really freshens up the working place. This could be difficult in desert areas where water is scarce but I would personally make it a priority.

2) Explosives - respect is still the word; pay your dues, put in the time required to study the subject. It is not a complex subject - so simple it is disarming. This is sort of a judgment thing - no second chance. I suggest setting boundaries and remaining strictly within them. It is possible to keep powder and primers separate until they go into the hole. This has always been a standard practice with me. The common practice, which I don't recommend is what is called "making them up". The powder and primers are made up, the primers put in the powder some time in advance and brought to the face when needed. What you have is a messy bomb with lots of fuses dangling. Often this bomb is placed on top of the remaining explosives. Anything could happen..a rock fall on the bomb or an air hose work loose and start battering everything and everybody. It's just a senseless practice because it is unnecessary.

Electrics are a good way to go, especially for the solo miner. Lightning is a hazard here. Read the literature. Don't use instruments not designed for the purpose to test the caps. Some electrical testers will set off a cap. Take the extra caps back out of the way and bring in your powder. As you complete each hole coil up the loose wire and be neat and professional. This pays off with results.

It's a bit more awkward to do this with fuse. The caps should be on the fuse at this point, connected in a good dry shop. Unroll the fuse and hang them to the side within reach and out of the water and mud. Have the powder on the opposite side, separate, and take one of each and prime each hole. Use professional care and keep the ends out of the mud and don't walk on the fuse. I suggest you don't use old fuse or fuse that has been abused. Be aware that you can get a runner, even with new fuse.

You've got to get air circulating. You can't use the same method deeper in that was fairly effective at the start. That is, you can't let the place breathe by itself. If you're looking for trouble, compressed air will get you by but it doesn't have the same amount of oxygen as air from the outside and you've got to be generous with it. Compressed air is usually dangerous to breathe. Using a blower and air ducting is safer. In some mines you can have an opening at both ends of a tunnel for air flow.

Simple and obvious as these things are, they aren't generally known or addressed properly, especially the dust problem. I share this with anybody I talk to that I think isn't aware of the effects. Miners are commonly indifferent to smoke from a blast, but this smoke has deadly carbon monoxide and accounts for many deaths in mining each year.

I've worked for numerous big companies, attending countless safety meetings and have only heard this problem addressed one time, told just as it is. This particular shifter happened to have a brother die from silicosis. This was less than 2 years ago and I repeat, only that one incident in all these years. Once I've told someone the facts what they do from then on is their business. I just want them to be aware of just one more hazard to be alert to.

I hope this goes over and rings a bell here and there. The Badger

> "the dust takes its time but is just as thorough. It just slowly cuts up the lungs...water is the best defense. Use lots of it."

NON-CYANIDE LEACHING

by Jim Humble

Only two leaching chemicals have ever been developed that are low enough in cost to leach head ore directly. All of the other leaching chemicals are expensive enough that they must be used only on concentrates. Those two chemicals mentioned in the first sentence are cyanide and chlorine. Cyanide still is the most important chemical leach for gold throughout the world. However, that is rapidly changing due to ecological considerations.

Chlorine does not work with as many ores as cyanide. There are many minerals within the ores that prevent chlorine from working. The fact is that much of the technology for using chlorine has been lost. Possibly it could be used just as effectively as cyanide. If so, the data remains buried in old books from a bygone time. Chlorine will probably be revived more and more as the ecologists make more and more noise.

New ways of using cyanide have been developed and thus we are not likely to ever see the end to cyanide. There are now methods of removing cyanide from ores that do not leave the ore contaminated. If one has a reasonable investment and if one knows where to go, probably the best investment possible in mining today is still in the cyanide process. If one has upwards of $400,000 to invest one can make a profit on .015 ounces of gold per ton of ore. This includes ecology considerations and all other expenses.

However, for the small miner today, cyanide is a thing of the past. Ecology considerations and other operating costs has made it practically impossible to use cyanide. There are a few that still do, but most do not. Cyanide operations are and have to be fairly large mostly because of the nature of cyanide. Cyanide is used and will work only in very small concentrations, about .25% to 1% solutions. This means a gallon of cyanide solution will only hold approximately 1 cent of gold. Thus to hold much gold there must be many gallons of cyanide solution. This means large tanks, big pumping equipment, much liquid handling capacity and thus lots of money invested. Chlorine has many of the same considerations.

On the other hand most of the non-cyanide leaching agents, and there are a number of them, will hold as high as an ounce of gold per gallon of liquid. Thus they are more suitable to be used on concentrates. The entire leaching operation can be 1/10th the size of the cyanide leaching operation. This means smaller tanks, pumps, and other liquid handling equipment. The cost of building and operating the leaching unit is a fraction of that for cyanide. Gold mining operations can be less costly now than in the past.

But because the non-cyanide leaching agents are much more expensive than cyanide they must be used on concentrates instead of directly on the ore. This means that ores that in the past could have been leached directly with cyanide must now be concentrated. Most miners when thinking of concentrating their ores think of many of the old ways of concentrating. Those old ways were necessary when one was going to sell his concentrates or smelt his concentrates. But with the invention of the new leaches there is a third much cheaper way of handling concentrates and that is leaching them with non-cyanide leaches in small leaching outfits.

Since one is not trying for perfect concentrates he can afford to use much cheaper methods of concentration. He can in many cases concentrate his ore 10 to1 or maybe even 5to1 instead of 100 or 200to1 that use to be required. Now if he winds up with one ton of concentrates for each 5 tons of ore that he concentrates he may find that he now has enough gold per ton to pay for the leach and make a profit. It is very easy and usually cheap to do a 10 to1 concentrate and very expensive to do a 100to1 concentrate. Thus the new non-cyanide leaches allow lower cost operations to be developed.

Non-cyanide leaching can be fairly simple for a small operation and can get very sophisticated for large operations. Costs can vary from $995 for a small 150 pound leaching unit with all of the associated equipment into hundreds of thousands for larger size leaching units. The basic procedures are the same as cyanide even to the point of using zinc to recover the gold from the liquid.

Non-cyanide leaching may be the way to go in many cases and is not the way to go in many other cases. Cyanide is still very useful. Concentrating on concentration tables for direct smelting of concentrates still works well under the proper circumstances with the proper ore. If one does use non-cyanide leaching of concentrates there are a number of such agents on the market. Thiourea, ammonium thiosulfate, sodium thiosulfate, aqua regia, CLS, and manganese oxide are a few. I will try to cover some of the advantages and disadvantages of some of these in a later issue. Meanwhile in some cases it might be a good idea to see if an ore can below concentrated instead of high concentrated. If one could change an ore from 1/10th ounce to even 1/2 ounce per ton by concentrating it may become payable. END.........

.......We were wrong. In the article on Non-cyanide leaching in last issue we stated that cyanide would only carry about one penny per gallon of liquid. That's not right. In approximate figures one gallon of liquid cyanide at i strength of .25% will have 2.5 grams of cyanide. This amount of cyanide will carry approximately twice as much gold or 5 grams of gold. In actuality cyanide solutions do often carry much less that this before they become saturated, but in other cases it seems that cyanide will carry much more gold than was stated. In any case CLS will carry up to 100 times more gold than cyanide.

To add a little more data, heap leaching solutions of cyanide often start as low as 1 pound of cyanide per one ton of water, about .05 percent.

PRECIPITATING GOLD WITH ZINC

by Jim Humble

Ken Horn, lab manager

Precipitating precious metals including platinum from solution can be very tricky. Many people haw thrown ounces of gold, silver or platinum away because nothing more came out of solution when they added zinc, or because they had already added zinc several times and then they were sure that there was nothing left. So one has to be very careful to make sure one has got the gold. What is even more important is not losing the results on a test that you have done.

Precipitation in chemistry or mining means making a metal that is in solution come out of solution and (usually) settle to the bottom of the container. One uses zinc or some other base metal (such as aluminum or iron) to cause this to happen. This is called the replacement process or the cementation process. The zinc will go into solution as the precious metal comes out of solution which is why it is called the replacement process. Cementation was an earlier word. It was called this because the precious metals seem to cement themselves onto the zinc. This is mainly true when zinc chunks are used.

So in the leaching of gold ores one uses zinc precipitation to get the gold out of the leach liquors. The old miners just filled up a box with zinc shavings and ran the cyanide across the shavings. The gold soon cemented itself onto the zinc shavings. At least that's what they hoped would happen. But it often did not happen. Cyanide must have oxygen in order to dissolve the gold, but the oxygen prevents the zinc from causing the gold to come out of solution. Thus many old zinc boxes failed. A lot of the old zinc boxes worked because the cyanide was so loaded that the gold just had to come out.

Many miners find that they still have trouble with zinc and cyanide, but technology has been developed to solve this problem. The Merrill Crowe process does this job. It is a process that is sold by the Merrill Crowe Company throughout the world. Basically they use a vacuum on the cyanide leach liquid first to remove the oxygen and then the system injects Just the right amount of zinc powder to replace the gold in solution. The gold then in small particles is caught in a filter. Since there is no oxygen in the water the cyanide will not re-dissolve the gold.

Other ways to remove oxygen from a cyanide solution are to heat the liquid for several hours without boiling it or to let it sit for several days in a closed container with Just a small opening. The zinc can then be added to get the gold to precipitate. It is best in this case to add powdered zinc, stir it for a while, filter it off, and wash the precipitate thoroughly. Then and only then add a weak (10 to 1) solution of HCl (hydrochloric acid and water) to the precipitate to remove the remaining zinc. The precipitate can then be cupelled or smelted. (Never allow the cyanide and acid to get together. They form a highly poisonous gas that kills with one breath).

The cyanide solution is highly basic, usually over pH 10. Using zinc with acid solutions to precipitate precious metals can be more tricky than with cyanide. This is true will all acid leaches including aqua regia. Now this may surprise some people, because many people believe that using zinc with acid is simple. All you've got to do is drop in the zinc to an acid solution and out comes the gold, or silver, or platinum. Well, not exactly. It does happen this way much of the time, but if you try it very often sooner or later it will get you. And you will lose some gold.

You see, precious metals are kind of finicky. They want to come out of solution only at certain pH's (when using zinc). (Please read the definition for pH in the glossary). The pH at which various acid solutions will give up their gold varies, but it is usually around 4 or 5 and sometimes higher. The thing that zinc does is start raising the pH. It will raise the pH right up through the point that the precious metals precipitate out of solution. And if you are lucky all of the gold will come out of solution (and drop to the bottom) as the pH raises upwards towards pH 7. (Note: zinc also causes cyanide solutions to to move towards a pH of 7. Zinc brings high pH solutions down towards 7 and low pH solutions up towards 7.) But if one uses powdered zinc sometimes he will find that the pH of the liquid has been raised too quickly and all of the gold did not have time to come out of solution. It's possible that if he waited several days that more gold would come out of solution, but few miners have that much patience. The zinc is replacing the gold in solution, but only at the right pH, and at a pH of 6 or 7 very little zinc is consumed into the solution so very little gold will come out of solution.

PRECIPITATING WITH A
ZINC CHUNK...

How do you overcome this problem? Well, one way is to add more acid taking the pH down to 1 or 2 and then add more zinc to bring the pH up through the critical pH precipitation zone again, maybe several times. But also remember that you must be sure that the oxygen is out of solution. And with the zinc powder you will always wind up with some in the precipitate. With zinc powder it is easy to make the mistake of not adding enough. You are not trying to add enough zinc powder to remove' the gold, you are adding enough powder to bring the pH up through the precipitation zone and thus remove the gold. So it takes quite a bit of zinc. Once the gold has precipitated, use weak HC1 as given above to dissolve the extra zinc, and cupel or smelt.

The preferable method of zinc precipitation is to use zinc chunks. Unlike powder chunks slowly dissolve bringing the pH up slowly. The pH then passes slowly through the precipitation zone and usually you get all the gold if you have removed the oxygen first. Do this same as above, by vacuum or heating or letting sit for several days. We usually add acid after 3 hours to allow the zinc to bring the solution up through the precipitation zone twice. Thus after six hours our precipitation is usually complete. The advantage of the zinc chunks is that you can physically remove the unused zinc and use it again next time and you do not get zinc in the gold.

So if you have had trouble with zinc powder and wondered why the gold didn't come out, it could be because you didn't use enough zinc, or you used too much too fast, or you didn't get the oxygen out of the liquid. Also you may not have enough zinc chunks for the amount of liquid present. Look at the zinc chunks, if they are still bubbling hard either you have not waited long enough or you did not add enough zinc chunks. When finished the zinc chunks will be bubbling very slowly. Take out a little bit of liquid and add some zinc powder. Anything other than a very slight bubbling indicates either more time or more zinc or both.

pH paper cones in small rolls. One tears off a short piece of paper and sticks it into the liquid to be measured. The color that the paper becomes where it is wetted with the liquid indicates the pH. There is a color chart on each roll of pH paper that one compares to determine the pH of the liquid.

Expanded Metal Sluice

by Torn Bryant

Editor's Note: This is the first of several articles by Tom Bryant on various aspects of placer mining. We believe that his articles will make it a little easier for everyone, especially the novice.

I suppose I should say right now that I don't expect everybody to go along with everything I say so....Tough. I'm safely hidden in the wilds of Canada so I'll call them as I see them. My background in the sciences and in placer mining makes me see things maybe a little differently from the "it looks good" gang. Some of the problems I'll be talking about may seem small or obvious but if I had a dime for every guy I've seen making these mistakes or wishing they had a solution to them I'd never have to look for that yellow stuff again.

For the first attack on the "it looks good to me" gang we'll discuss the use of expanded metal mesh under the Hungarian riffles in a sluicebox. Many manufacturers and miners set up their sluice by laying a carpeting, then a layer of expanded metal mesh, then Hungarian riffles on top it all. But this actually leaves holes under the Hungarian riffle that water can pass through. The system works and there's lots of people who can show you the gold to prove it, but I wonder how many people know that the system actually works against itself in saving fine gold.

(Fig 1)
Blow Out

The riffles, both Hungarian and mesh, are there to create cavitation. (When water passes over or past something especially the edge of something a low pressure area is generated on the downstream side. When the water is moving fast enough this low pressure area becomes so strong that it sucks air and other particles into it. This is called cavitation). By causing an area of low pressure to be created at the downstream side of a riffle the gold is knocked out of suspension. Now you've got it but can you keep it? Fine gold needs a place to hide. It hits the bottom of the sluice and that's where the problem that I'm discussing occurs.

Take a close look at the system. When a diamond of mesh is under the Hungarian riffle it makes a hole that can let water shoot through and blow stuff out. Actually one of two things can happen depending on the angle of the sluice and the water flow. You could get blowout as I've said or perhaps even worse you could get an upstream cavitation that could cause "suckout". Take a look at Fig. 1 and 2 and you'll see what I mean.

Blowout is obvious but suckout might surprise you. If a strong cavitation were to be created on the upstream side of that hole under the Hungarian riffle it can suck out the concentrate from the pile on the downstream side. Since the stuff that is sucked out is from the bottom of the concentrate pile where the heaviest of the heavies are, you're losing gold! To compound the problem a rotational movement (for the steam engineers) can be set up by the weight of the new concentrate piling up on the existing pile on the downstream side of the riffle. This movement actually aids the suck out by pushing the stuff being sucked out from behind.

(Fig. 2)
Suck Out

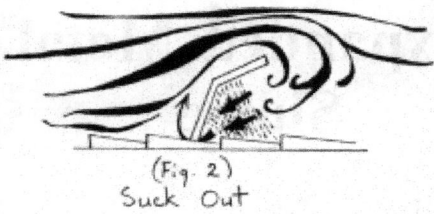

Oh my gosh - this is worse than you thought. Blowout, suckout and now pushout!! What can we do?!! Simple, close the holes. You may say, there are only a few holes under the riffle and all the rest of the expanded metal will hold most of the fine gold. But not so, most of the fine gold winds up right down at the area we are talking about, under the Hungarian riffle. So close those holes.

If you don't want to make major changes to your Hungarian riffles, the easiest way that I've found is to fill in the offending mesh diamonds with something like silicone rubber seal.

Assemble your sluice dry and mark off those diamonds that pass under the Hungarian riffles with a magic marker. I simply run my marker along both sides of the Hungarian riffle leaving a stripe across the mesh the width of the bottom of the Hungarian riffle. I then mark off (depending on the size of the diamonds) at least 1/2 inch on either side of that stripe. Lay a line of masking tape to the outside of the area you're going to close off.

With the mesh out of the sluice on a flat surface and wax paper underneath lay a couple of 2X4's on either side of the work area and put some weight on them to push the mesh down tight to the wax paper. (See Fig. 3) You should use a filler compound that dries non-tacky and stays flexible. I've been successful with the silicon rubber bought at any hardware store.

Run a bead of the stuff across the mesh and by using a wide spreader of cardboard or flexible plastic moosh the stuff down into the mesh. Moving the spreader from the downstream end of the area to be filled to the upstream end will fill it in very well. Make sure that each diamond is filled right to the top - add more filler as needed. When you're done you'll have stripes of filled in diamonds that should line up under your Hungarian riffles to stop blowout suckout, and pushout.

Your mesh should be placed into the sluice the same way every time. If you have problems mark the top edges of the sluice in line with the stripes of filler. Assemble your sluice dry to check out your handiwork. Line up the marks and your stripes every time you set up.

Some miners might say that the fine gold they would save by doing this wouldn't be worth the effort. I say an ounce of 100 mesh gold weighs the same as an ounce of 10 mesh gold. If you have gone through the effort of getting it into your sluice you should save all you can! No matter what you say an inefficient system costs you money - you wouldn't leave your windows open in the winter no matter how good your furnace was so why would you give the yellow stuff an excuse to get away.

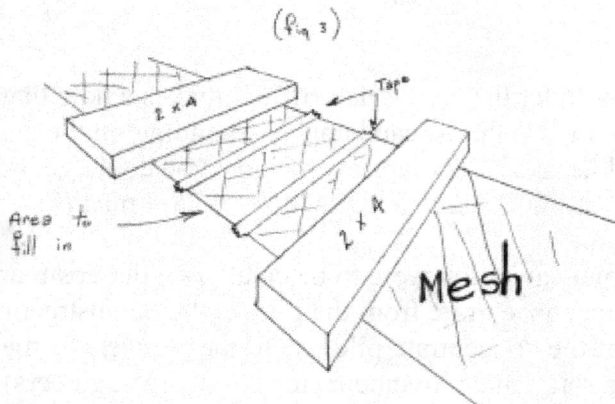

(Fig. 3)

Tape

2 x 4

Area to
fill in

2 x 4

Mesh

One last point about holes. I've seen all sorts of guys out there that don't pull their riffles tight against the carpeting. Quite often the mesh gets bows in it that can't be held down with side clamps or wide spread Hungarian riffles.

Deeper pile rug will solve a lot of problems so one might give way to astro turf style of indoor/outdoor carpet. If you just have a "loose" system add an under pad or buy a thicker base carpet. That way the carpet will act as a cushion from the bottom to tighten everything up. You should always do the best you can with what you've got. Closing those holes is certainly a step in the right direction. END

Learning About Flotation

by Paul H. Skinner

This series of articles, over the next few months, will be for the purpose of educating and helping small miners and prospectors in the flotation process. We will cover the history, scope and mechanisms, and get into the different types of flotation along with testing of ores and the different types of flotation machines going into the positives and negatives of each make. Hopefully when we are finished you will be able to build and operate your own flotation plant or at least be able to tell if a custom mill is doing it right.

FLOTATION DEFINED

Flotation, or more specifically froth flotation, is a physio-chemical method of concentrating finely ground ores. The process involves chemical treatment of an ore pulp to create conditions favorable for the attachment of certain mineral particles to air bubbles. The air bubbles carry the selected minerals to the surface of the pulp and form a stabilized froth which is skimmed off while the other minerals remain submerged in the pulp. Although flotation was originally developed in the mineral industry the process has been gradually extended to other fields.

In general, sulfide mineral particles coarser than 48 mesh (about 295 microns or .295 mm in diameter) cannot be effectively recovered; consequently, an ore that is to be floated must first be ground fine enough so that the desired mineral is all, or substantially all, smaller than the limiting size and small enough to get liberation. Liberation means that the mineral you want is broken away or liberated from the gangue or waste rock. The size of this grind will vary greatly, each ore being very individual.

HISTORY OF DEVELOPMENT

The earliest patent which may be considered as relating to the flotation process is that of Haynes in 1860. His recognition of the differences in wettability of various minerals by water and oil formed the basis for a number of "oil" flotation processes. During the following fifty years, flotation passed through three principle stages of development: (1) Bulk oil flotation, (2) Skin flotation, (3) Froth flotation.

Bulk oil flotation was based on the fact that minerals of metallic luster are preferentially wetted by oil in the presence of water and consequently pass into the interface between the oil and water while the water wetted gangue drop out. This process required large amounts of oil, usually about one part for each part of ore.

Skin flotation, on the other hand, depended on the fact that when finely ground dry ore was gently brought into contact with still water the metallic particles tended to float more than the gangue. This process was developed in the period of 1890 - 1915, but both it and the bulk oil flotation were made obsolete by the froth flotation process.

As early as 1902, Froment in Italy and Ballot in Australia recognized gas bubbles as the ideal buoyant medium for carrying the oiled sulfide mineral particles to the surface of the ore pulp. They and other investigators generated bubbles by chemical action or by applying a partial vacuum above the ore-water pulp. It remained, however, for Ballot, Skumand and Picard (1905) to utilize a rising stream of air bubbles and to materially reduce the quantity of oil required. This process, when in acid pulps and with non-selective oils used as collectors) provided an economic method of concentrating many of the sulfide ores during the period of 1906 to 1925. However, it proved inapplicable in many complex ores, such as those of copper-iron, copper-lead, zinc-iron, lead-zinc-iron, copper-zinc-iron, and copper-nickle-iron.

The need for treatment of such complex sulfide ores as well as the non-sulfide ores has inspired continual progress in the development of more selective reagent combinations and more efficient flotation equipment.

Modern flotation is usually considered to have begun in 1923, when C.H. Keller discovered the use of Xanthates as collectors for sulfide minerals. It should be mentioned that a very large number of patents have been granted within recent years. These patents have generally been for more specific reagents and reagent combinations. As such, they are of limited application and can hardly be considered milestones, despite the fact that each adds to the knowledge of froth flotation and expands it scope.

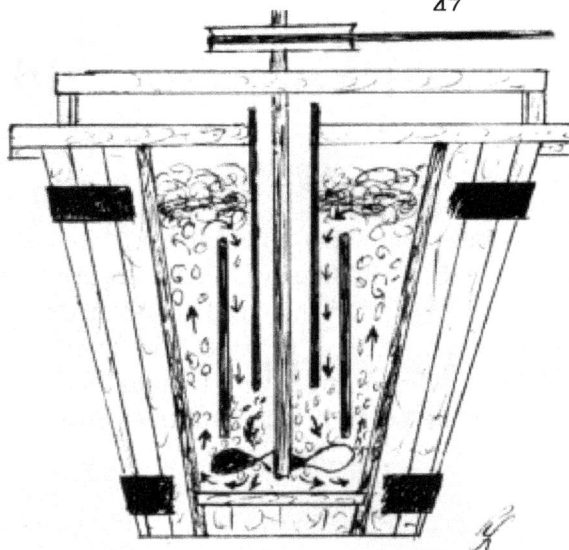

Exploring Old Mines

by Robert Louis Desmarais

You never know what you might find in a old cave or mine. I am sure you have read of some great lost treasure someone had found or that lost vein of gold left in an old abandoned mine. I heard of an older couple who found a stack of silver bars while exploring some old mines in Montana. What a find! Wish I could come across something like that. I have found a few relics and antiques during my explorations but nothing of real value until this trip.

This trip takes place in the White Mountains near Bishop, California in the High Sierras area. This trip really turned out to be a jewel. Five of us were working on cleaning out a debris-laden tunnel. We had several tons of rock, dirt, and old rotten timbers to remove before we could start drilling and blasting to resume our ore gathering. We were sorting out the tailings as we dumped each load. These particular mines were founded around the 1880's and I was always finding some old relics. The mines were again worked in the late 1920's and the late 1950's. These claims were re-staked in 1967 and further exploration and plans were made to start reproduction of the silver ore removal. What held up production at that time was the fact that the prices of gold and silver were quite low and the cost of mining exceeded the precious metal in the ore. As the prices of metals finally started to rise, a company expressed interest in our ore, and we set up a contract to deliver some ore to their mill. They gave me the position of foreman and a crew of four men. They supplied us with enough money for reasonable expenses and what equipment we basically needed to get the mine operational. This was my miner's dream come true. Here was my chance to make my first million. Well, back to work.

We were gathering up a good pile of high grade galena ore and sacking it up for delivery. We had to use mules to get the ore up the one mile steep narrow trail to the waiting truck. Our capacities on delivery were quite small, but with high grading (taking only the best ores available) it was worth the efforts. We were trying to keep costs down so the profits could be higher.

While continuing our digging, we came upon some old mule shoes, hammer handles, a few rusty chisels, a length of twisted ore car track, and many old tobacco cans. One old can had some 1927 claim papers in it. As I was watching one of the workers swinging a pick into a large pile of loose rock, I heard a metallic sound. I yelled to him to hold up a minute thinking it might be another tobacco can. I dug around the area he was picking at and felt some canvas material. I moved enough dirt to see that it was an old ore bag. I grabbed part of it and pulled it from its long time grave. It was partially rotten and had crystal-like flakes on the bottom side of it. We were all getting excited as to what was in the bag; since, when the bag was shaken it made a jingling noise like coins in a can. We all gathered for the unveiling of the bag's content. I was hoping to find someone's lost gold poke or a can of old silver dollars. Better yet a can of twenty dollar gold pieces. You could feel the anxieties as I untied the bag and peered in. It was full of some real Jewels all right. Inside the bag were thirty sticks of old oozing dynamite all primed with electrical caps!

There was also a box of one hundred fuse-type blasting caps in a coffee can along with a hundred foot roll of fuse. What a boom it would have made if it would have been hit by the pick directly on a cap or maybe even one of those old sticks of dynamite. We would have all been killed and buried alive in the mine. Many old mines have these kind of dangers in them, so be careful on your digging and picking into rubble. Old dynamite and especially damp oozing crystallized sticks are extremely dangerous. It will go off as easily as nitro glycerine. The moral of this story is be careful when exploring old mines and remember old dynamite laying around is more dangerous than new dynamite.

A Vibrating Backpack Sluice

by Ron Lukawitski

Here are some pictures of a "vibrating sluice box" that I built. The water pump transmits enough vibration to the sluice box which sits on four springs. There is another spring underneath which can be adjusted to equalize the downward pull of the box.

The upper part of the sluice has two "L" shaped riffles which are bolted down onto the box. They save the heavier material quite efficiently and are simply cleaned out with a bulb sniffer. The fine gold is caught in the plastic gratings (which are fluorescent light gratings) The gratings are removable.

Because there is a lot of clay on my claim, the hopper is divided into three sections to give the material more distance to break up. These sections fold down into each other and the whole unit, although a bit bulky, can be carried on a pack frame.

The base is an old hospital stretcher cut down. The poles are from an old tent and the grizzly is a refrigerator grate. The whole thing minus the water pump and plastic grate cost about $50 to make. The plastic grating cost $20 and the water pump $200.

1 hope this can be of use to some miners.   Ron

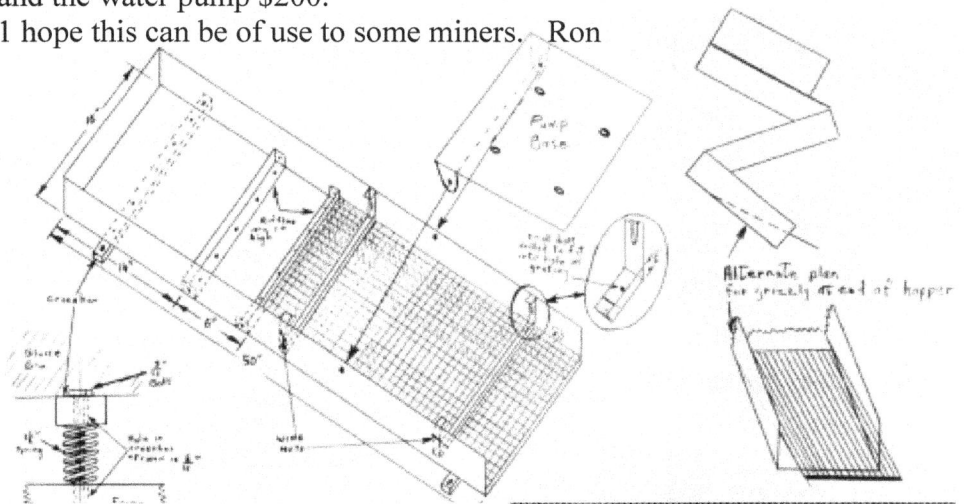

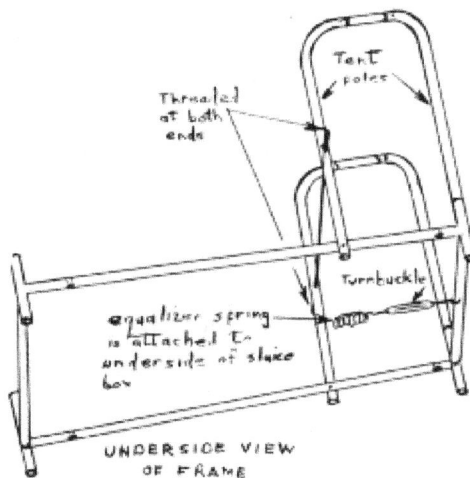

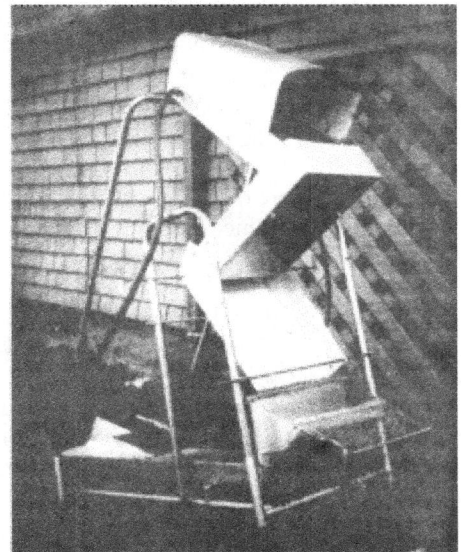

Dredging for Gold in North Carolina

by Ellsworth Boyd

ROBERT PRATT offensive lineman for Seattle Seahawks. KEN HUFF offensive lineman for the Washington Redskins, and JIM CANTER guide, test the dredge before operations. Photo by Ellsworth Boyd.

Gold fever isn't likely to be fatal but there is little hope for a cure. Ken Huff and Robert Pratt had the disease. While Pratt tended the dredge. Huff grabbed the suction tube and disappeared beneath the brown waters of Badin Lake. They both skipped lunch and didn't return to camp until sunset. Their plan had called for much preparation and hard work and they intended to lose little time.

Not everybody finds gold in North Carolina but these guys figured they had a better than average chance because they were searching where the gold panners had never tried. They were beneath the water, while panners are limited to the shoreline.

Using one of Jerry Keene's (Keene Engineering, Northridge, Ca.) most efficient dredges, Huff and Pratt moved over 20 cubic yards of overburden an hour. A panner averages only one cubic yard per hour. The dredge sucked the overburden across three riffle boxes, then dumped most of it back into the water. The same engine that powers the dredge also runs the compressor which pumps a continuous supply of compressed air to the divers.

For Pratt and Huff the plan at Badin Lake was simple, but the work complicated. With the help and advice of guide Jim Canter, from Asheboro, North Carolina, the linemen chose a small U shaped cove to begin their quest. A steep V shaped trough ran down King's Mountain and into the cove. This was evidence of a commercial gold operation begun in the early 1800s. It was here that men dug quartz rock to be milled and then flushed the milled rock down a wooden sluice built into the mountainside. Gold tailings that overflowed from the sluice might be lodged in the cracks and fissures of the cove where a number of obstacles such as boulders, trees, tree trunks and other debris blocked their way.

Most of the work was in five to ten feet of water where one diver always dredged the overburden after the obstacle was removed. The cove was too much for one day... the area was too big. The diving prospectors decided to return the next morning and continue their attack at the foot of King's Mountain.

That evening, by the camp fire, Jim Canter, the guide, told all about glory holes, Colonial gold and the Carolina] Slate Belt. "Everybody dreams of finding a glory hole" Canter says, "a crevice or fissure in the bedrock of a lake or stream where countless nuggets have lodged. They might be as small as a pinhead and as large as a marble, but when you find 'em you pluck 'em up like grapes from a vine." Canter told about in 1799 a 12-year-old boy named Conrad Reed was playing in Little Meadow Creek, on his father's farm in Cabarrus County, North Carolina. He found a 17-pound, wedge-shaped, yellow rock that he carried back to the house where it was used as a kitchen doorstop for three years. One day Conrad's father took it with him to Fayetteville where he sold it to a jeweler for $3.50. It was worth about $3,500 then, but that rock of gold would command a small fortune today. That, by the way, was the first authenticated discovery of gold in the United States. This is all on record in the North Carolina state archives.

"We're sitting on a mountain of gold," Canter continued. "Most of the mines were abandoned by 1930, but by then over $60 million had been taken out of 650 mines spread throughout 50 of North Carolina's 100 counties. About $500 million worth still remains in this state, a lot of it right here beneath us. There is talk of restoring some of the mines again, but it takes a lot of capitol to start a commercial mining operation. That leaves more for placer prospectors like us. King's Mountain gold has been washing into the basin of Badin Lake for the past 80 years."

Pratt and Huff were out of touch with reality now, overcome by pyrexia, a fever that Jim Canter says is incurable. It is accompanied by tachycardia, the medical term for fast heartbeat, and delirium, something they fell victim to the next day in the cove beneath King's Mountain.

Starting early the next day, they moved some small rocks from the outer perimeter of the dive site, then used the long monstrous suction tube to suck up lots of overburden. It was doing a fine job until a flat rock jammed it, forcing the divers to cut the engine and shake the obstruction loose. It was time to check the riffle box anyway. Pratt lifted the riffle trays and dumped the fine overburden into a large can whose contents would be carefully panned later in the day. Disgruntled at seeing no immediate signs of nugget, he routinely lifted the ozite mats from the floor of the trays and rinsed them in the cloudy water of the large can. That's when he spotted the pinhead nuggets laying in the bottom of one of the trays. Somehow they had worked themselves beneath the mats instead of lodging in the thick napping on top. The burnished stainless steel tray intensified the gilded treasures, six in all, shimmering in the morning sun. Delirium set in.

There was no more nuggets that day, but plenty of fines or flour gold turned up when panning the cans of overburden collected by the divers. They were picking up the overflow from the old commercial sluice operation. There was plenty of hope and expectation in camp that evening. Pratt hummed the theme song from "Gold-finger," while Huff sat back and toasted the day's finds with another cold one.

Badin Lake was a wise choice. Each day just enough gold was found to feed the fever. Pratt, the gourmet chef of King's Mountain, fed the group royally. He brought 10 pounds of the thickest New York strips I have ever seen and proceeded to hickory smoke them over an open fire. Huff was afraid to drink the water so he brought Budweiser, Michelob, Schlitz and Miller Lite.

Camp was bustling the next morning. It was raining, but that didn't deter anybody. The group huddled together in the cove, deciding on a plan. Huff descended first, in 10 feet, with the compressed air hose strapped to his back. Pratt followed and they took turns buddy breathing from the regulator, while sizing up the dredging from the previous day. Nobody had checked the gas tank and neither diver heard the engine quit. The ensuing comedy was like something from the Keystone Kops. The situation could have been very dangerous.

The diving buddies handed the regulator back and forth like a fumble neither could get a handle on. They were sucking rubber... but no air, and nobody was heading up. Finally the light dawned and they broke the surface together, like missiles from a submarine, gasping and choking for air.

Pratt filled the gas tank and remained with the dredge while his partner continued where he left off the day before. Later, Pratt took over and manipulated intake hose into a shallow water embankment. It was only four feet deep and the red clay fell away in big sheets, bringing roots and rocks with it, clouding the water around the diver. It took Pratt about 30 minutes to dredge a hole large enough to get his head and shoulders into, but realizing the danger involved, he withdrew and started a new excavation. Seconds later, the old one caved in!

But when the veil of mud and clay lifted, we saw a four foot chunk of quartz rock protruding from the embankment. It had a stringer in it, something Jim told them to watch for. A series of criss crossing lines in the gray quartz, it is indicative of volcanic eruptions that used to push gold toward the surface. Sometimes the gold becomes crystallized and lodges in fractures in the quartz.

Pratt surfaced long enough to grab a crowbar. Then, like a man possessed, he attacked the stringer and split the rock down the middle, spilling tiny fragments of fractured quartz at his feet. He then sucked up everything in sight with the dredge intake. Then they checked the "goodie trays" on the dredge.

There were no nuggets visible, but that didn't mean anything. Sometimes they are hidden in the small mounds of sand and gravel or discolored by the red clay. My job was to pan down about 30 pounds of overburden. It was slow work, but I got good color. I also got a few pinhead nuggets and lots of fish hooks, sinkers, lures and a hand-forged nail. I was nearly finished when I heard a "growler" in my pan. A growler is a large nugget... you hear it before you see it. As I moved the pan in a circular motion, if broke free suddenly from the overburden and rolled across the top of the pan making a slow, rumbling noise. There was no mistaking its identity. It growled and it glistened. This one was crystalline, more valuable for jewelry than its weight in gold.

"I've got him - a growler!", I shouted. Huff thought I was referring to Pratt until he spotted the thumbnail-sized nugget gleaming in my gold pan. Pratt thought we had found a glory hole and went into another frenzy.

But the trail of gold has many detours. It rained all night and the forecast on Robert's "BDR" - "Big Damn Radio," called for three more days of it. We could dredge in the rain, but life in camp was miserable. The tents leaked and everything was sopping wet; there was no dry place to cook and the road leading down the mountain was nearly impassable. Reluctantly, we packed for the eight-hour return trek to Huff's home in Baltimore.

We never found Pratt's elusive glory hole, but we dredged enough gold to pay for our trip. For Pratt and Huff it wasn't solely the lust for gold that drove them to Badin Lake campground in search of gold, it was also their love for the outdoors... a return to nature. It was an ideal way to relax and unwind before the start of another gruelling training camp at the NFL.

KNOWN PLACER GOLD DEPOSITS

Will we return? You bet. After the season.

The dredge we used was Model 5508, an older model. It will suck up rocks as large as your fist from a maximum depth of 30 feet. We sold the gold to a jeweler for near $500. There were no local regulations concerning weekend dredgers. Panning is required as a supplement to the dredge. It must be done fast. Most of the flour and fine gold is too tedious to keep.

Donna Sundstrand, assayer

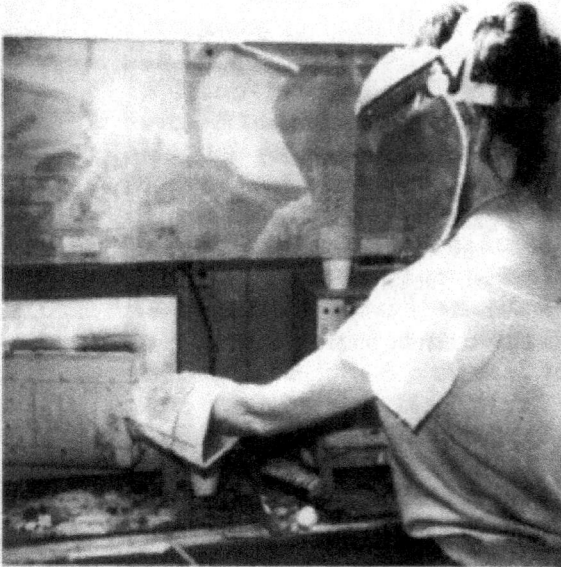

How to Fire Liquids

It is often necessary to determine the gold, silver, or platinum content of a liquid. There are several ways of doing this kind of an assay. One way would be to precipitate the gold from the liquid and then fire the precipitate. There are many methods of precipitation. One very good method is to use hydrazine. One merely adds hydrazine to the liquid, filters the resulting precipitate from the liquid and fires the precipitate.

Hydrazine may be bought from industrial chemical supply houses and from chemical supply stores that supply chemicals to water treatment plants. Hydrazine is a nice chemical to have around the lab; because it can be used often for just this purpose, determining the content of precious metals in a liquid. Hydrazine is a very poisonous chemical so do not breath the fumes or get it on your skin. This is true of many chemicals in the assay lab.

On the other hand if you do not have hydrazine easily available there is another simpler easier method of fire assaying liquid to determine the precious metal content. We do it this way: (1) Measure out 90 grams of standard fire assay flux. If you do not have a standard flux available you can purchase some from GAME CO (an advertiser in this magazine) or you can mix your own standard flux from any fire assay manual. (2) Measure out 100 milliliters of liquid that you wish to use. (3) Add your 90 grams of standard flux and 100 ML of liquid to a plastic cup and mix well. (4) Set the mixture in a drying oven or in any warm area of not more than 150 degrees F. Wait until the mixture is dry. (5) Put the dried mixture in a 30 gram crucible and fire as any standard fire assay.

The resultant bead that you get should be multiplied by 37.85 to give you the amount of gold per gallon of liquid. This is because there is 3785 ML in a gallon of liquid.

Do not pour your liquid into a crucible with the flux. All crucibles are porous and will absorb some of the gold. Be sure your liquid flux mixture is dry before adding to the crucible. If you would prefer an answer of gold per ton of liquid then use 29.16 ML of liquid instead of 100 ML. Then multiply your assay bead by one ounce times each milligram obtained in the assay. This will give you ounces of gold per ton of liquid.

Improving Your Sluice

by Ivan Carlson

Here is a way to make a very efficient sluice or to triple the time between cleanup on your present sluice and also save much more flour gold. Or if you have a good sized sluice already, use this idea in the top area to catch more flour gold.

This sluice would be three or four feet long, (i) Cut burlap to cover the last three or four feet of sluice. (2) Lay the burlap in place and then cover the entire sluice length with expanded metal. This metal also covers the first piece of burlap. (3) Cut a second piece of burlap the same size as the first piece. Lay it directly above the first piece but over the first piece of expanded metal. (4) Lay the second piece of expanded metal on top of the layers of burlap to form the fourth layer. The expanded metal is now clamped in place by side clamps. The burlap is held in place by friction and gravity. The sluice is ready to use. When it is full of gold it is cleaned up, or maybe a little before then. END

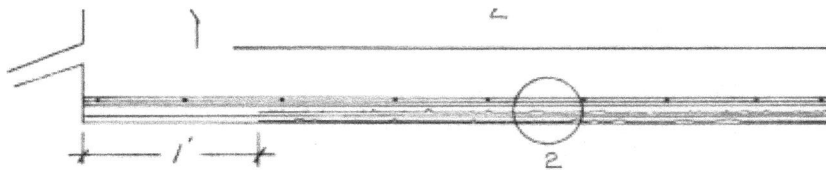

UNDER CURRENT SLUICE

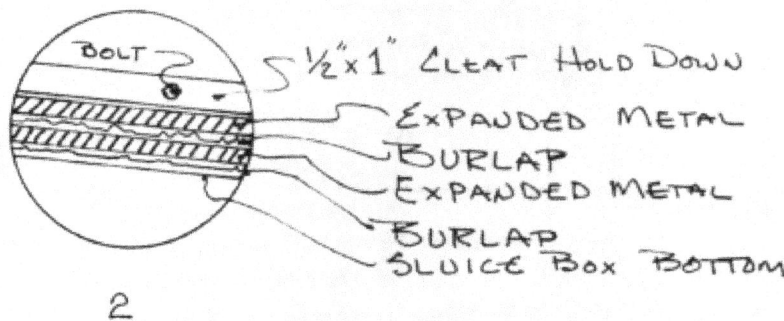

BOLT

S ½"x1" CLEAT HOLD DOWN
EXPANDED METAL
BURLAP
EXPANDED METAL
BURLAP
SLUICE BOX BOTTOM

2

A Mercury Retort from Pipe

by D. G. Piper

Here is a retort made totally from pipe that you can buy at the hardware store. I. use a weed burner to heat the retort. It will retort about 300 ML of mercury in about 20 minutes. When mercury quits running out of the retort I used a water hose to cool it off before removing the mercury residue.

The picture is pretty much explanatory. Be sure to use pipe teflon tape wrap on all joints. Use extra amounts of the pipe teflon tape wrap on the cap that you must screw on and off each time you retort. If you use extra tape here you will have little trouble getting the cap off. Otherwise it will bind. So use teflon pipe tape wrap every time that you use this retort.

One bit of construction that you must do in making this retort is to drill out the 1/2" tube adapter so that a 1/2" stainless steel tube will go through it. All the rest is simple assembly of pipe.

When using the retort follow all of the rules listed in Issue 1 of this magazine concerning mercury safety. Allow a small trickle of water to run continuously through the water jacket while you are retorting. If you don't want to use your hose then use a small reservoir and a water pump from a swamp cooler to pump water through the water jacket.

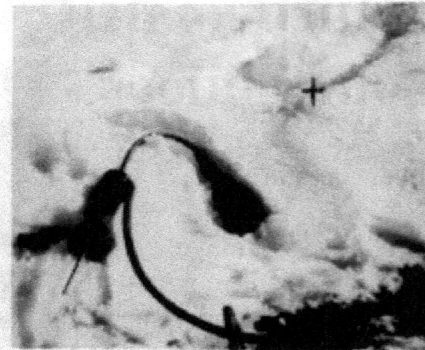

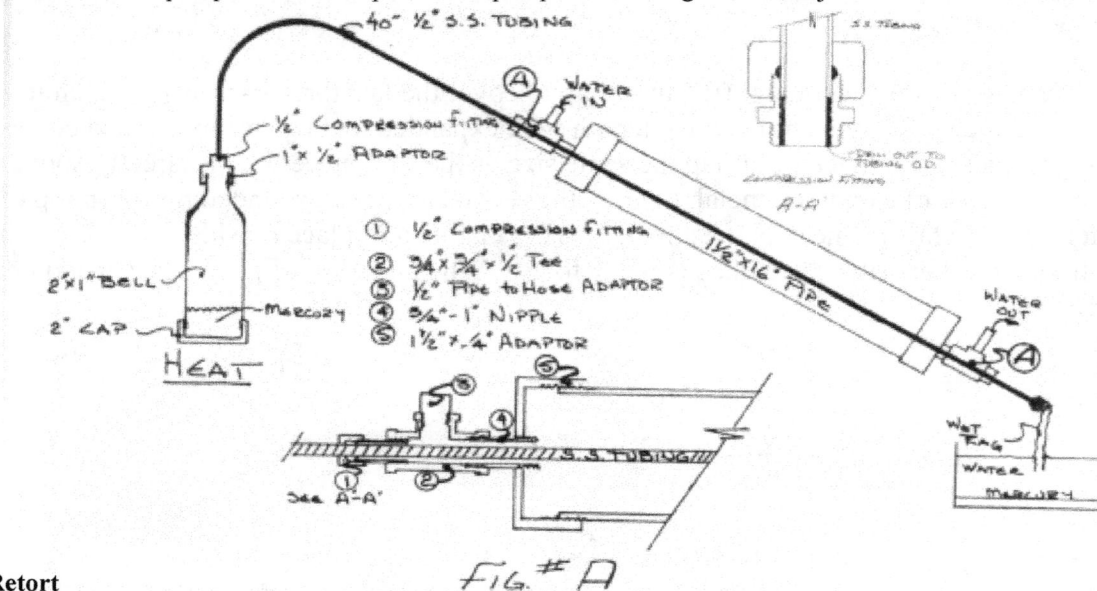

FIG. # A

Mercury Retort

Editor: While looking back in the April/May issue of POPULAR MINING, on page 22 the article on a "Mercury Retort from Pipe" is sound, but remember galvanized and copper pipe are not totally compatible with mercury, as the copper and other alloys tend to "kill" or deaden the mercury. The mercury can be rejuvenated with other chemicals at an additional cost. Try to construct the retort with stainless steel, a one time larger expense, but is better than buying additional chemicals later. Remember, SAFETY FIRST and double check all operations for SAFETY. Stainless is only needed wherever mercury comes into contact with the metal, the outer jacket can be any 1.5 pipe as it does not contact the mercury or the fumes. The exit pipe should be submerged underwater a little to prevent the mercury fumes from escaping into the air YOU BREATHE.

I consider myself a recreational-weekend miner, it is a lot of fun to find a little spot of color in your pan or sluice. It is something like trout fishing, "there is a bigger one just in the next pool, or it got away", or "you should see the gravel bar I could not work as we had to come home". Does that sound familiar? .....Keep up the good work. Yours truly, Stephen B. Barnett - San Carlos, California

More on Mercury

Dear Sir: I've read your magazine since its inception and noted the articles on mercury and retorts. So, I thought I would add my two-bits worth.

My operation requires the retorting of 30 to 60 Ibs. of mercury at time and I use a cast iron retort with a clamp-type lid. The old timers used red clay to accomplish a seal and this works fairly well, but I use asbestos ground up in water to make a thick paste. When the paste is placed on the seat of the retort and the lid clamped down, it just doesn't leak.

I enclose my retort in a cairn (a cone-shaped mound) of fire bricks that form a bee hive shape that encloses the whole thing and its burner. The gas burner is never turned up high, but is allowed to slowly come up to heat - it usually takes 6 hours to run 30 Ibs.

In addition, I place a sheet of aluminum foil on top of the cairn, and if it leaks even a little bit, it causes minute perforations in the sheet that are easily seen during the daylight. If I have to run at night, I use a UV lamp shining on the cairn until I've recovered the first pound or so.

I enjoy your magazine -keep up the good work. Leonard Whaley-Sacramento, California

More Information On Mercury Retorts

(The following information was sent in by Ernie Wells in Elgin Oregon. He received it from Walt Lashley of the American Society for Applied Technology. The correspondence is in regards to a Letter to the Editor in Issue #5 and the original retort article in Issue #2.....) by Walter Lashley, American Society for Applied Technology, Silver City, NM

Walt wrote to Ernie: "In regards to the plans for a mercury retort written in the April/May 1984 issue of PM, "a wet rag is ok, and so is the stainless steel. If the tube is immersed in water, there must be a surge tank between the water and the retort. For if a drop of water were drawn into the kettle, it would become a bomb."

Additional information regarding the retort...all the parts should be of black iron, including the condenser tube. Just bend 1/8" pipe to a gentle curve which slopes down at about 45 degrees, and let the end protrude into cold water. Watch your temperature, for if the mercury boils violently, it will carry off all of the fine gold with the vapors.

The mercury dissolves the zinc from the galvanized fitting within minutes and forms a stiff, silver/grey mass which could be crumbly if any other metal is present. Blue coloration in nitric acid denotes the presence of cupric copper ions, and the brown fumes are nitrous oxide, which is always given off under these circumstances. Your light black residue is. the gold. Finely divided gold has a color dictated by its particle size, and it ranges from light brown to purple black.

If you will put your black powder in a scorifying dish, with 3:1 lead and 5 to 7 grams of borax glass, and fire it at 1000 degrees C. for 30 minutes, you may then cupel the resulting button in a bone ash cupel and recover your gold in one piece."

The American Society for Applied Technology is a nonprofit scientific foundation located at Box 1705, Silver City, NM 88062 (505) 538-3849. It would be worth your while to check it out about joining. They share technology with members.

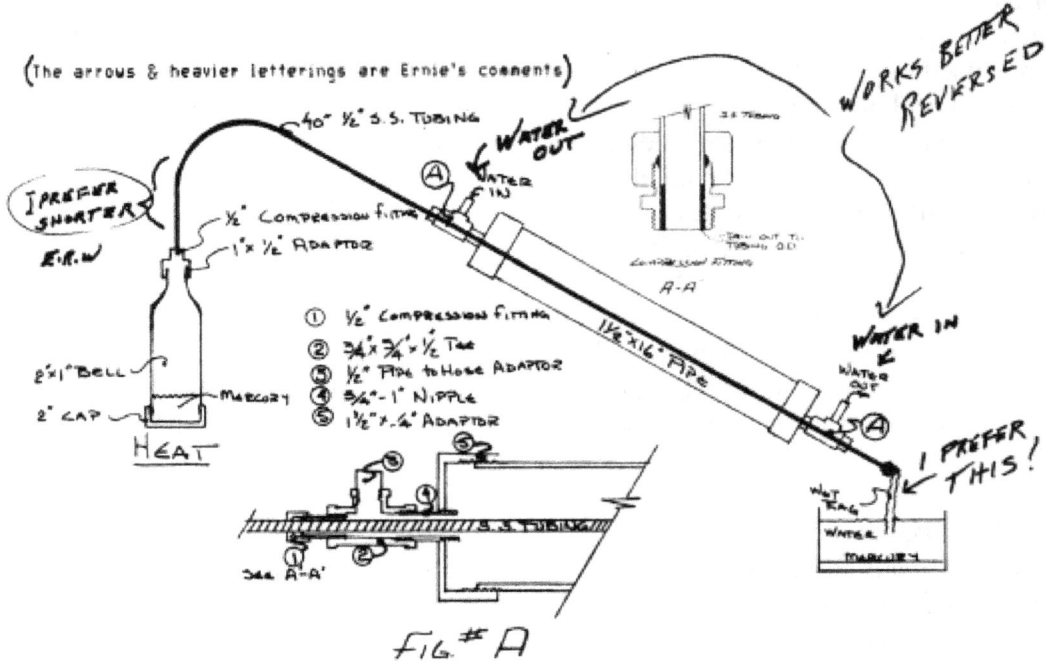

Fig. # A

Ernie's comment's on retorts follows:
Dear Jim, the retort you showed in Popular Mining (April May issue) is a good one. I read the comments about it being better to discharge underwater (Issue *5 Letters to the Editor). Never, never discharge underwater unless there is a lot of expansion space on the discharge line. This is the why for the surge tank. Many people advocate the use of black iron only. This is because galvanized pipe or fittings raise hell with the works. Stainless steel is fine . I wrote to Walt to be certain as I do so many things I can not be an expert in all (see above) and I wished to be sure this was not my error.

I would caution your readers not to discharge their retort underwater but to use a cloth tube hanging into the water. (See original article with drawing) If the whole works was hot and you made a quick cool an unwanted explosion could occur.

I made one similar to the one shown in PM except I heated the tubing and used a "flare fitting" on top and did not let the tube rise so high from the retort tank before it started down and put the water in the bottom and make overflow at the top of cooling pipe. *Use extra amounts of the pipe teflon tape wrap on the cap*

EDITOR NOTE: See original write up on previous page and the notes that Ernie made on it for reference A Mercury Retort from Pipe

Making and Maintaining BABBIT BEARINGS

By M.D.(Doc) Isely

A small mining or milling operation can conserve capital by rebuilding old, discarded, machinery. Such machines often have babbit bearings. Bearing failure was a common cause of the final breakdown which caused the machine to be discarded. Major plants with their lubrication crews rarely suffer shut down because of a failed bearing but small operators often ignore lubrication with disastrous results. So, if you expect a plain bearing to run with no more attention than a sealed anti-friction bearing: forget it. The old machinery is not for you.

Babbit bearings are often lubricated with grease. Grease cups must be periodically screwed up until they go no further and then refilled. Always Keep all dirt out of lubricants. This may be difficult in a dusty place but it's very important. If a grease gun is used keep it filled and handy and clean. Wipe all dust off the fittings before attaching the gun. Your bearing will probably have no dust seals except the surplus grease which runs out the ends, so don't wipe it off. If oil cups are used keep them filled above the minimum level. The same applies to wick oil founts. (Drip oilers must be started before the machine is started and as the temperature rises and changes oil viscosity the rate of flow must be reduced or all the oil will quickly run out.) Look at the transparent reservoirs occasionally to be sure that they are always filled.

The exact amount of lubrication will depend upon your machine and its uses and the temperature. A general rule is to feel each bearing and if it is much warmer than the
adjacent metal, lubrication is needed. If it is smoking lubricate at once. If the added lubricant increases the smoke, shut down immediately, or lose the bearing and perhaps score or break the shaft. It is better to over-lubricate than to under-lubricate.

BABBIT BEARING SPECIFICATIONS -
Babbit metals are formulated to bear loads from 15 to 5,000 PSI. Your machine probably does not impose loads over 300 PSI and most mine and mill equipment is so over-designed that the figure is much less. Thus you can use a low priced lead-antimony babbit. A babbit containing graphite will compensate somewhat for underlubrication but, of course, will not exclude dust. Dust seals form at the ends of unshielded bearings even if oil is used, since the dust adheres to the oil.

MAKING A BABBIT BEARING -
Clean all rust and other foreign matter from the bearing cavity so that bare metal shows. Tin the inside of the cavity using solder with an appropriate flux. If you can not completely tin cast iron, drill holes in the bearing cap and counter-bore the outside of the holes. When you pour, babbit will enter these holes, harden and act to hold the bearing in place. Place little sheet steel dams around the outside of the holes to prevent babbit run off. Complete tinning is the best method if it can be accomplished.

Pouring the babbit

If a bearing cap is lost or broken make a new one. Be sure it is stress relieved before installation. (You can stress relieve by heating to cherry red and let air

Checking the temperature

cool). The shaft must be perfectly round and free of rust or any other roughness. The babbit is softand will quickly be torn away by a rough shaft. Make a mandrel slightly smaller than the shaft if the bearing is to be bored. Otherwise make it the size of the shaft or don't even use a mandrel but pour the bearing right on the shaft. Install about four .005 shims (from the automobile parts store) between the bearing cap and the rest of the bearing housing. (These can be removed as the bearing wears).

If two bearings are to be aligned place one above the other and connect the mandrels with a rigid rod (if a mandrel is to be used). To keep babbit from running out the bottom of the bearing during the pouring, the lower plate on the bearing housing must fit the rod very closely. A press fit is best.

MAKING THE BEARING HOUSING READY TO POUR THE BABBIT -

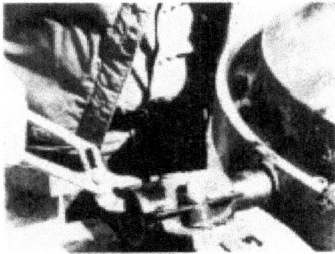
Removing the bearing to put in grooves

To keep babbit from running out the bottom of the bearing during the pouring, the lower plate on the bearing housing must fit the rod very closely. It can be made of steel and packed in place with asbestos and graphite pump packing bought at a pump supply house. It can also be packed in place with babbit bearing packing. Both of these packings will stand the heat of the pouring. Any combination of packings that will prevent hot babbit from leaking out before cooling is OK. Buy babbit packing called Babbit-rite from Marquette Industrial Materials Co. Inc. 329 E. Lake Ave, Peoria, IL 61614. (5lbs for about $10)

Use a pour temperature recommended by the babbit maker. These can run about 625 F to above 800 F but they are not critical. If you lack a pyrometer, a rule of thumb is to insert a dry stick from a piece of pine lumber into the molten babbit. If it turns brown but does not ignite the temperature is right for babbit which is mostly lead. If it bursts into flame it is right for a tin based babbit. As babbit heats, oxides form on the surface. This can be prevented by putting charcoal dust on the surface. If charcoal dust is not available the oxide itself pretty much blocks the air and retards further oxidation. Since the babbit is much heavier than either the oxides or charcoal and since it is a liquid while they are solids, it is possible to pour babbit through the lip of a ladle without getting oxides or charcoal into the bearing. Have the bearing cavity and the mandrel or shaft at about 200 F. You can use regular carpenter's chalk on the mandrel or shaft to prevent sticking or you can wrap them with two layers of newspaper. Do not use more that two wraps with the newspaper. It will burn by the heat from the melted babbit but will prevent sticking and will leave enough room between the babbit and the shaft to allow free turning.

While it is not the best way, you may have to finish the bearing by scraping the inside if the shaft will not turn or if you have used a mandrel. Scrapers are hand tools and are not high priced. You can make your own from old files. In scraping a bearing, insert the shaft or mandrel around which it was poured and turn it by hand. Add or remove shims from the bearing housing until it can be turned by hand with some resistance. Remove the mandrel or shaft by removing the bearing housing cap.   Next paint the inside of the bearing with machinist bluing. Insert the mandrel or shaft and reinstall the cap.   Turn the shaft and again remove the shaft and cap and observe the parts of the bearing from which the blue has been wiped away. These are the high spots. Remove them carefully with a scraper. Remember babbit is soft so don't overdo it.   Leave a smooth surface. That's easy to do with babbit.   Remove one shim from each side of the cap if the bearing becomes loose. It is not absolutely necessary to do a perfect job.   When the bearing fits, drill a grease hole at a convenient place in the bearing housing. This hole can be drilled with a 7/16" drill. It should go all the way through the bearing. This hole can then be tapped with a 1/4" pipe tap which is a standard size of grease fittings.   Then cut lubrication grooves all the way from one side of the bearing through the drill hole to the other side of the bearing. (This groove would be parallel with the shaft). Cut three more such grooves evenly spaced to make a groove at each 1/4 distance of the radius of the inside of the bearing.   The grooves should be about .08" wide by .04" deep.   Reinstall the shaft and clamp in place with the cap.   It must now turn without wobbling.   Next add lubrication and run the machine without load.   If the bearing heats or squeals take it apart, make sure every-thing is smooth and lubricated and add .002" shim on each side of the bearing cap.

As the bearing wears it will become somewhat loose.   If it runs out of true or knocks, it is time to remove more shims. Bearings of this type run for years, 24 hours a day, 365 days per year* without trouble when they are well adjusted and aligned to begin with and adequately lubricated after they have been installed.

Note: We know those who have succeeded in making good bearings on the first try with less data than is in this article. It is often necessary to redo the old bearings on equipment because new bearings cannot be bought. So get out those old pieces of mining and milling machinery and get'em going.

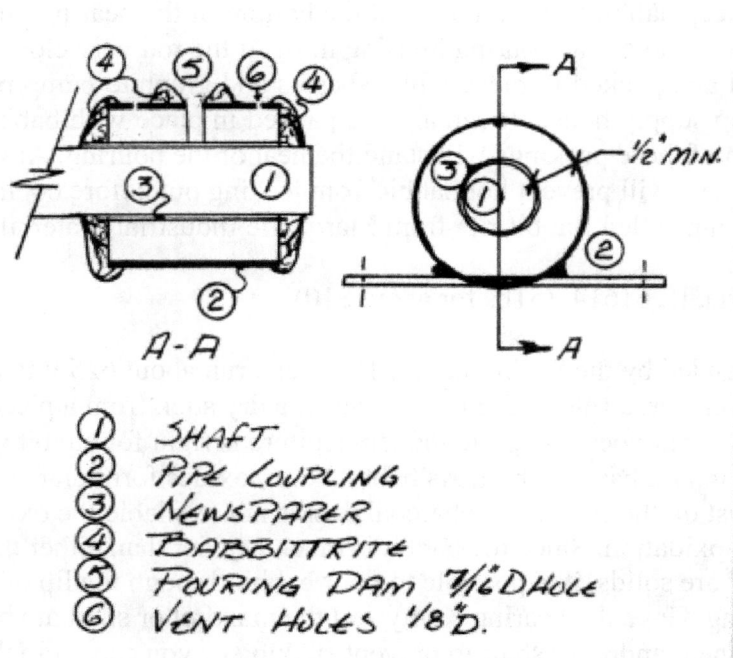

A-A

① SHAFT
② PIPE COUPLING
③ NEWSPAPER
④ BABBITTRITE
⑤ POURING DAM ⁷⁄₁₆ D HOLE
⑥ VENT HOLES ⅛ D.

How to do Sampling

by Pete H. McLaughlin

So there you are. You have found this outcrop of mineral and you crushed it and panned it and the values are staring at you from the bottom of the pan. You pick up several pieces of the mineral and send them to your assayer and back comes the report showing $900.00 per ton and you whip out your calculator and start computing the riches as though the whole mountain were ore.

Your next stop is the smelter with a 40 ton load and the owner tells you to get that worthless rock off his property. What went wrong?

Well, as an assayer I have seen this happen a couple of times. I'll get a call and the phone will smoke while the fellow tells me about what a rotten assayer I am and after he cools down a little I ask about his sampling methods and
get a dead silence that ends with "........sampling method?"

In prospecting and then exploring a mineral find, there are two different sampling techniques with very different purposes and methods. The "Prospector's sample" is a piece or pieces of the high grade material from the vein or deposit. It contains the highest concentration of the mineral that the prospector can find at the scene. This sample is sent to the assayer and should receive a broad spectrum assay. You are trying to find out what minerals are present in the vein that may be of commercial value and you really are not interested in the quantity indicated by the assay report at this point. As best you can, be sure that you have identified all the minerals present as the history of mining is riddled with stories of fabulous mineral finds that prospectors walked over for years because they did not recognize them and did not test for the right mineral. The Kelly Mine in Red Mountain, California is a case in point. The discovery point is on the path that existed between the tungsten mining town of Atolia and the bars and stores in Randsburg. The miners walked right over the outcrop for years and the ore was not discovered until someone looking for pigment for red paint had a piece assayed. The assay came back at 336 ounces of silver and 36 ounces of gold per ton.

If you have gone to all the trouble to prospect for and have located a mineralized outcrop you would be wasting your money to test for only one or two elements. Find an assayer that does spectroscopic or spectrographic assays or a wet qualitative analysis and reports the major constituents in the sample.

Now that you know what mineral or combinations of minerals you have you are no longer prospecting, you are now evaluating a mineral vein or deposit.

Grinding the ore

Vein or deposit exploration consists of a number of samples taken from the volume of the deposit to determine: (a) The location and extent of the commercially valuable material in the deposit, (b) The tonnage of the commercially valuable ore present.

Since we are talking about a large number of samples taken at the surface and below the surface this is going to cost money. You must therefore sit down and figure out what is the minimum grade of ore that would recover with a profit and how many tons of ore will you need to pay for the mining and processing investment and begin to show a profit on the operation. Your exploration must

503 826-9330

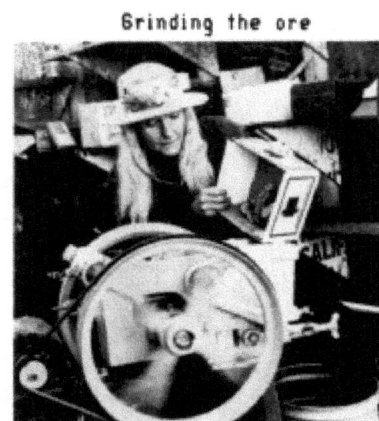

prove up enough tonnage to pay these costs or you will not be justified in starting any mining operation.   The exploration then needs to continue until you are satisfied that you have this tonnage and hopefully more.

Now we are back to the prospector in the scenario at the beginning of this article. Let's assume that he is pretty experienced at mineral identification and the outcrop he has found does not have a mineral he is not familiar with. At this point he has done the prospector's sample himself and he should begin the exploration. If he is looking for gold, the crushing and panning showed some values so there is the possibility of commercial interest here. Then the exploration begins with samples taken across the width of the vein including enough width to allow a shaft or tunnel to be bored into the deposit. (You pay the cost of removing the rock from the full width of the tunnel whether it has value or not.) The cut to remove the sample will be a "V shaped cut or even depth and width across the width of vein and gang. This material is crushed to at least 1/4 inch minus but better to 50 mesh minus and quartered and sent for assay. Send your assayer about 1 to 2 coffee cups of material so he has sufficient to do the analysis. Quartering can be very important. It is best to put the entire amount of ore in a cement mixer and let it mix for an hour or so. Then take the ore out and quarter it down to several coffee cups.

Splitting the ore

You need to make a series of samples along the surface of the vein at regular intervals of 5 or 10 feet and then get a core drill or sink shafts at the locations of these samples and get assays on the material from each 10 feet of depth until you have analyzed enough tonnage to prove the claim can be worked profitably.

There is one special case which avoids this expense of exploration before starting to mine the outcrop. When the values in the outcrop are so rich that any ton taken is capable of being processed profitably with equipment immediately available (as is the case when there is a mill within shipping distance and the prospector has enough equipment to start mining without the need of obtaining capital). Then the exploration can be considered complete when the prospector (now miner) can see that ore of commercial grade exists and is coming out of the mine as he works it. This assumes that the values are very visible or easily determined as by panning a sample of the ore after crushing it or by some other test at the site.

So there you have it. Sampling is a technology that is very important to the prospector and miner and it must be done intelligently or the resulting assays will mislead you more than they will help you.

Editor's Note: Pete McLaughlin is an Assayer. He resides in Ridgecrest, CA.

Lab Centrifugal Concentrator

Here is a simple centrifugal concentrator that you can build and it may save you thousands. It allows you to test your ore in the laboratory, or in your own backyard, to determine the feasibility of using centrifugal concentrators. In any case it can be a "quick and dirty" method of panning or concentrating from two or three pounds to 40 pounds of ore.

It's made of several skate boards (we paid $11 each skate board), an electric motor, 4 feet of 2x8 board, and plastic plumbing odds and ends. It could cost you from $22 to $80 depending on how much junk you have laying around.

ABOUT CENTRIFUGAL CONCENTRATORS

There are two basic types of centrifugal concentrators. (A) vertical models and (B) horizontal models. Obviously this Laboratory Model is a horizontal unit. We have tried both types of centrifugal concentrators and we feel that there is a lot to be said about either kind. We are convinced that both kinds will work and that there can be bad designs of both kinds.

The basic advantages of the vertical kind, such as the Knelson or the Hy-G is the ability to inject water at the bottom of the riffles creating a fluid bed to catch the micro gold. On the one hand the ore must be washed uphill to get out of the centrifugal bowl and the length of the centrifugal bowl is only 18" on the largest model with each riffle being only about 1 1/2" wide. On the other hand the horizontal kind has a length of 12 feet on the large model with riffles varying in width from a few inches to several feet depending on who makes it. So instead of 18" you have 12 feet to catch the gold. The horizontal units do not have fluid beds. We think that this is a disadvantage. The horizontal units do not have ore flowing uphill against gravity. It may be that the horizontal units, because they do not have a fluid bed, will not catch gold that is as fine as the gold caught by the vertical units. But this has yet to be proven. We did a very good job of concentrating gold in a sulfide ore using a 8 foot length horizontal centrifugal concentrator.

follows:
In building this unit be sure to refer to the drawings as well as the pictures. It is easily

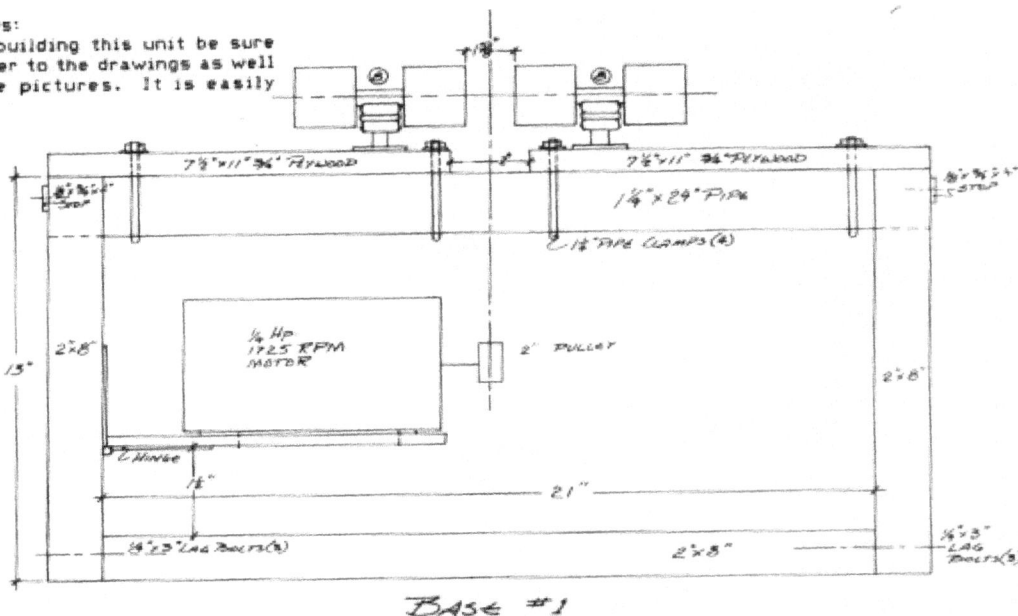

The main advantage of the horizontal unit is its ability to hold concentrates. A large horizontal unit can hold several tons of black sand. The 18" depth of a vertical unit will never hold more than a few pounds of black sand. This may or may not be an advantage. Some people do not want to keep their black sand and thus don't care if most of the black sand escapes so long as they keep the free gold. But many people have in mind the processing of black sand itself for extra

BASE #1

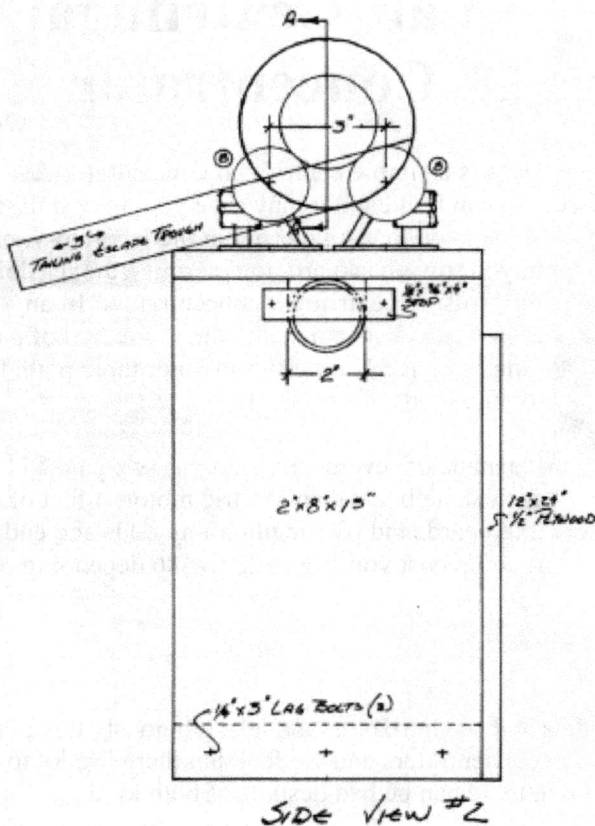

gold and thus need to keep the black sand. In such a case the vertical units are almost useless. And the horizontal units do the job. They can run for hours before they need to be cleaned out.

Finally as far as differences are concerned the horizontal units can be cleaned out in minutes by simply throwing a lever, while the vertical units must be stopped, opened up, connected at the bottom, and washed carefully out. So, as you can see the two different kinds of centrifugal concentrators have somewhat different uses as well as a different operation. I wish to state again that there are good and bad horizontal units being sold now. So if one intends to buy such a unit he definitely should see it in actual operation on his ore or similar ore before buying.

Our laboratory model of the horizontal kind does not have the lever mechanism for clean-out; since, the centrifugal riffle tube can be lifted out of the unit and washed out by hand. We saw no point in adding this feature. There are several improvements that can be made to our unit and we will report on them next issue.

STEP BY STEP INSTRUCTIONS
In building this unit be sure to refer to the drawings as well as the pictures. It is easily
built and should not take more than two days construction time. It is best to just follow the step by step instructions one at a time.

BUILDING THE BASE
(1) Cut three pieces of 2x8, two pieces to 13" and one piece to 21". The two 13" pieces must have a 2"x2" notch cut in the top end. The notch should have a rounded bottom. Do this by drilling a 2" hole at the top of each piece and then cut the hole open with a small hole saw. (See Side View #2)
(2) Glue and bolt these three pieces together as shown on the drawing. Do this by holding two pieces together and drilling 3/16" holes through one piece
and into the connecting piece. Use three holes at each end of the 21" piece. Drill the end pieces with 1/4" holes and drill the 21" piece with 3/16" holes about two inches deep. Add the white wood glue to these Joints and bolt them together with 1/4"x3" lag bolts.
(3) Cut a 1/2" plywood piece to 12"x24". See view A-A for mounting details. Glue and nail this piece in place.
(4) Coat the entire base with waterproofing. Use either a 50% mixture of Varathane 93 clear satin and alcohol or use Thompson's Waterseal. Use two coats as suggested on the directions of Thompson's Waterseal can.
(5) Mount the motor in place. The top of the MOTOR MOUNT BOARD should be 1 1/2" up from the top of the bottom of the BASE board. Use a hinge to mount the MOTOR MOUNT BOARD. We used one of the skate boards cut up as the MOTOR MOUNT BOARD. Mount the motor so that the pulley that is on the motor shaft winds up directly in the center of the BASE.
(6) Make the STOPS from a strip of 1/8"x3/4" steel bought from the hardware store. These STOPS should be cut to 4". Fasten them in place with a 1" wood screw. (The STOPS prevent the ROLLER ASSEMBLY from shifting out of place). (See Side View #2 for correct positioning).
(7) Paint the BASE some suitable color. We used blue.
This completes the base.

The ROLLER ASSEMBLY is next. The ROLLER ASSEMBLY sets in place on the BASE and is held in position by gravity and the STOPS that were installed in STEP (6) The tension on the belt also helps hold the ROLLER ASSEMBLY in place.

ROLLER ASSEMBLY CONSTRUCTION
The ROLLER ASSEMBLY is made from a 1 1/4"x24" iron pipe, two pieces of 3/4" plywood the rollers, and various assembly pieces.
(8) Cut a piece of 1 1/4" pipe to 24". Cut two pieces of 3/4" plywood to 7 1/2"x11".

(9) Using four 1 1/4" pipe U clamps clamp the two pieces of plywood to the 24" pipe. There should be 2 inches between the two plywood pieces directly in the center of the pipe. When the pipe is laid in the notches of the two end pieces the two plywood pieces should fit flush with the outside of the two end pieces and rest on top of the two end pieces. The pipe should be clamped to the center of the plywood pieces.

(10) Mount the skate wheels in place. Note that the two plywood pieces have 2" between them. One set of skate wheels should be mounted at the edge of one plywood piece so that there is 2" from the edge of one set of skate wheels to the edge of the other set on the other piece of plywood. The wheels in either set of skates should be 3" from wheel axle to wheel axle and directly over the iron pipe. (See Drawings for details). Use 10/32x1" screws with a nut and large washer on the underside of the plywood. Do not try to fasten the skate wheels in place with wood screws.

(11) Construct and install the **SPLASH GUARD**:

(A) Make the SPLASH GUARD from two 4" plastic gutter pipe caps and a 3" long piece of 4" plastic gutter pipe. Use the 3" piece inside of the caps to hold the caps together. This forms the SPLASH GUARD ASSEMBLY. Cut open each end of the SPLASH GUARD ASSEMBLY with a 3" round hole. This then allows the CENTRIFUGAL RIFFLE TUBE to pass through. Then cut a 2"x1/2" hole in the SPLASH GUARD ASSEMBLY to allow the water and tailings to drop down into the TAILINGS ESCAPE TROUGH.

(B) Mount the SPLASH GUARD ASSEMBLY with the 2"x1/2" hole at the bottom at the right end of the ROLLER ASSEMBLY so that the end of the CENTRIFUGAL RIFFLE TUBE will set inside of the SPLASH GUARD ASSEMBLY.

(C) Use a piece of 1/16"X3/4" rod bent into an L shape to mount the SPLASH GUARD ASSEMBLY. Use 8/32 screws and nuts. Hold the SPLASH GUARD ASSEMBLY in place and mark the L shape piece and the SPLASH GUARD ASSEMBLY then drill and mount both pieces as needed.

Roller assembly

(12) Mount the **TAILINGS ESCAPE TROUGH**:

(A) Make the TAILINGS ESCAPE TROUGH from 3' of white (PVC) plastic gutter. You will need to bend the gutter as shown in the pictures. Heat some sand up to about 300 F. You can do this in a frying pan on your kitchen stove. The sand should be clean and dry. Use a cooking thermometer to determine temperature.

(B) Cut your plastic gutter at the proper places for bending. Then pour the hot sand on the gutter at the point where you wish to bend it. Be sure to use gloves. Wait several minutes and then bend the plastic at the point you wish it bent.

(C) Cut pieces of gutter to fill all gaps and glue them in place with PVC pipe cement. Then fill corners and holes with silicon rubber caulking.

(D) Mount the TAILINGS ESCAPE TROUGH that you have just made on the ROLLER ASSEMBLY underneath the SPLASH ASSEMBLY. Use two screws. Drill holes in the TAILINGS ESCAPE TROUGH for the screws or you will crack it. Then silicon rubber caulk the heads of the screws.

This completes the ROLLER ASSEMBLY.

There are two assemblies left to make. The FEEDER FUNNEL and the CENTRIFUGAL RIFFLE TUBE. The FEEDER FUNNEL is next.

(13) Construct and mount the **FEEDER FUNNEL**:

(A) As you can see from the pictures the FEEDER FUNNEL is made from an oil change funnel bought from an automobile parts store. To bend the neck of this funnel again heat the clean sand to 300 F and then pour into the funnel after stopping up the bottom. After a couple of minutes, wearing gloves, bend the neck of the funnel until it is at a 90 degree angle to the mouth of the funnel. Pour the sand out and run cool water in it while holding the neck bent.

(B) At the opposite end of the BASE from the SPLASH GUARD, screw a piece of 1/8"x3/4"x24" steel rod to the 2"x8" BASE end piece. This rod will stick up higher than the CENTRIFUGAL RIFFLE TUBE and the FEEDER FUNNEL will be mounted to this rod. Mount this steel rod with two 2 1/4" 8/32 screws, flat washers and nuts. Bolt the bottom four inches of the rod leaving the upper part of the rod free to bend. This is necessary when removing or replacing the CENTRIFUGAL RIFFLE TUBE.

(C) Mount the FEEDER FUNNEL to the rod. Hold it in place and mark both rod and FEEDER FUNNEL. Then drill 2 holes and using 8/32 screws, flat washers, and nuts, mount the FEEDER FUNNEL.

(C) Mount the FEEDER FUNNEL
to the rod. Hold it in place and
mark both rod and FEEDER

① FUNNEL
② 2" PVC CAPS
③ 2" PVC COUPLING (CUT)
④ V-BELT
⑤ 4" ABS CAPS
⑥ 1½" PVC COUPLING (CUT & SIZED X 10)
⑦ 4" X 3" ABS PIPE
⑧ ROLLER ASSEMBLY

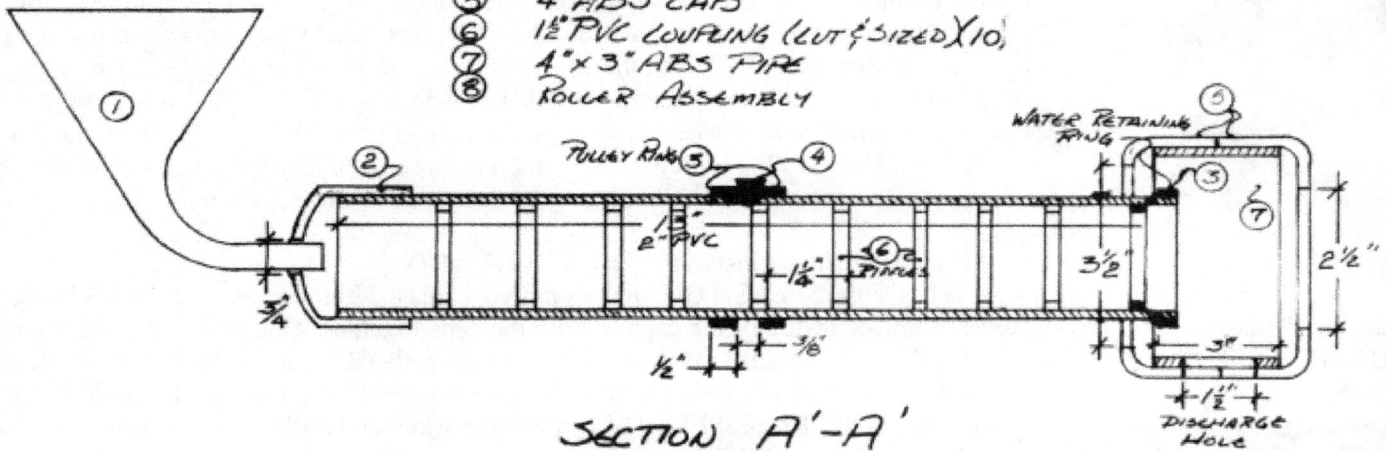

SECTION A'-A'

MAKING THE **CENTRIFUGAL RIFFLE TUBE**:

(14) Cut a piece of white 2" PVC water pipe to 13" length.

(15) Cut and install the riffles within the CENTRIFUGAL RIFFLE TUBE as follows: (The riffles are narrow rings cut from pipe that fit snugly into the 13" PVC TUBE which will become the CENTRIFUGAL RIFFLE TUBE)

(A) When cutting PVC pipe it is easiest and fastest to use a radial saw. If you do not have a radial saw then a table saw will do. If you do not have a table saw then a band saw will do. If you have none of these a plain old hack saw will do the job.

(B) After you have bought your 2" white PVC pipe you must find a piece of pipe that will fit snugly into the 2" pipe to form the riffles. This may take a little searching. A 1 1/2" PVC pipe is just a little too small to fit snug and a 1 1/4" pipe is definitely too small. We found a 1 1/2" PVC pipe cap that is somewhat bigger than PVC pipe. It fit snugly into our 2" PVC pipe. We bought five of these pipe caps and cut two 1/4" rings out of each cap. Do this by pushing the cap on a piece of 1 1/2" PVC pipe and then cut it twice with a saw to make two rings. You may find that pipe caps do not work because of different manufacturers. Such a case just keep trying various fittings; you are bound to find one that will fit snugly.

(C) When you have cut 5 caps up to make 10 rings you are ready to cement the rings into place which will form the riffles. It is easiest to put in the middle riffle first. Take a long narrow wooden stick, dip it into the PVC glue, carefully push the end of the stick half way down the 13" PVC pipe TUBE cut in STEP (14), and make an inside ring of wet PVC glue. Do this 3 or 4 times in the same place to get enough glue in one ring.

(D) Take the 13" PVC TUBE into the light and while looking down the inside use a 1 1/2" pipe to push the riffle ring down into the ring of glue. The riffle ring should set right in the center of the glue.

(E) Do this 6 more times with each ring 1/4" to 1/2" apart. The distance apart is not critical. The seventh ring from the center should be about 3/4" from the discharge end. The remaining three rings should be installed one inch apart from the center towards the input end. This should leave a couple of inches as a wash out area as the ore enters the TUBE.

(16) Install the Pulley rings: Cut two rings from a 2" PVC pipe coupling. Each ring should be about 1/2" wide. These rings will fit snugly over the outside of the 13" PVC TUBE.

(A) Lay the 13" PVC TUBE in place on the rollers and position the pulley rings in place between the two sets of rollers. Be sure the end of the 13" PVC TUBE extends into the SPLASH GUARD ASSEMBLY. Position the pulley rings 3/8" apart and mark their position with a pen.

(B) Move a pulley ring and liberally apply PVC glue to the position marked in STEP (A) above. Reposition the pulley ring back in position and on the PVC glue. Do this to both rings.

(17) Install the rear end cap.
Use a 2" PVC pipe cap. Cut a 7/8" hole in the center of this cap first and then glue the cap in place on the 13" PVC TUBE at the end where the ore will enter the tube. Be sure to lay the TUBE in place on the rollers before doing the final gluing.

(18) Install the water retaining ring. Cut one additional ring similar to the pulley rings. This is the water retaining ring. It serves to prevent the water from running back up the outside of the tube. Use PVC glue to glue this ring at the very edge of the discharge end. This completes the assembly of the CENTRIFUGAL RIFFLE TUBE.

OPERATION OF THE CENTRIFUGAL CONCENTRATOR

The operation of this LABORATORY CENTRIFUGAL CONCENTRATOR (the LCC) like all concentrators takes a bit of practice and experimenting to make it work. It seldom works the first time. It just plain takes a bit of adjustment.

The LCC should be set up so that you can try it with the CENTRIFUGAL RIFFLE TUBE at various angles downhill. We liked about a 20 degree angle but we suspect that various angles will give different results.

IMPORTANT -- We found it necessary to put a TURBULATION GENERATOR in the CENTRIFUGAL RIF-FLE TUBE in order to make the LCC work. The TURBULATION GENERATOR is simply a piece of wire stuck up into the CENTRIFUGAL RIFFLE TUBE. We used a size 12 electrical wire without removing the insulation. This wire does not spin with the CENTRIFUGAL RIFFLE TUBE. It slides against the bottom of the tube causing enough turbulation to keep the ore in the riffles fluid and thus prevent packing.
INSTALLATION of the TURBULATION GENERATOR (wire) is simple. Use a Number 12 electrical wire. Pull it out straight, push it up into the tube about 11", then bend it out from the discharge of the Roller assembly CENTRIFUGAL RIFFLE TUBE several inches so that you can fasten it to the wood BASE back. Several wood screws extending out far enough to bend this wire around several times will make a good tie down. Now this wire should lay lightly against the bottom of the CENTRIFUGAL RIFFLE TUBE. It will remain in position as the TUBE spins.

TO RUN YOUR FIRST ORE.
Set the LCC on a 20 degree slant. Fasten a water hose in the FEEDER FUNNEL. Turn on the LCC motor and then turn on the water. Adjust for a flow of water about as big around as your little finger. The ore should be fed on a continuous flow basis. When you have fed in 10 to 50 pounds of ore, turn off the water first then the motor. Remove the CENTRIFUGAL RIFFLE TUBE and wash the contents into a bowl or pan. Pan or assay the concentrates to determine the effectiveness of the LCC.

An alternate method to emptying the CENTRIFUGAL RIFFLE TUBE is to not remove it from the machine. Just turn off the motor and leave the water run. Turn the TUBE slowly by hand. It will clean itself out. You can catch the concentrates as they come out of the TAILINGS ESCAPE TROUGH.

A SUPER AMALGAMATOR?
We have added mercury to the LCC as it was running. The mercury catches in the riffles and does not come out. This means that all of the ore passes over the mercury at 10 to 40 gravities (it's 10 to 40 times heavier while spinning in the TUBE). As this article goes to press we have not had time to check out this possibility, but it appears good. We will update you next month on this feature.

When using a 2 " CENTRIFUGAL RIFFLE TUBE mercury will not spin with the tube and thus will probably not work if you use a 1 1/2" pulley on the motor. If you use a 2" pulley the mercury will spin and stay in the TUBE if you turn the water on before adding the mercury. If you use a 2 1/2" pulley it seems to us the speed is just slightly too fast.

A 1 1/2" pulley gives approximately 56 gravities. (That means increases the weight of the ore 56 times). A 2" pulley gives approximately 58 gravities and a 2 1/2" pulley gives approximately 67 gravities. We suggest that you do not even try a 3" pulley.

We built this machine so that if you wished you could easily make a larger size CENTRIFUGAL RIFFLE TUBE. The ROLLER ASSEMBLY will accommodate up to a six inch CENTRIFUGAL RIFFLE TUBE. You will have to modify the FEEDER FUNNEL and the SPLASH GUARD, but by doing this you could make a small pilot production model. Keep in mind as the CENTRIFUGAL RIFFLE TUBE gets bigger in diameter it must be spun more slowly. A six inch TUBE should only be spun at about 300 to 500 RPM.

ACKNOWLEDGMENTS: Built by Ken Horn and Tracy Sarrett. Designed and written by Jim Humble

Lab Centrifugal Concentrator Modified

by J. Warr, Riverton, Ut.

I have modified the centrifugal concentrator, as follows. It works much better for me, especially with flour gold.

I have used a variable speed motor, and modified the centrifugal riffle tube. The main problem with this type of concentrator is the tendency for packing in the riffles, resulting in a loss of heavies. The "Knelson-type" concentrator has overcome this problem by injecting high pressure water under the riffles to fluidify the bed. However, this is a difficult and expensive process.

A simpler solution for the horizontal centrifugal concentrator is to eliminate all the concentric riffles except the last one, and replace them with 2 or 4 longitudinal riffles running parallel to the length of the tube.

PROPER SPEED IMPORTANT

When material is fed through at the proper speed, it will be caught by the longitudinal riffle and slowly flow down it toward the discharge. As it flows, it has a chance to classify without packing. By the time it reaches the remaining concentric riffle at the discharge, the gold has settled to the bottom of the feed, and is caught.

This concentrator is run somewhat slower than recommended in your original article, in order to allow all the material to flow. However, speed must be adjusted for different types of feed, thus requiring an adjustable speed motor.

Also it works much better if the feed is fine and uniform (true of any concentrator). I screen mine to -100 mesh since I am working with flour gold. Other refinements can be made by adjusting feed, water flow, and angle of descent.

CLEAR TUBING

In order to facilitate speed adjustment, I have constructed the riffle tube out of clear plastic tubing rather than PVC. As a result, I can readily see if the feed is packing or flowing, and adjust the speed accordingly.

The ends of the tube are sanded flat, and 2 (or 4) 1/4" plastic rods are glued to the inside, being careful to space them evenly. The rods run from the top of the tube to within 1/4" of the bottom, leaving a space for the concentrate to evenly fill the concentric riffle.

The end caps are cut from 3/16" or 1/4" plastic sheet. Before cutting them out, drill a 1" hole in the center of each with an electric wood bit. The caps are then glued on each end with regular plastic cement.

The materials can be purchased from hobby shops, or plastic supply houses.
The only other modification is to make sure that the turbulation generator extends clear through, and out the intake opening, so that it doesn't strike the riffles.
The concentrate is flushed from the tube periodically the same as your original model.

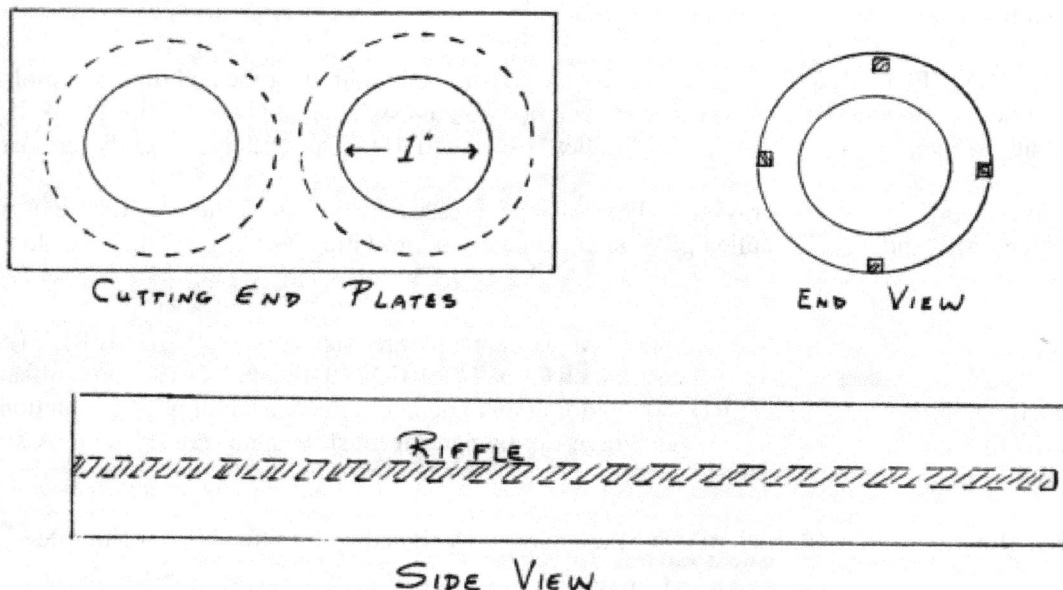

CUTTING END PLATES

END VIEW

SIDE VIEW

A Poor Man's Rock Crusher

Author unknown

The picture shows a rock crusher made from a piece of six inch pipe welded to 1/2" steel plate. The pestle is made from a pre-49 ford rear axle. The face of the bevel gear is ground flat. It will reduce goose egg size rocks real quick. Use it with a screen to turn out assay samples or samples for panning.

a

■-■

Letters to the Editor

Dear Editor, I enjoy your magazine when it comes to the office. As a chemical lab tech/assayer I must comment on your article "A Mercury Retort from Pipe". A more effective condensation of mercury vapor would result from having the water inlet at the lower end (instead of the top end). This would fill the pipe much better. Sincerely, Robert E. Johnson, Steamboat Springs, CO

Thanks for the correction.   It would indeed be best that way.   Jim.

Dear Editor;

We are one of your newest subscribers. You've no doubt heard this before, but it won't hurt to mention it once again. WHAT A FANTASTIC MAGAZINE!!! It was long overdue and we wish you all many successful years of publication. We have a question, the answer to which would no doubt provide more than enough copy for a whole series of the subject. We feel that it would also create a tremendous amount of interest.

In the March 1984 issue of the California Mining Journal was an article entitled, "Geochemical Exploration for Gold and Other Minerals". It was authored by "Evaldo L. Kothny PhD of the Soil Search Lab. The gist of the article was the assay, analysis, location of gold and of "duff" materials (mulch, leaves, pine needles, humus, etc.). The article made a good case for this type of prospecting, but it does not go into the meat of the question. Our question to you folks is: How is this material prepared and tested and evaluated?   Jim Scott Coeur d'Alene. ID.

PREPARING LEAVES, PINE NEEDLES, HUMUS, ETC. FOR ASSAYING.

We will have many more complete articles in the future on how to do Geochemical Exploration for Gold and Silver, but for those of you who can assay and want to try a few tricks here is the basic idea. (1) As you get closer and closer to a gold vein the plants will have more and more gold in then. (2) You can prospect by assaying the plants in the area of a suspected gold vein or veins. (3) The plants directly above the vein on flat land or directly below the vein on a mountain side will have the highest gold readings. (4) sometimes it is most practical to assay the plants in an area in a set pattern so that you can plot your results on a map to get the direction of the gold flow.

To assay a plant take the leaves and dry then out in the sun or in a drying oven (not over 180 F). When the leaves (or pine needles or any green part of the plant) are dry then weigh oat a given quantity, usually several ounces. One pound is more accurate but a good deal harder to handle. Whatever you start out using you will have to use every time. Your evaluation must contain the amount of gold in a certain amount of weight of a certain of certain plant. It must be the same part every time. Once the leaves are dried and weighed you must next ash them. Do this by placing them in a furnace in an open container, leave a small crack in the door, and very slowly, begin to raise the temperature. Do this so that there is no great fire and smoke. The temperature should be raised to 1500 F over a period of possibly 8 hours or more. A great deal of smoke could carry off some of the gold. When the leaves have turned to complete ashes there should be less that one ounce left. If this is not so, continue to roast until you have less than one ounce of ashes.

The ashes can then be assayed in a standard assay. If you have trouble with the carbon not burning up in the assay or sometimes taking several hours too long to complete the assay just add several teaspoonsful of sodium nitrate. This will burn up the carbon quickly. And that's it. Plot the quantities of gold that you have found on your map and you may have a vein.

Idea Corner

Idea by Jack Cota

To increase your gold recovery add strong magnets directly beneath your sluice box. As any piece of metal moves through a magnetic field it generates a small current in the metal which in turn makes the metal put off a small magnetic field. This magnetic field is then attracted towards the magnet. This then slows the movement of the gold particles helping them to come to rest against the bottom of the sluice. - Jack

Improve your mercury amalgamator by hanging a strong magnet over your moving mercury. At any metal moves through a magnetic field a voltage is generated in the metal (that's the principle of all electric generators). This voltage can, under most circumstances, help you recover your gold. Before starting mark each magnet with a red marking pen. Hang them the same way. You will find that one way will buck the amalgamation and one way will aid it. If you should be bucking, reverse the magnetics.— Jack Cota.

The pictures below show several more whirlpool concentrators that people have built. Everyone seems to be happy with them. For very finely ground ore they really do the job. A 55 gallon barrel would probably do a hundred tons a day if adjusted and handled correctly. If anyone uses one of these units for production Iet us Know. If anyone wants to buy one Action Mining can provide one ready for use for $33.00.

Running 3 Phase Motors On Single Phase Power

THE THREEPHASER UNIT

If you have three phase motors lying around but do not have three phase power it is usually more economical to use the 3 phase motors than it is to buy single phase motors, even if you can find used single phase motors. Often single phase motors will not even fit into the places where three phase motors are required. You may find it very useful indeed to buy one or two "Threephasers" for the purpose of using your three phase motors.

What is a "Threephaser"? A threephaser is a device that converts 208-240 volt single phase power to 208-240 volt three phase power. The only real disadvantage is that you will only be able to operate any such three phase motors at 80% of their ratings. This is because single phase power is less efficient than three phase power. If you bought a single phase motor it would be somewhat bigger to do the same job. So when you are using three phase motors only single phase power they will only produce 80% of their rated horse power.

Three phase power consists of three lines of electrical voltages that are always varying in respect to one another. The way these voltages vary is very precise, of course. This varying voltage is Known as phase. These voltages are varying at a 60 cycles per second rate. The Threephaser takes one of the single phase wires and makes two wires from it. This is done by adding a capacitance in one of these new wires and some inductance in the other wire. The inherent nature of capacitance and inductance creates the varying voltage that we call three phase. The third wire is one of the original single phase wires.

The Threephaser will also work with single phase 440 volts changing it to three phase. All the same rules still apply. The schematic shows the actual wiring of the Threephaser. The starting capacitance and running capacitance must be bought separately. They do not come with the Threephaser. You will need 100 MFD starting capacitance for each horse power of the motor that you are going to use. You will need 20 MFD running capacitance for each horse power of the motor that you are going to use. Both capacitors are not exactly the same kind in that starting capacitors may be much smaller and cheaper than running capacitors. This is because they are in the circuit only for short periods of time. The running capacitor is the more expensive capacitor. If your running capacitor overheats during running, it is built physically too small. Running capacitors must be oil filled.

Only Y connected three phase motors will work with the Threephaser. Three phase motors will have a Junction box with wires numbered T1 through T9. To make a Y connection of these motors connect T1 and T7 together. This becomes line one. Connect T2 and T8 together; this becomes line 2. Connect T3 and T9 together; this becomes line 3. The remaining lines T4, T5, and T4 are connected together and taped up; this does not connect to anything.

If you have an air compressor that was originally loaded more than 80%, it may not run. Just change the motor pulley to a slightly smaller size, which will give slightly less air, but will usually be satisfactory. As far as air conditioners are concerned some will work at 80% and some will not. Care must be taken to ensure that nothing is overheating in the air conditioner.

The Threephaser can be purchased from Reliance Electric Mfg. Co., Route 2, 1024 West Lake Brantley Rd. Maitland, Florida 32751 for (69.95 each for motors up to 7 1/2 H.P. If you buy 12 or more they are $59.95. You must buy capacitors from your local wholesale electrical supply. If you wish to operate a motor larger than 7 1/2 H.P. two Threephasers can be hooked in parallel. Units come with complete instructions.

Vacuum Dredge Nozzle
For Less than $15.00

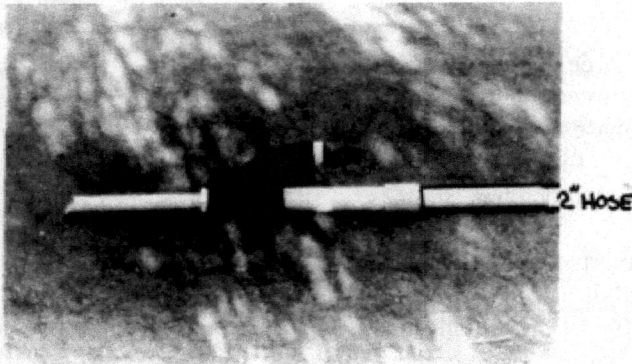

The picture and drawings show a simple vacuum head. It, like all vacuum nozzles, works on the principle of a venturi. We did not have a lot of time to experiment with this nozzle, but we did try it and it seem to work OK. We were able to pick up 1" rocks and larger with no difficulty. We were feeding the nozzle with 60 GPH of water with about 25 PSI or 50 foot of head pressure. Of course, it will take a great deal more pressure if one should wish to deliver the vacuumed material at a greater distance.

PARTS LIST
☐ 2" ABS "Y" sewer pipe piece
☐ 22"x 1-1/4" PVC water pipe
☐ 2" to 1-1/4" reducing bushing
☐ 10"x2" PVC water pipe
☐ 2"x2" PVC water Pipe
☐ 2" ABS 45 Degree elbow
☐ 2" slip to 1-1/2" thread bushing
☐ 1-1/2" hose adapter

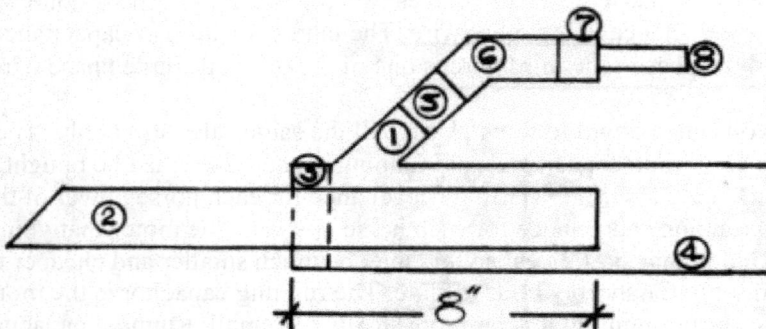

Only one piece was modified to make the complete unit. The 2" to 1-1/4" reducing bushing had to be shortened by cutting with a hacksaw in order to allow the 1-1/4" pipe to slide all the way through. It didn't exactly slip on; it was pounded on. There should be about 8" of the 1-1/4" pipe sticking through the 2" ABS "Y". This is what creates the venturi.

The 1-1/4" PVC pipe was sliced on a slant to help pickup. All pieces were glued in with PVC glue even though ABS was involved. We have found that PVC glue has always worked fine for this purpose. All the rest of the parts were simply glued together as you see in the photo. The output of the nozzle should go into a 2" hose. We did try reducing the hose down to 1-1/2" and it did work. But the suction at the nozzle was greatly reduced. Higher water pressure might make this work better but we did not have the water pressure to make the test. So we would have to suggest 2" hose on the output for proper operation and 60 to 100 GPM with a pressure of at least 25 PSI.

The nozzle will suck any size rock that will go through a 1-1/4" PVC pipe. You may have to do a bit of experimenting for your operation. Good luck. Jim

Handling Dynamite

Special Box for Transporting Dynamite

The first thing that you should know about handling dynamite is that there are permits involved. The procedures for obtaining permits are different in each state as well as different in each county in each state. However, in most places it is relatively easy to obtain a permit and it usually doesn't take more than 7 days. Of course, in some areas it could be a little longer than that.

(Please note that in the following article the word dynamite is used loosely and should be considered to cover explosives of all kinds for all mining use.)

Hopefully we can give you enough data for you to have some idea what it will require to use dynamite. Do not try to circumvent the red tape and use dynamite without the proper permits. It could leave you open to prosecution. This could be trouble far beyond the red tape involved. It is nothing like, say, building a spare room on your house without a permit. That is something you would probably never be prosecuted for, but possessing dynamite could be a felony in some cases. That means, state prison.

In many counties a small miner does not need a permit to use dynamite, but in every case he must have a permit to transport dynamite. The state of California does not require the small miner to have a permit, but some county ordinances might. However you will have to have a transportation permit from the Sheriff in order to buy the dynamite. In Los Angeles County and some other highly populated counties this is hard to get, but in many areas it is very easy to get.

Even though a small miner may use dynamite on his own claim without a license or permit the storage of dynamite requires a number of approvals. If you do not have an approval for your powder magazine it is assumed that if you buy a box of explosives that you are going to use it all up in several days time.

On the other hand if you have even one employee or you are working for someone else or if someone has hired you on a contract (in California you then must have a license issued by CAL OSHA (CALIFORNIA OCCUPATIONAL SAFETY AND HEALTH Office It requires that you have minimum of 3 years experience in explosives and that you pass written and oral examinations
In some areas you will have to construct a special metal box with a wooden liner with no metal showing on the inside. The box will have to be paint red and have the words "explosives" written on the outside. We bought a metal box from a army surplus store and lined it with 1/2" plywood and painted it red.

To sum it up: If you are miner or a mining partner but with no employees at all, you are in the state of California and you wish to hand: dynamite you would usual proceed as follows: (1) apply to the Sheriff for transportation permit. The Sheriff will usually want check your vehicle. He may require that you get a perm from the fire department. This usually takes about 7 day This time is required because they will do a check 1 determine if you have ever committed a felony. You may I required to build a special transportation box. (2) you can then buy your dynamite. (3) you cannot store the dynamite for more than several days unless you have an approved magazine for that use. Approval must come from CAL OSHA and possibly the local fire department too.

If you have one or more employees you must do all of the above plus you must have a state license issued by CAL OSHA, if you are operating in California.

Some counties may have additional requirements. Check with the Sheriff. If you are in another state check with the Sheriff as the rules will probably not be the same.

If you have not used dynamite before, or even if you have, there are many safety rules that you should know about before jumping in and doing it. Do not assume that you can do it without reading a book of rules on dynamite. We believe that anyone can do almost anything for himself, but there is no reason for being stupid. People are killed every year by dynamite. Many lose their lives from breathing carbon monoxide fumes from exploded dynamite. Be sure to read up on what you are doing.

One good explosive handbook is the "Blaster's Handbook". It can be ordered from most book stores.

The following is an incomplete list of do's and don'ts when around dynamite. (1) Don't try to use dynamite without reading a handbook first. (2) Don't store explosives, fuse, or fuse lighters in a wet or damp place, or near oil, gasoline, cleaning solutions or solvents, or near radiators, steam pipes, stoves, or other sources of heat. (3) Don't use explosive or accessory equipment that is obviously deteriorated or damaged. (4) Don't attempt to reclaim or use fuse, blasting caps, electric blasting caps or any other explosives that have been water soaked, even if they have dried out. (5) Do handle fuses carefully to avoid damaging the covering. In cold weather, warm slightly before using to avoid cracking the waterproofing. (6) Don't use short fuses. Know the burning speed of the fuse and make sure you have time to reach a place of safety. Never use less than three feet. (?) Don't cut the fuse until you are ready to insert it into the blasting cap. To insure dry fuse ends, cut off an inch or two from each end. Do not twist the fuse after inserting it into the blasting cap. (8) Don't insert anything into a blasting cap but fuse. (9) Don't crimp blasting caps by any means except a cap crimper designed for the purpose. Crimping with your teeth is utter stupidity and has caused death. (10) Do make up primers in accordance with proven methods. Use adhesive tape to hold the cap in place in holes in the cartridge. Do not make such a primer that the cap can fall out.

Further data will be written on explosives in future magazines. END

Building a Furnace Part III

Designed and built by Mike Glenn

FURNACE IN OPERATION

Last month we built a very cheap pot furnace from a 5 gallon bucket. We have used that furnace and it works well. This issue we have built a larger furnace from 55 gallon drum. We have tested this new furnace with a number of smelts using a rger crucible. It works surprisingly well. In fact we were more than pleased with the results. The mechanism for lifting and lowering the 100 pound lid is simple and works with no effort at all.

We did use a new furnace material that has better specifications than the smaller rnace last issue which we made from standard fire clay and sand. Since we were building a bigger and more expensive unit we felt that it might be worth the trouble and expense to use a better material. On the other hand our experience gained from the first unit indicates that we could have used fire clay and sand on this one too. What we art trying to say is that you could probably get by with $25 for fire clay and sand instead of $130 for special cement. However there is no real question, the special cement Super Refaete 2800 is more durable, stronger, and will take somewhat more heat than fire clay and sand.

a
la

fu

sp
cr

FURNACE IN OPERATION

We used the same torch as given in PM issue #1. It fired this larger unit with no problem and it gave plenty of heat. I'm sure that we could get 2500 F if we tried. It might take a little larger torch to get a full 3000 Ft but then most people don't need to take their furnace that high and the cement is rated for only 2800 F anyway.

This furnace has a reasonable capacity for someone wanting to smelt concentrates, or other smelting jobs. The maximum size crucible that this furnace will handle would be a number 40. That would hold 140 pounds of brass or about 40 pounds of ore and flux. But you couldn't handle that much weight by yourself. You would need either a two man pair of tongs or some other lifting method. Next issue we will show a lifting and pouring mechanism by which one man can easily handle a number 40 crucible. Meanwhile we suggest that the maximum size that any one man can lift with difficulty would be a size 14 crucible which would handle about 15 pounds of ore or 50 pounds of brass.

CRUCIBLE SIZES:

Height	bilge
#1 3-5/8"	3-5/16"
#2 4-9/16"	4-1/16"
#6 6-9/16"	5-3/8"
#10 8-1/2"	6-9/16"
#14 9"	7-14"
#40 12-1/2"	10-1/4"
#80 16"	12-7/8"

Crucibles can be ordered from Baker Supply or there are usually suppliers of crucibles in most areas. Look in the yellow pages.

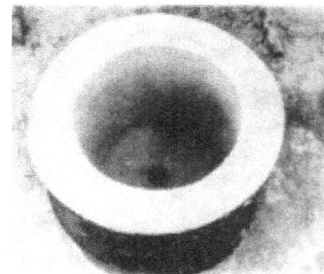

To get to the subject of building a furnace the first thing you will need is a clean metal 55 gallon barrel. Get a barrel that i« open at the top. You will be using the top portion of the barrel. There are only a few parts that need welding. If you cut everything out and get it ready you should be able to get a welding shop to do all of the welding for less than $25.00.

To make the barrel part of the furnace you will need the following parts:
1- 55 Gal. steel barrel
1- 1' piece of 2" pipe
500 lbs. Super Refacrete 2800
1- 15 gallon barrel or a section of 15" flue pipe
The Super Refacrete 2800 can be ordered from North American Refractory, 7831 Paramount Blvd, Paramount, Ca. (213) 723-3816.

(1) Measure down from the top of the barrel 21" and make a mark all around the barrel at 21" inches down. Remember you art going to be using this top portion of the barrel.

(2) Using a cutting torch cut the barrel along the lint made in step #1 above. This leaves you with a 21" barrel with no bottom. But don't throw the bottom away, you are going to use it as the top.

(3) Cut a 2-1/2" hole so that the bottom portion of the hole is 5" up from the bottom of your barrel. This hole is for the torch.

(4) Place the barrel (with no bottom) on a flat piece of cement or boards. Cover the cement or boards with a plastic trash bag first.

(5) Mix up one bag of cement (100 LBS). Use 8 quarts of water for the entire bag. Dump the cement into your bottomless barrel up to level with the bottom portion of the 2" hole. That should give you a 5" thick bottom. Pack this in place with a 2x4. You have now made the bottom. IMPORTANT: exactly in the center of your bottom make a 2" hole in the cement. This is to allow spills to run out of the furnace. You can make this hole with a pipe when you art pouring the cement.

Note: Use your fingers to form the bottom so that there are no sharp corners and the bottom slopes somewhat toward the center hole.

(6) Place the 1'x2" pipe in place. This is where you will put the torch to fire the furnace. Put this 2" pipe in place so that the fire will go into the furnace on a tangent. Put the pipe through the holt and bend it to the side so that the flame will flow in a circle. Look at our picture. This pipe should not stick out into the furnace at all, but should be set just into the furnace wall. It is best to cut this pipe off on an angle so that the corner of the pipe will not project into the furnace. Set the pipe level.

(7) At this step either use your 15 gallon barrel or a flue pipe that is 15" in diameter. (The 15 gallon barrel must be about 15" in diameter; a plastic barrel would be best.) If you use a flue pipe you will need an extra several hundred pounds of sand. Put the 15 gallon barrel or flue pipe in place in your 55 gallon barrel right on top of the 5" of cement in the bottom. This is what you use to form the sides. Be sure the flue pipe or barrel is exactly in the center.

(8) Mix another 100 pounds of cement and begin pouring around the sides of the unit. Use 7 to 9 quarts of water per sack of cement. It must not be too runny or it will seep out from under the flue that you are using to form the sides.

(9) Continue to mix cement and pour into the sides. The entire amount of cement used should be approximately 400 pounds.

(10) When you have reached the top, add Just a little extra cement at the top. You want the cement to stick 1/8" to 1/2" above the metal of the 55 gallon barrel. The cement will have a tendency to shrink when drying. So keep a little cement on hand to add if this area should start to shrink below the metal level.

(11) One problem remains. That is to remove the 15 gallon barrel or flue pipe before the cement entirely hardens and then open up the 2" pipe hole so that the flame can come into the furnace. This must be done before the cement hardens or you will have much difficulty chipping the hole out properly. This cement becomes very hard.

This completes the furnace barrel.

MAKING THE FURNACE LID
PARTS REQUIRED:
100 lbs Super Refacrete 2800
1) 16"x 1-1/4" pipe
2) 3"x 8" x 1/4" steel
2) 1-1/2" pipe couplings
10' 3/8" rebar
2) 2"x 1/2"x 1/4" steel

(1) The bottom of the 55 gallon barrel is used for the lid. Again mark and cut the barrel. The lid will be 4" thick so mark the barrel 4" from the bottom and cut. This will give you a bottom and 4" sides.

(2) Cut a 4" hole in the center of the bottom. This hole is a must. The heated and spent gasses from the torch must exit from this hole and you will also use the hole to view the condition of your smelt.

(3) Cut a second hole 1-5/8" in diameter in the bottom up against the side. The side must be reinforced at this point. See drawing. Use two pieces of 1/4" steel 3"x8". Use a hammer to bend these pieces of steel in a radius that will fit inside against the curvature of the barrel and outside against the curvature of the barrel. Clamp them in place one on either side of the barrel side wall with a "C" clamp and drill four holes for 3/8' bolts and bolt them in place.

(4) Directly opposite of the hole cut in step 3 above and at least two inches beyond the middle 4" hole run a 3/8" rebar (concrete reinforcing bar) from one side of the lid to the other. Make holes in the side of the lid for this rebar. The rebar should be 2" deep in the concrete when it is poured. Let this rebar stick out at least 2" on each side then bend the 2" piece upward. You will need to weld a brace to this rebar later.

(5) The lid should have several more rebar pieces from one side to the other. Be sure to miss the 4" hole in the middle. Each rebar should stick through the side at least 1/4". If you are using the fire clay and sand instead of the Super Refacrete 2800 cement you should put in extra rebar. In the cast of fire clay we would suggest obtaining several pounds of baling wire (from most any hardware store). Just wad the wire up in balls and stuff into the lid. That will help hold it together and prevent cracking.

The next three steps must be done by a welder. If you cannot weld then we suggest that you get all pieces ready and take them to a welding shop. The cost should be very low.

(6) Refer to drawing. Piece #2 (the 1-1/4"x16" iron pipe) should be welded in place next. This pipe sticks out the top and rear of the lid. This pipe must be welded to the inside reinforcing piece that you bolted in place in step 3 directly above.

(7) Weld pieces 3a and 3b (1-1/2" pipe couplings) in place. One piece (3a) is welded to the lid directly out from piece 2 (the piece in step six directly above). The weld will be on the outside reinforcing piece that was bolted in place in step 3 directly above. The other piece 3b is to be welded in place with the bottom end of 3b 8" above the top of the furnace lid. There must be two small pieces of 1 / 4" thick metal (1/4"x1/2"x2") between piece 2 and piece 3b. This is to keep pieces 3a and 3b directly lined up with 3b directly over 3a.

(8) Weld two rebar braces (No. 8) in place. These braces will weld to the rebar pieces installed in (4) directly above and to piece 2 9" above the top of the furnace lid.

(9) Drill a 3/8" hole through both sides of piece 2 down 1" from the top. The direction of the drill must be parallel to the lid. (Refer to the pictures and drawings if there is any question).

(10) Finally you are ready to pour the cement (or fire clay). Brace the lid upside down on a small flat surface. (The 15 gallon barrel will do). Mix one sack of cement (100 pounds) with about 8 quarts of water and pour the lid full. The cement should extend above the sides about 1/4". You will have to form the center hole with your hands. (Wear rubber gloves). To keep the lid from cracking, put several thicknesses of damp cloth over the lid and keep the cloth damp for 12 hours. This completes the lid.

THE LID LIFTING MECHANISM:

The LID LIFTING MECHANISM consists of the BASE upon which the FURNACE rests the PIVOT PIPE, and the LIFTING MECHANISM. (And next issue the CRUCIBLE LIFTING MECHANISM will be attached to the PIVOT PIPE.) Some welding is required, but the cost should be very low if you cut all the pieces first.

PARTS REQUIRED:
4) 18" x 2" x 2" x 1/4" angle
3) 4" x 2" x 2" x 1/4" angle
1) 7' x 1" pipe (PIVOT PIPE)
2) 18" x 1-1/2" x 1-1/2" x 1/4" angle
l) 24" x 1-1/4" pipe
1) 5" x 1-1/4" pipe
1) 14" x 1-1/4" pipe
1) 36" x l" x 1/8" flat steel
1) 2" x 1-1/4" pipe

(1) Build the BASE. Cut all of the pieces and weld the BASE together. Use the 7'x1" PIVOT PIPE as one leg of the base.

(2) Measuring up from the bottom of the 7' x1" PIVOT PIPE 22" make a mark. Slip the 2"x 1-1/4" pipe piece over the and of the 7' x 1" pipe down to the mark and weld it in place making a weld all the way around the pipe. Be sure to make a good weld but do not weaken the 1" pipe in the process.

(3) The last piece to be welded is the 5" x 1-1/4" pipe. You must weld two 1/2" nuts in place to be used as the pivot bearing for lifting the furnace lid. Do this as follows: Take two 1/2" nuts and weld them on either side of the pipe about half way from one and of the pipe. Then grind the corners of the nuts to make them somewhat round. Grind off the welds to make a smooth pipe with short round pieces sticking out. (You could use two short pieces of 5/8" rod in place of the nuts if you wish.) No more welds should be required.

(4) The FURNACE should be set in place on the BASE. Set the FURNACE so that the TORCH hole is at one side or the other out of your way.

(5) The one inch pipe must now be bolted to the FURNACE at the top of the FURNACE. Use the 36"x1"x1/8" piece of flat iron. Bend this piece in the center across the 2"x 1-1/4" piece of pipe that was welded in place in (2) directly above. Drill through the flat iron into the pipe and thread the pipe with 1/4"-20 threads. Drill a hole in the flat iron large enough to pass a 1/4"-20 bolt and bolt it in place.

(6) At each end of the 36" x 1" x 1/8" flat iron drill 3 holes that will pass a #12 sheet metal screw. Through those 3 holes drill small holes that will accommodate the threads of a #12 sheet metal screw into the sheet metal of the FURNACE. Install the sheet metal screws (use a large screw driver).

(7) Mount the FURNACE LID in place. Get several men to help you lift the lid up and slip it down over the 7' x l" PIVOT PIPE.

(8) Make the LEVER ARM. Use two 18" x 1-1/2" x 1-1/2" angle irons and one 24" x 1-1/4" pipe. Do this as follows:

(a) With the two 18" angle irons clamped back to back with the flat portion on top, drill at 3" from one end a hole directly through the down portions of the angles. These holes must be large enough to easily fit over the 1/2" ground nuts welded in place in (3) (c) above.

(b) Mount the two 18" angle irons onto the 24"x1-1/4" pipe so that the flat portion of the angle it on top and so that the end of the angle iron that has the hole is sticking out beyond the pipe approximately 6". You can do this with three 1/4" bolts and nuts or by welding. When bolting or welding this LEVER ARM together be sure to put the piece made in (3) directly above in place in the holes drilled into the 18" angle irons.

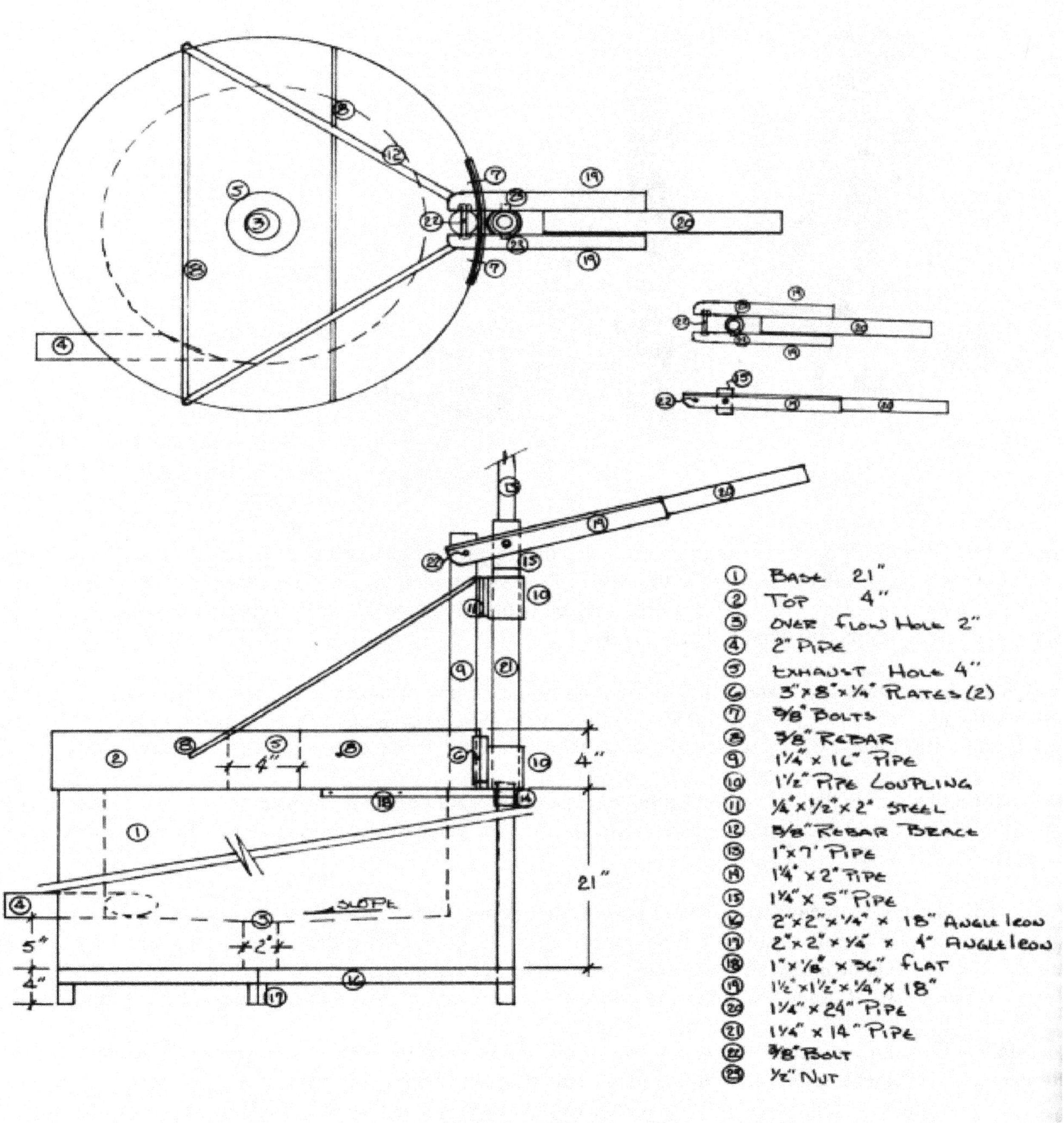

4"
21"
5"
4"
2"
SLOPE
(1) BASE 21"
(2) TOP 4"
(3) OVER FLOW HOLE 2"
(4) 2" PIPE
(5) EXHAUST HOLE 4"
(6) 3"x8"x1/4" PLATES (2)
(7) 3/8" BOLTS
(8) 3/8" REBAR
(9) 1 1/4" x 16" PIPE
(10) 1 1/2" PIPE COUPLING
(11) 1/4"x1/2"x2" STEEL
(12) 3/8" REBAR BRACE
(13) 1"x7' PIPE
(14) 1 1/4" x 2" PIPE
(15) 1 1/4" x 5" PIPE
(16) 2"x2"x1/4" x 18" ANGLE IRON
(17) 2"x2"x1/4" x 4" ANGLE IRON
(18) 1"x1/8" x 36" FLAT
(19) 1 1/2"x1 1/2"x1/4"x 18"
(20) 1 1/4" x 24" PIPE
(21) 1 1/4" x 14" PIPE
(22) 3/8" BOLT
(23) 1/2" NUT

(9) Slip the 14" x 1-1/4" pipe in place. Do this by slipping this pipe over the 7' x l" PIVOT PIPE and lowering it down through the two 1-1/2" couplings welded to the FURNACE LID until it rests against the welded 2" x 1-1/4" pipe. (Welded in place in (2) directly above).

(10) Slip the LEVER ARM in place by slipping the piece made in (3) directly above down over the 7' x l" PIVOT PIPE. It should rest directly on top of the 14"x 1-1/4" pipe put in place in (9) directly above. Now check the rounded off 1/2" nuts. They should be about 1/2" higher than the 1/4" holes in the top of the 1-1/4" pipe attached to the FURNACE LID. If these rounded nuts are more than 1/2" higher then remove the LEVER ARM and cut a piece off of the piece made in (3) so that the rounded nuts are 1/2" higher than the 3/8" holes mentioned above.

(11) Drill 3/8" holes in the LEVER ARM so that you can put a 3/8* bolt through the end of the LEVER ARM and the top of the 1-1/4" pipe attached to the FURNACE LID.

At this point you should be able to use the LEVER ARM to lift the lid high enough to swing it out of the way. You are ready to use the FURNACE. Next issue will cover a lifting mechanism for a large crucible and in the future we will have a TILT and POUR furnace for you to build. Any operation that does smelting should have all of these furnaces available to them for various different jobs. So why not build them all and send us some pictures.

Understanding Sluices Part II

by Tom Bryant

Riffle Boil vs. Riffle Splash

Further in our search for the perfect sluice box we will discuss too much of a good thing. I don't know how many times I've come across people running their sluice box in such a way as to defeat the whole purpose. A common mistake is <u>too much</u> cavitation. It is simple enough to do. Here you are working away and your sluice keeps loading up way too fast with concentrate. What do you do? With most systems you can either increase your water which will have some effect or if you want to have a major effect you change the angle of the sluice. Velocity is what moves the material through the sluice and since angle effects velocity to a much greater degree than volume, your angle adjustments will be the major control on what your sluice keeps and what it throws away.

With a dredge or with any system already running to maximum volume you are stuck with increasing the angle as the only easily adjustable variable. If increased angle equals increased velocity that means more water units per second can be run through the sluice. So... increased angle equals increased velocity equals increased volume capacity.

Now all that means is that if we don't increase our water volume when we increase our sluice angle, our water flow through the sluice will get shallower and shallower. If your water flow becomes too shallow you will have too strong a cavitation (riffle boil).

Figure 1 illustrates good cavitation. There is sufficient depth over the top of the riffle. The lightweight material stays high and moves out of the sluice without getting caught up in the cavitation eddies. Rocks are also transported in a skipping motion down the sluice and there is less chance of rocks getting hung up on the riffles. Meanwhile your heavy material is lower in the current, if not at the bottom of the sluice, and is fair game for the cavitation traps behind each riffle.

Now take a look at Figure 2. Here we have too shallow a water flow. The water pours over the riffle like a waterfall and gums up the cavitation to the point that gold is being dug out of the concentrate pile by waterfall currents as fast as the cavitation current can grab it. This action allows everything to contact the riffles. You will see rocks and light sand sucked into the cavitation traps which can only make the system inefficient if not useless.

Here are a few "rules of the thumb" you can use when you start adjusting the variables of velocity and volume:
(1) If you make a substantial change in angle you will have to adjust the water volume.
(2) If you see a pronounced bump or wave over each riffle your water is too shallow. Don't confuse riffle boil with riffle splash. Just because you don't see the water over each riffle jumping around making whoopee doesn't mean your sluice isn't working.
(3) Remember deeper water doesn't move gold - higher velocity does.
(4) If you are in a situation where you are using all the water you've got and still getting too much cavitation you might consider getting lower riffles. Lower riffles do not need as much water to "drive" them.
(5) If your riffles are as low as you want them and you still have too much cavitation you will have to narrow your sluice if you can't increase your water volume.
(6) The only time deeper water (higher volume) is a problem is if it's so deep and fast compared to the length of the sluice that fine gold can't settle through the depth of water before it reaches the end of the sluice. A longer sluice would help solve the problem.

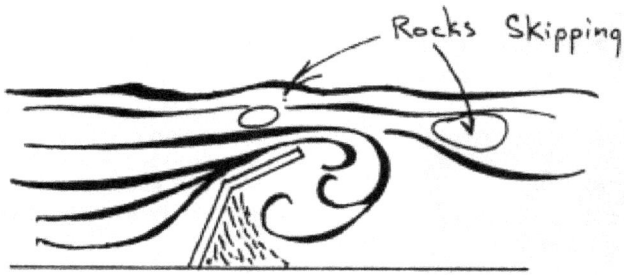

Fig. 1
Good Cavitation

(7) If the riffles have at least one inch of water running over them when running 1/2" minus material - use deeper water for coarser material -then you will be close to the needed volume. A dredge running coarse gravel would need deeper faster water which makes it harder to save fine gold.

The triple sluice dredge sold by Keene is the best way to go if you want to save fine and coarse gold with a dredge. The center sluice runs 1/2" plus material with deeper, faster water while the two outside sluices process material that's 1/2" minus using slower, shallower water.

(8) The most efficient gold separation can only take place with classified material. Trying to save 200 mesh gold while running 4" rocks through a sluice is not the most efficient way to go. The better your classification the easier it is to adjust your separation system.

(9) Can you see the top halves of your riffles as the water clears between gravel feedings to the sluice? If you can't you're burying your riffles too deeply in the rocks and dirt - back to adjusting sluice angle and water volume.

(10) Volume keeps the solids in your sluice suspended, velocity moves them out. You may have enough velocity to move light material out but if you don't have enough volume to keep your solids in a highly liquid state your sluice won't work worth a darn. You will have a loading problem you wouldn't believe, as the sluice is processing wet dirt instead of dirty water. A sluice is designed to process dirty water.

(11) Compromise!! -You'll reach a balance between volume and angle compared to riffle loading. Always keep in mind that your aim is to save gold. Running a nice, neat, unloaded sluice that only has to be cleaned out once a week may be the worst way to save gold. That beautiful system may be blowing gold off the end like crazy. Always pan your tailings as you adjust your sluice.

(12) Once you've compromised, consider the deal you've made. Have you adjusted your sluice to save every piece of flour gold but have to clean out the sluice every 15 minutes? If you have to go through $5.00 worth of labor to save 25 cents worth of gold that you might lose from the ounce of gold you would recover, you don't have a compromise, do you!?

Fig. 2   Rocks caught
Poor Cavitation

TEST YOUR TAILINGS -test your TAILINGS! - if your sluice box isn't loading up too quickly and your tailings aren't showing an unacceptable loss in gold then you're off to the races.

*****

SAVING YOUR FUEL

by Ernest R. Wells

The unit shown in the drawings eliminates any possibility of a spark from the exhaust setting a fire in a dry environment and it acts as an extremely efficient muffler. It reduces octane demand of the engine and increases the amount of power for the amount of fuel burned. The vapor mixes with the burnable gases which tend to stratify within this unit which are returned to the engine causing a more complete and cooler combustion with better upper cylinder wall lubrication which lengthens engine life. This unit may be used with any internal combustion engine that uses spark plugs that has a stationary or semi-stationary application.

Exhaust from engine to tank through steel flexible exhaust pipe.
Vapor and unburned gases back to engine from top of tank to carburetor.
Water level indicator -nylon tube on end.

A magnesium slug should be used in the crankcase to eliminate the - formation of sulphuric acid. These can be bought from most auto parts stores.

Water is poured into this unit through the exhaust outlet and a level is maintained between indicator marks. The water quiets the exhaust and as soon as it becomes heated (from the engine exhaust), the engine exhaust bubbles up through the water carrying a certain amount of water vapor with it. This vapor tends to stratify with the heavier gases which do not burn well going to a lower plane and the lighter gases rising. This causes many of the unburned gases to go back to the engine for returning along with the water vapor which lowers octane demand and allows a lower grade of fuel to be used.

While this unit is very simple a few precautions must be taken. The exhaust distributor tube in the unit must run along the bottom or floor beneath the water level. Small holes are drilled evenly spaced along the topside of this tube which is underwater. The size of these holes and the number of them must equal the area of the engine exhaust outlet. The size of the container used with an engine is determined as follows......

Use a foot or so on the unit discharge outlet to prevent outside air currents from disturbing the stratification of gases inside. This unit will cause your generator to run very quietly. I am sure that you will enjoy the silence. Ernie * *

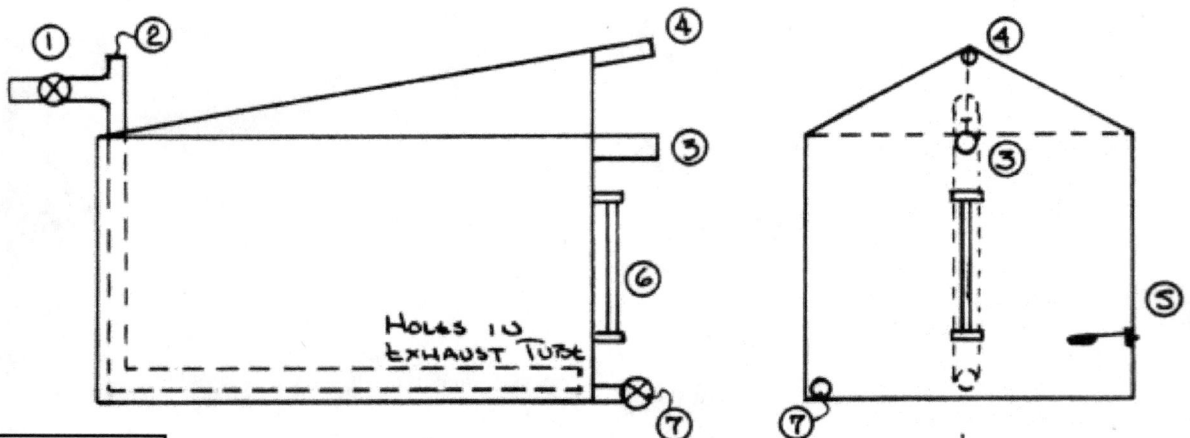

HOLES IN EXHAUST TUBE

SIDE VIEW   END VIEW

1	Exhaust Inlet Valve
2	Vacuum Break
3	Exhaust Outlet
4	Carburetor Outlet
5	Safety Shut off Switch
6	Water Level Indicator
7	Drain Valve

Construction drawings next
issue. Designed and built by
Mike Glenn.

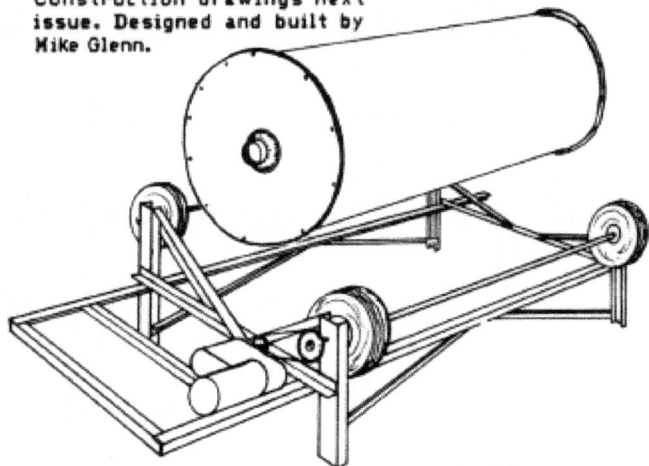

BALL MILL

BUILDING A BALL MILL

For our first ball mill we have chosen to build a simple unit 24" in diameter and 24" long. This ball mill under ideal conditions will grind 700 pounds an hour from 1/2" minus to 60 mesh minus. Under normal conditions you could expect about 500 pounds per hour and under bad conditions about 200 pounds per hour.

The cost of building this ball mill if you buy all the parts new should be less than $250 not including the driving gear motor. The gear motor should cost $250 new from a surplus house or about (150 used. If you can do the welding you will obviously save yourself money and time. We had some of the things cut and built for us by the company where we bought the steel. We will give you all the details concerning theory and how and where to buy the parts in this issue, and we will give exact construction details in the next issue. If you have experience building such equipment there will probably be enough data in this issue for you to build. If not, you should get the parts together and wait for the next issue.

THEORY

It is important to get your ball mill spinning at the proper speed. If it turns too slowly it will not grind the proper amount of ore. If it turns too fast it will grind very little ore. Most people error on the side of turning their ball mills too slowly. The proper speed is where the balls are riding up almost to the center of the ball mill before beginning to drop.

Obviously the faster the ball mill turns the more ore it is going to grind in any given amount of time. Well, that's true until you pass the speed where the balls are dropping off of the top of the ball mill. Any faster than that the balls just stick where they are. This is called critical speed and little or no grinding takes place. The proper speed for our ball mill (24" diameter) is 48 RPM.

If you want to build a larger ball mill we have included a chart for you to determine the proper speed. This chart shows various percentages of critical speed for any given ball mill diameter. For steel ball grinding wet, use 75% of critical speed. For steel ball grinding dry, try 50% to 70% of critical speed. That is also true when using ceramic balls instead of steel.

As a general rule of thumb many ball mill companies use 1/2 horse power for each cubic foot of inside ball mill space. And we have found that is not a bad figure. However, on the ball mill shown here we have used less than that. We have used about .4 HP per cubic foot of inside ball mill space.

If you are going to use a chain as we have to drive the tires that drive the ball mill then be sure to buy sprockets that have odd and even numbers of teeth. If you put an even number of teeth on your large sprocket then the small one must have an odd number. That works vice versa as well. That's the way it is.

A ball mill should have 45% of its area filled with steel balls. The size of the balls should be 3/4" to 2" for fine grinding (that's 100 mesh minus). You can use 1" to 3" balls if you only want to go to 60 mesh minus. This ball mill (24" by 24") holds 540 pounds of balls at the 45% level.

This ball mill is made from a piece of 24" steel pipe. We have lined it with rubber. The rubber that we used was 1/2" thick used rubber belting. The rubber has many advantages even though it does reduce the amount or ore ground by a slight amount. First it reduces the noise factor. But mainly it prevents the thin ball mill walls (3/8") from wearing out. The rubber itself wears very slowly and it is easy to replace.

The flanges on both ends of the ball mill were ordered cut from the metal company. Many metal companies will cut such flanges to order and the cost is very small (about $20 for the cutting).
Most of the materials for this ball mill are available in any area, however, we have listed below a number of sources of materials that we believe to be very good.

Pauley Brothers Junkyard in Rosemond, California sells steel balls for $.15 a pound. Phone (805) 256-2464 Rosemond is just outside of Mojave. Lincoln Machinery sells used mining machinery, steel balls (somewhat higher price), and rubber belting for lining the ball mill. 12920 E. Imperial Hwy, Santa Fe Springs, CA (213) 582-5491. C&H Sales sells used and surplus gear motors. You should call them for their catalog. PO Box 5356, Pasadena, CA 91107 (213) 681-4925.

If you are going to build this ball mill or any number of other things you will want a catalog from both of the following companies. Give them a call. Surplus Center, PO Box 82209, Lincoln, Neb. 68501 (402) 435-4366 and Northern Hydraulics PO Box 1219, Burnsville, MN 55337 (612) 894-8310. The last address is for the wheels and hubs that you will need.

BILL OF MATERIALS
(From Northern Hydraulics):
1) #1387 I' shaft 36" long
2) #1805 1" pillow block bearing
2) #13829 ATV adaptor hub
2) #1211 480 x 8 Tire and wheel
2) #1330 410 x 4 tire

OTHER MATERIALS:
24" of 24"x3/8" steel pipe
2) 26" x 1/4" disk plate
2) 26" OD x 24" ID x 1/4" ring
11' of 24" rubber conveyer belt
3' of 1-1/4" pipe
3' of 1" pipe

CAPACITY: A ball mill of this size is rated at 300 to 500 pounds per hour ground to 60 mesh by Denver Equipment if it is turned at the correct speed (and this one is), and if it is being fed with 3/4" rocks. We suggest that you feed it with 1/4" rocks, and thus you can get a much better return even if you are pulverizing to 100 mesh. From Denver Equipment specifications and from past experience we would say that you could get about 500 pounds an hour pulverized to 100 mesh using this mill.

If you wish to have a greater capacity, you could increase the size from 24"x24" to 24"x36" or even 30"x36", which would increase output to 750 and 1000 pounds per hour respectively. The cost increase would be less than $100.

Those people who expect to pulverize their ore to 400 mesh will need a second ball mill filled with 3/4" and 1/2" steel balls. The steel balls should be mostly 1/2" and certainly nothing more than 3/4". When the ore is pulverized to 100 mesh by the first ball mill it is then routed through the second ball mill where it is pulverized the rest of the way.

BUILDING THE BALL MILL
There are a number of things that must be welded when building this ball mill. You can get everything cut to shape and then have the welding done at a welding shop. It is* of course* much easier to just do all the welding yourself if you have a welding machine.
(1) We have used, for the barrel, a piece of pipe 24"x24" with 1/2" thick side walls. You can have such a piece of pipe cut for you at any large pipe supply yard. We suggest that you try a used pipe supply yard. One could use any size pipe up to 36" (we think) on this frame. Anyway, we are sure that a 30" pipe would work fine. But ii you use anything different than our 24"x24" you will have to change a number of the dimensions as you go along. It would probably be easier to follow our plans on the first one and then build your next one on a larger scale. When buying your pipe take a tape measure along and check the various pipes that you look at. Some pipes are not very round and you need one as round as you can get it. Just check the diameter one way and then go 90 degrees and check again. If it's within 1/2" both ways it should be ok for your purpose.
(2) Next is to buy the two END PIECES. You actually need two END PIECES and a FLANGE. The FLANGE is a ring that is as big as the END PIECE on the outside diameter, and inside diameter of the FLANGE is the same as the pipe diameter. This ring is used as a FLANGE for mounting the END PIECE that is not welded in place. The END PIECE that is at the input to the BALL MILL should be welded in place on the 24" metal pipe. (See print for details of construction of the END PIECE). The opposite end of the 24" pipe will have the FLANGE welded in place. The END PIECES and FLANGE can be bought from any large steel supply yard. They will be able to cut the pieces for you. See the print for details.
(3) The discharge END PIECE should be laid in place on the end FLANGE and a notch made in both pieces (the END PIECE and the FLANGE) so that they can be put back together in exactly the same position. Then six holes should be drilled and tapped for the mounting bolts. Use 1/2" bolts minimum. (Don't forget to notch the two pieces, otherwise you will always have trouble putting them back together).
(4) The inside of the barrel must be covered with the 24" rubber conveyer belt. Cut a piece 3.14 times the exact inside diameter of the barrel and fit it in place to make sure it fits properly. Then use Devcon cement to glue the rubber in place. Follow directions on the can very carefully. Be sure to buy Devcon. Most hardware stores carry it. Devcon makes a number of products so buy the product for gluing down rubber.

BUILDING THE BALL HILL FRAME
(5) The first step to building the BALL MILL FRAME is to build the two END FRAMES. They are shown on the print under the headings "FEED END" and "DISCHARGE END". The print shows which parts are 4" channel iron and which parts are 2"x2"x l/4" angle iron. Those dimensions marked with an asterisk (*) on the print are the important dimensions. As long as you get those dimensions correct, the position of the other bars and welds can vary slightly without affecting the operation of the BALL MILL. So, cut the various pieces of angle and channel iron, lay them out, and weld them together as shown to make the END FRAMES. Use a flat surface to lay them on.
(6) Drill and mount bearings for the drive axle and the angle iron pieces (No. 14) for the idler axle.
(?) Stand the two END FRAMES up in position and lay the interconnecting 4" channel irons in place, but do not weld them. Then mount both axles in place at they must be when in operation. Make sure measurements are correct. Make sure the axles of each wheel are the correct distance from one another at both ends of the axles. That's one of the asterisk measurements on the print. Make sure the

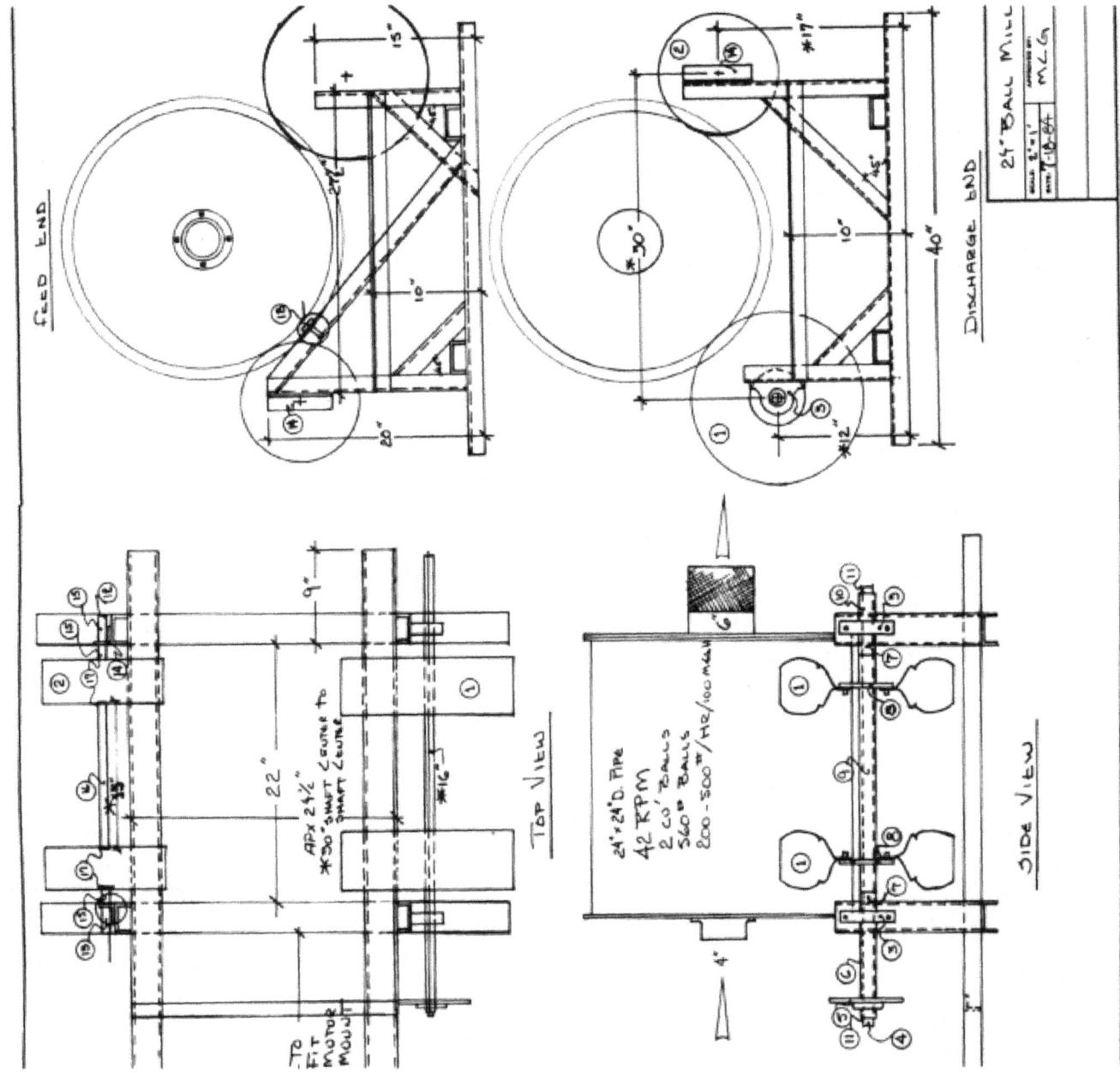

FEED END

DISCHARGE END

24" BALL MILL

TOP VIEW

SIDE VIEW

24"x24" D. Pipe
42 RPM
2 cu' BALLS
560# BALLS
200-500 #/HR/100 mesh

DRUM GUIDE

PRINT KEYS

1. (2) # 1211 480x8 tire and wheel
2. (2) # 1330 410x4 tire and wheel
3. (2) Pillow bearings
4. (1) 1"x36" shaft
5. (1) power sprocket
6. (1) 1"x?" Black pipe (adjust to your motor)
7. (1) 1"x2" Black pipe
8. (2) # 13829 ATV adaptor hub
9. (1) 1"x16" Black pipe
10. (1) 1"x1" Black pipe
11. (2) 3/4" nut & washer
12. (1) 3/4"x30" shaft (rod)
13. (1) 3/4"x3" Blk pipe
14. (2) 2"x2"x1/4"x6" angle iron
15. (3) 3/4"x1" Blk pipe
16. (1) 3/4"x15" Blk pipe
17. (4) 3/4" washer
18. (2) D.S. bearing 15mm I.D.x35mm O.D.

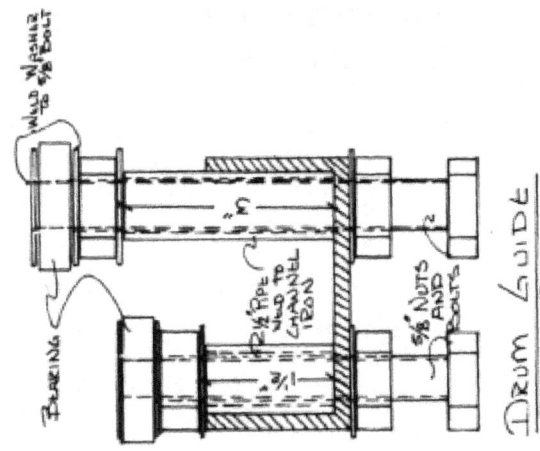

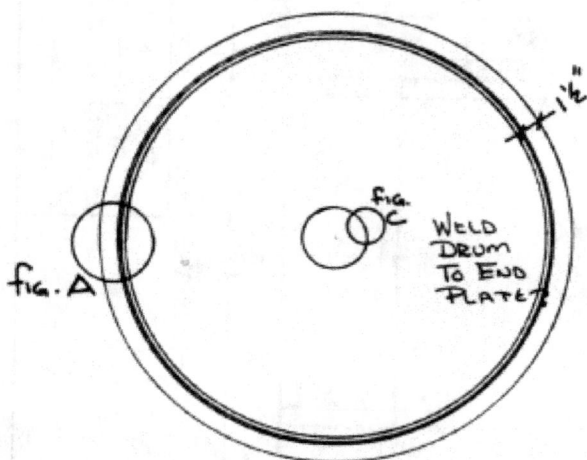

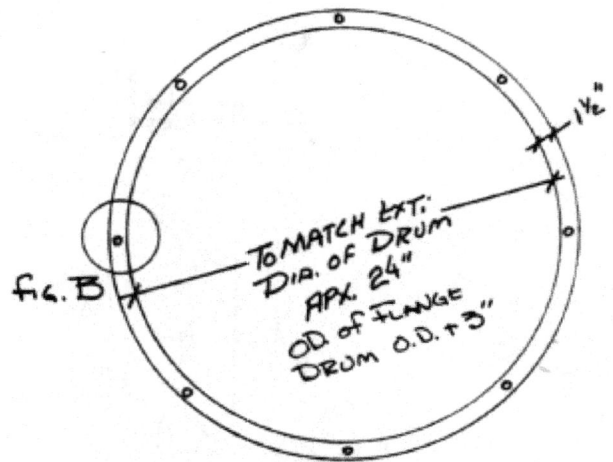

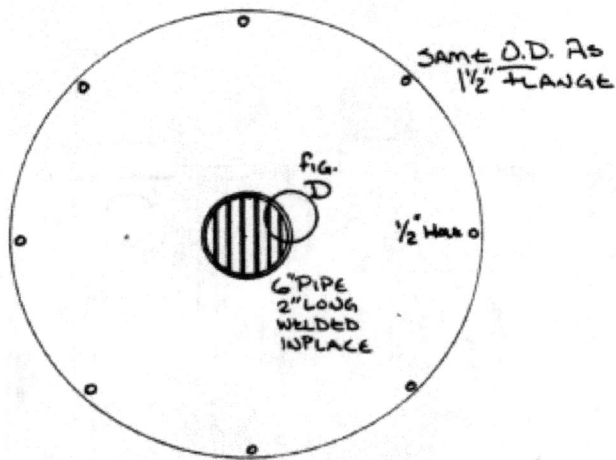

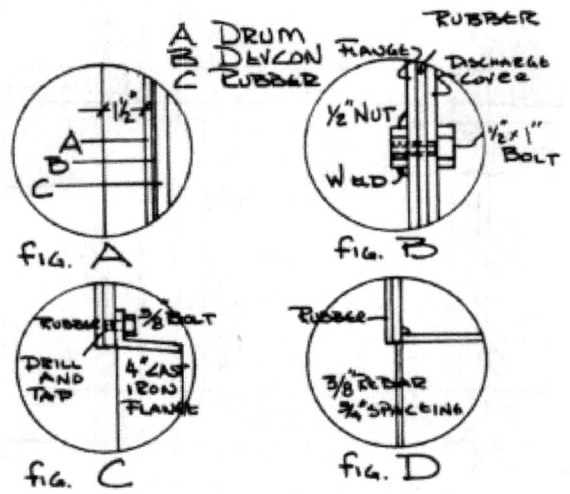

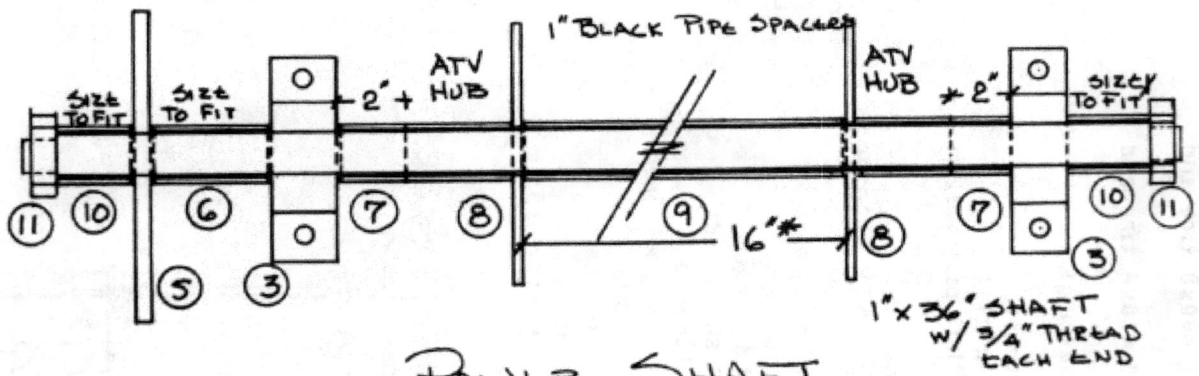

axles art not binding. When all measurements are correct, weld the interconnecting 4" channel irons to the END FRAMES. When welding the channel iron in place, keep in mind the motor that you expect to use. We allowed the channel iron to extend 18" beyond the FRAME but some motors may require more room. The motor mounting plate will set on the two extended channel irons.

(8) THE ADAPTER HUB (No. 13829) requires a bit of special attention. It is manufactured to fit on the end of the axle. You will have to modify it slightly. Use a 1" metal hole drill. Mount the 1" drill in an electric hand drill and drill into the end of the ADAPTER HUB that is smaller than 1" until you have bored through the spot welds that hold the 3/4" insert in place. Then use a bar to knock this insert out. Now you have an ADAPTER HUB that will fit onto a 1" shaft. Be sure to use a 1/4" key in the keyway provided in the axle and the ADAPTER HUB. Use the black pipe as shown on the print to space the wheels properly.

(9) Install the idler axle and two idler wheels. Use the 3/4" shaft and the 1" black pipes. Set the print for details.

(10) Weld on the motor mounting plate. It should be about 18"x24". Use 1/4" steel. If your motor is larger, use a larger plate.

(11) Mount the BALL MILL barrel in place.

(12) The D.S. Bearings 15mm I.D. by 35mm O.D. are used to prevent the BALL MILL barrel from moving sideways and falling off. These must be installed correctly. See pictures and prints for details.

USING THE BALL MILL

The first thing is getting the proper motor. A three horse power motor should work OK. But you will need a type C motor. That is a type of motor that gives 2.5 times the output while starting, in other words a very high starting torque. (That's 7.5 HP during start up until the motor has reached running speed). Otherwise you should use a 5 HP motor.

Install the END PLATE against the FLANGE on the output end of the BALL MILL. We used a heavy piece of weather stripping as the gasket. Put the gasket on the inside of the bolts, and if you install it carefully, it will not leak. Let the weather stripping run side by side for several inches where the two ends join.

You should have 560 pounds of steel balls for this mill. Use 3" balls, 2" balls, and 1" balls. Buy an equal number of each.

You can feed this mill by running a pipe into the inlet side. Wash the ore into the mill with a stream of water. You can adjust this stream going in to determine the size of particles that will be coming out. The more water flow, the larger the output particles will be.

If you don't cover the inside of the barrel with rubber don't expect it to last much longer than 100 tons of ore. Of course, the barrel is fairly cheap and you could replace it for less than $100 if you build it yourself, but there is one other factor. This BALL MILL will make tremendous noise if there is no rubber on the inside.

So, there you go. A Cheap BALL MILL for use as a pilot mill or minor production. We like ours. And let us know how yours is running. - Jim

GRINDING BALL DATA

Denver Equipment Division
Joy Manufacturing Company
621 South Sierra Madre
P.O. Box 340
Colorado Springs, CO 80901
Phone: 303/471-3443
TWX: 910-920-4999
Telex: 45-2442

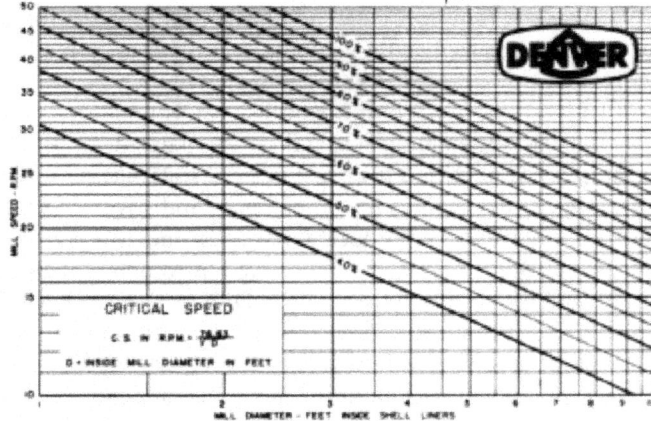

SELECTING THE PROPER INITIAL MILL CHARGE:
Many factors control the proper ball charge for a given grinding problem. These include mill size, feed size, product size, hardness and grindability of the ore. No established formula can be applied to all problems. Since the most efficient initial charge of grinding media varies for each mill application, recommendations should be obtained from Denver Equipment Company. Experienced engineers will recommend an initial charge of both the number and size of balls to approximate a seasoned charge. Operating experience will determine the optimum range of grinding media sizes needed for each application.

SELECTING SIZE OF GRINDING BALLS:
A seasoned ball charge will contain grinding balls of various sizes ranging from large replacement size down to those sizes discharged with the product. Usual practice is to initially charge a mill with selected diameter grinding balls calculated to approximate the seasoned charge.

A coarse feed or product normally requires predominantly larger diameter grinding balls. A finer feed and product require predominantly smaller balls. For either case, replacement grinding media is usually of the largest size ball unless various sizes are required to maintain a specific size gradation.

In general, the smaller the size of the grinding media the more efficient and economical the grinding operation. This is because smaller media provides more surface area for grinding. Therefore in selecting the maximum size ball, the size chosen should be just large enough to break the maximum size particle present in the feed. In selecting

the smaller size ball, consideration should be given to the fact that smaller balls wear out faster.

Primary grinding usually requires a graded charge from 4" to 2" diameter balls. Secondary grinding normally requires from 2" to ¾" sizes. Regrind circuits with fine feeds permit the use of 1" diameter balls for most efficient grinding.

VOLUME OF THE INITIAL MILL CHARGE:
Initial charges usually range between 40% to 45% of the internal volume of a ball mill. The ball charge will on the average weigh about 280 pounds per cubic foot and have about 42% void space between the balls. Volume of an initial charge should be carefully regulated so as to avoid overcharging the mill with grinding media. Overcharging results in inefficient grinding and increased liner and media consumption.

GRINDING BALLS—QUALITY:
Use a good grade of forged steel grinding balls. Experience has shown that in most cases forged steel grinding balls, uniform in roundness, hardness, roughness and density give the most efficient service. The absence of any one of these physical properties can impair grinding efficiency.

GRINDING MEDIA CONSUMPTION:
Rate of grinding ball consumption varies considerably with each application depending on factors such as hardness of material, feed and product size involved. Steel ball consumption will normally vary between 0.25 pounds to as much as 2 pounds of steel per ton of new feed depending upon the specific application.

Ball mill operating speed is expressed in percent of mill critical speed. Critical speed is obtained when centrifugal force compels material within the mill to cling to and rotate with the mill lining. This condition prevents continuous cascading of grinding media upon which effective grinding is dependent.

Critical Speed is determined by the equation:

$$C.S. = \frac{76.63}{\sqrt{D}}$$

C.S. - Critical Speed in revolutions per minute.

D - Mill diameter, in feet, inside shell liners.

The Critical Speed chart above is based on this equation and makes it possible to quickly determine the Critical Speed for various size ball mills.

Depending upon the application, normal operating speeds for wet grinding with steel balls generally range between 65% and 85% of Critical Speed, average being 74% to 76% Critical Speed. Slower operating speeds are suggested for dry grinding with steel balls, and for grinding with less dense media such as ceramic or porcelain balls. Whenever a scoop type feeder is used, caution must be exercised to keep speed of tip of scoop lip below approximately 95% of its Critical Speed.

How to Do Fire Assays

Putting crucible into furnace

Anyone can do a fire assay with a furnace and a few tools. For thousands of years fire assaying was pretty much kept secret. There are still a few assayers here and there that will tell you that you need ten years of experience and four years of college, but for the most part fire assaying has given way to the "do-it-yourselfer" just as everything else has, including childbirth. The steps are easy and just like anything else if you are careful you will get just as good results as the big laboratories. Ready?

(1) Be sure that your ore is a representative sample of your ore vein. Take chips and rocks from the entire surface of the vein and drill the vein if possible. If your mine is a placer, fire assay the concentrates only, never straight placer gravel.

(2) Pulverize the ore into a powder and stir the powdered ore thoroughly. It is ok to pulverize ore to 80 to 100 minus mesh for most standard assays, but some ores will not assay properly unless it is pulverized to 200 mesh minus. 200 mesh is about like face powder. And some ores that are complex will not fire assay properly unless pulverized to 400 mesh and finer.

(3) Measure out 29.16 grams of ore (this is an Assay Ton). One ounce is 28 grams. So on an ounce scale you would add just a touch more.

(4) Mix three ounces of the proper assay formula with the ore and place into a 30 gram crucible. A 30 gram crucible is a standard size crucible. Use standard assay formula for most assays, or use the sulfide formula for sulfide assays. You can mix your own formulas. They are given in most assay books or you can buy your assay formulas premixed from GAME CO. (We have never noticed any advantage to capping the assay, that is, covering the mixed ore and formula with salt or other chemicals).

(5) Furnace the assay for 1/2 hour beyond the time that the crucible reaches 1850 F. When starting with a hot furnace this should be about 45 minutes.

(6) Using tongs, remove the crucible from the furnace and pour its entire contents into a pouring mold. This should be some kind of a metal container with cone shaped bottom or V shaped bottom. (The GAME Co. catalog has a picture of all the equipment mentioned here).

(7) When the contents in the pouring mold has cooled, there will be a lead button at the bottom of the slag. Break this loose from the slag.

(8) Pound the lead button with a hammer until the slag is all broken off. Most assayers pound the lead into a cube shape. A small amount of powdered slag will adhere to the lead; this will not be a problem.

(9) Place the lead on a cupel and place back into the furnace at 1700 F until the lead has all been absorbed into the cupel or volatilized into the air leaving the gold and silver as a small bead in the cupel. Leave a small crack in the furnace door when doing this step.

(10) Weigh the bead, or measure its size. If you have a laboratory balance, that's not a problem. If you have no balance, a comparator will measure its size and then you can determine the weight from the chart given. Comparators are available from machinist tool stores and from GAME CO ($22.50).

The weight determines how much gold and silver was in your sample and that tells us how much gold and silver is in your ore if the sample was taken properly. For each one milligram that your bead weighs there is one ounce of metal (gold and silver) in your ore. But we must yet determine how much of the bead is gold and how much is silver.

(11) DETERMINATION OF GOLD IN ASSAY BEADS. If you can see any yellow or yellowish green color in your bead you Know that it is at least 77% gold and the rest silver. If your bead is bright yellow it will be around 85% gold and the rest silver. If your bead is dark yellow it is probably 90% to 99% gold. If your bead is a silvery color, then there is still a bit more work for you to do. You must determine the gold content of the bead and that is usually referred to as parting.

(12) PARTING - Well now that we said it, we are not really going to do it. We are going to do something that is somewhat less accurate, but is much faster. (We'll tell you how to do proper parting next time.) We are going to show you how to do COLOR DETERMINATIONS for the content of gold in gold beads in this issue.

COLOR DETERMINATIONS

(A) Prepare a small drying dish with a small amount of nitric acid solution. The solution should be two parts distilled water to one part nitric acid. You need about one teaspoonful of this solution.

(B) Heat this solution until you see fumes rising from the solution.

(C) Drop the bead into the heated solution. (Never add. the bead to a cold solution, the results will be different).

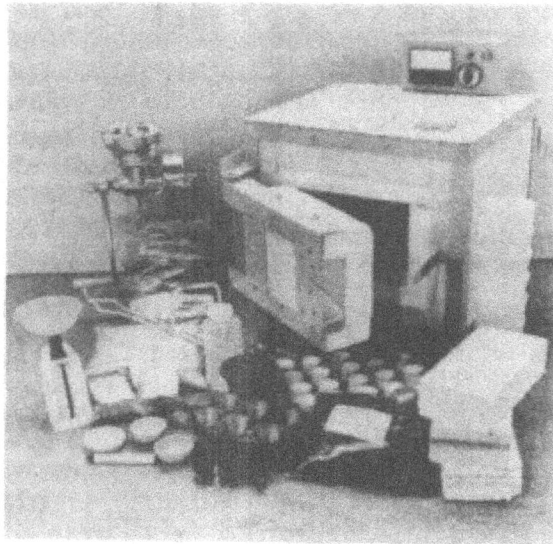

Various Assay Equipment

WEIGHT OF ASSAY BEADS

BEAD SIZE	100% GOLD	50% GOLD	100% SILVER
0.005"	.02 MG	.015 MG	.01 MG
0.008"	.08 MG	.06 MG	.04 MG
0.010"	.16 MG	.125 MG	.09 MG
0.012"	.28 MG	.215 MG	.15 MG
0.015"	.55 MG	.425 MG	.30 MG
0.017"	.81 MG	.625 MG	.44 MG
0.018"	.96 MG	.74 MG	.52 MG
0.020"	1.23 MG	1.62 MG	.72 MG
0.023"	2.01 MG	1.55 MG	1.09 MG
0.025	2.58 MG	1.99 MG	1.40 MG
0.027"	3.25 MG	2.51 MG	1.77 MG
0.030"	4.47 MG	3.45 MG	2.43 MG
0.033"	5.59 MG	4.59 MG	3.24 MG
0.035"	7.10 MG	5.48 MG	3.87 MG
0.040"	10.50 MG	8.13 MG	5.76 MG
0.045"	15.00 MG	11.61 MG	8.23 MG
0.050"	20.70 MG	16.00 MG	11.30 MG
0.060"	35.70 MG	27.60 MG	19.50 MG

(D) If the bead turns any color other than black within 5 seconds, then quickly add extra water to stop the nitric action. Or if the bead does not change color add water to stop the nitric acid action. (Remember the nitric must be hot).

(E) If the bead turned black, look at it with a magnifying glass while it is in the nitric. If the bead is black but shiny like metal, then add water to stop the action. If the bead is not shiny (flat black) and is not bubbling, also add water to stop the action. If the (black flat) bead is bubbling, then continue to heat until the bubbling stops. Then add water to stop the action.

(F) GOLD CONTENT OF ASSAY BEADS AFTER NITRIC TREATMENT.

COLOR	GOLD
no color change	74% to 77%
Yellow	65% to 74%
Bright bronze	50% to 65%
Dark bronze	45% to 55%
Brown	35% to 50%
Black shiny	25% to 35%
Black dull	10% to 25%
Black crumbly	1% to 10%

Now that you know the gold content of your bead and you have weighed it, you simply multiply the weight of the bead times the percentage found above, and that will give you milligrams. Once again, the number of milligrams will equal the number of ounces of gold per ton of ore. If you have not weighed the bead, then you have measured its size.

(13) DETERMINING THE WEIGHT AND AMOUNT OF GOLD OF AN ASSAY BEAD FROM ITS SIZE.

(a) Use the chart to first determine the weight of the bead. The chart shows how much the bead would weigh if it were 100% gold, how much it would weigh if it were 50% gold, and how much it would weigh if it were 100% silver. Your bead will lie somewhere within the two 100% figures for any given size. Your estimate will be good enough for standard assaying for gold on your own claims. After all, you just need to know about how much gold is there. Knowing exactly isn't going to do you much good; since, the next assay from the same area is going to vary anyway.

(b) Now that you have a fairly good estimate of how much your bead weighs, the next step is to multiply that weight by the figure that you got in step (F) above and that will be how much gold is in the bead in milligrams which will equal ounces per ton of ore.

MAKING CUPELS

by Pete H. McLaughlin

I have been assaying for a couple of years now and of late my bills for material started to catch my attention. It appeared to me that things had been less expensive when I started.

I checked my records and sure enough the cost of doing assaying has been going up. Cupels in particular have risen and now are some $.62 each retail.

Since I was sure that I had seen a write-up on making cupels in one of my books I did some research and found that making cupels is a simple procedure that the old time assayer sometimes had to do himself. The mold is simply made and the materials are cheap. Just how cheap depends on the quantity you buy.

The "Manual of Fire Assaying" by Charles H. Fulton and William J. Sharwood, third edition, 1929, proved to be the best source of information. It is the finest book on assaying that I have been able to find.

The Cupel
The "Manual of Fire Assaying" discusses 3 materials for making cupels. Bone Ash is the most accepted material for making cupels. Chemically Bone Ash is tricalcium phosphate and this chemical has the property that at 1800 degrees F. will absorb liquid lead oxide like a sponge keeping the molten lead in the depression of the cupel free to continue oxidizing.

MAKING CUPELS
Two other materials are discussed in the book as doing a good job of absorbing lead oxide. Magnesia which is magnesium oxide is said to work well and Portland cement also will absorb lead oxide.

I did not try to find a source of magnesia, as Bone Ash and Portland cement are both readily available. I priced them both and found that Denver Fire Clay, Ceramics Inc., Box 110, Canyon City, Co 81212 is a supplier of both ready-made cupels and Bone Ash. Bone Ash comes in premium grade and regular grade, the difference being that premium grade is pure white and regular grade is discolored. This coloration has no effect on cupellation. For cupels you will want 3X mesh Bone Ash.

A cupel that is $.62 retail has as little as $.08 worth of Bone Ash in it. This sounded good to me until I checked the price of Portland cement. A bag of cement at the hardware store is about $9.00 or $.10 per pound at 12 cupels per pound. This works out to $.008 or 8/10ths of a cent per cupel. Now that's a bargain!

The cost comparison, of course, lead to the question of losses during cupellation. Would Portland cement cupels compare to Bone Ash? In the "Manual of Fire Assaying" test results on Portland cement, Bone Ash, and 50-50 cement and Bone Ash cupels are reported. The Portland cement cupels had losses on the order of 1.3% to 1.34%. The 50-50 mix had a loss of 1.21% and pure Bone Ash had a loss of 1.26%. Portland cement is therefore equally as efficient as the pure Bone Ash material.

The Mold
Since the price savings looked promising, I checked around for a mold and found that Denver Fire Clay has a foot-operated mold available with various sized cupel dies. The unit starts at $1000.00 though, so I decided not to get one.

The only pictures of a cupel mold I could find are, again, in the "Manual of Fire Assaying" where both a foot-operated and hand-operated mold are shown. Since the hand mold was simply constructed and the dimensions given, I had a local machinist make one for me. Mr John Adam made the cupel mold for me that is shown pictured with this article. This mold varies from the one shown in the "Manual" in that the cylinder is 3" tall and the lip for the cup is 3/16" wide which makes a stronger edge on the cupel. Mr. Adams made the mold for $45.00. The price of the mold was paid for with the first hundred cupels I made.

Molding Your Cupels
I have tried Portland cement, mixed cement and Bone Ash, and pure Bone Ash cupels. All 3 are easy to make and work well. The only difficulty I had, occurred when I allowed cement to compact on the sides of the cylinder. It made it difficult to get the cupel out of the mold. This is easily handled by wiping the cylinder with your fingers or a rag after each cupel is cast. If you damage a cupel as you make it, simply crush it in your hand and put the residue back in the can for remolding.

I work with 2 pounds of material at a time. This will easily fit into a 3 pound coffee can which I use to mix the dry ingredients and the water.

The amount of water is critical. Use 1.28 ounces of water per pound of dry material. This makes 8% by weight. The "Manual of Fire Assaying" states that as the quantity of water approaches 5%, the cupel will be dry and crumbly and at 15% the cupel is difficult to remove from the mold. At either extreme of moisture content, the cupel will crack when used in the furnace and this could lose your precious metal bead. You will need several items to make cupels:

the mold
8 oz plastic mallet
4" x 5" x 1/2" steel anvil
block or any smooth, hard surface
14 mesh sieve
3 lb coffee can with plastic lid
scale to measure 2 pounds

scale to measure 1.5 oz avoir or 43 grams

The procedure is to:
1. Weigh the dry ingredients into the coffee can, add the water and mix again for 2 or 3 minutes.
2. Force the mix through the sieve to break up the balls of damper material and make a more homogenous mix.
3. Weigh the material out 43 grams at a time and pour it into the mold.
4. Place the handle into the cylinder and press out the air in the mold.
5. Strike the handle with the mallet 4 or 5 times to compact the material.
6. Place the top of the handle against the floor and hold the cylinder in your hands and press the cupel out of the mold.
7. Rub the inside wall of the cylinder to remove any compacted material.

If you have followed the procedure correctly you have a perfect cupel in your hands. Set it on a piece of cardboard and when you have made the number of cupels you wish, date the batch and place in a dry place to set. It is best to leave cupels for 2 or 3 weeks to dry before use. In the desert climate where I live I have found that I can use cupels sooner than this, but if the cupel is too wet for use it will crack in the furnace and let me know I've made a mistake. Always preheat the cupel to a dull red color before use. This will drive out any moisture or gases in the cupel and prevent losses due to spattering.

In making a number of cupels I find that the 50-50 mix of Portland cement and Bone Ash is the easiest to make and use. The cement in the mix sets up stiff and forms up stiff right from the mold making it easier to handle. 100% Portland cement cakes on the side of the cylinder and has to be cleaned more often than the other 2 mixes. Bone Ash by itself is soft and somewhat brittle and has to be treated gently both when it's formed and after it is dry.

1 hope that those of you who are doing your own assaying can use this information. I will certainly be using it in my own assaying. Good luck and good prospecting.

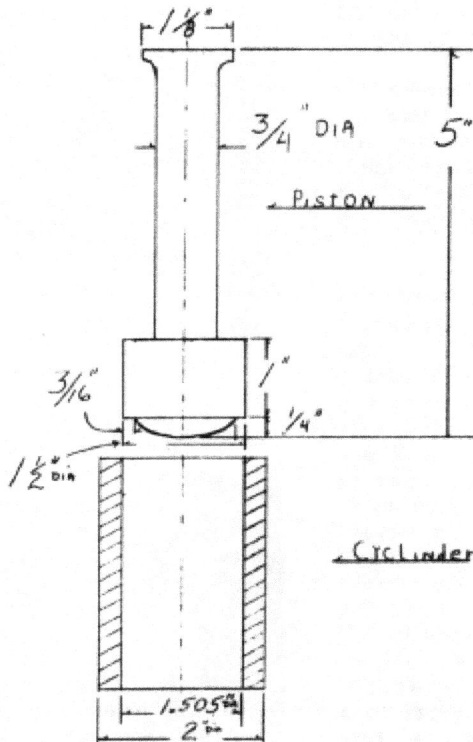

Learning About Flotation
Part II

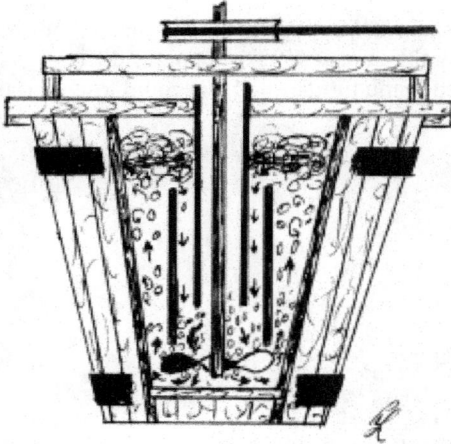

Scope and Application by Paul Skinner

The flotation process is today the most efficient, most widely applicable, and most complex of all ore concentration methods in use by the mining industry. Fully 95% of all metallic minerals processed today are processed by flotation. Theoretically, it can be applied to any mixture of particles that are essentially free from one another and small enough to be lifted by rising gas bubbles. I once had a professor who said, "Any mineral can be concentrated by flotation if the correct suite of reagents and conditions are applied". This has proven to be true, because in recent years minerals have been concentrated by flotation that were untouchable twenty years ago.

Particles ranging in size from minus 20 mesh down to a few microns are responsive to flotation. This quote is from Dow Chemical Metallurgists, I personally have had a very difficult time getting anything larger than 65 mesh to float.

Actually however, limited knowledge of the flotation mechanism has restricted the use of this technique to a relatively small but growing list of minerals and elements. Some of the more common are listed below, a number of which will be more fully discussed in a later chapter.

SULFIDES:
Copper, Iron, Cobalt, Lead, Molybdenum, Nickel, Zinc, Arsenic, Copper-nickel-platinum
NON-SULFIDES:
Phosphates, Iron oxides, Fluorite, Salt, Limestone, Chromite, Potassium salt, Feldspar, Tungstates, Beryl, Tellurides, Silica, Coal, Rhodochrosite, Manganese oxide, Lead oxide, Zinc oxide, Barite, Silver minerals
METALLICS:
Gold, Silver, Copper

I only recently learned that flotation equipment is being used to separate crude oil from water in the mixture that comes from steam-injected stripper wells.

MECHANISMS INVOLVED
The essential mechanism of flotation involves the attachment of mineral particles to air bubbles in such a manner that the particles are carried to the surface of the ore pulp where they can be removed.
1. Grinding the ore to a size sufficiently fine to liberate the valuable minerals from one another and from the adhering gangue minerals.
2. Making conditions favorable for the adherence of the desired minerals to air bubbles.
3. Creating a rising current of air bubbles in the ore pulp.
4. Forming a mineral laden froth on the surface of the ore pulp.
5. Removing the mineral laden froth.

Although grinding of the ore is not, strictly speaking, a part of flotation, it does have an important bearing on the process. For optimum flotation results, the valuable minerals should be separated completely from the waste rock (gangue) and from one another in the grinding step. In practice, however, this is not often economically feasible, and even when complete separation is attained other complicating factors may arise. Thus, with the ordinarily used ball mill or rod mill grinding, considerable gangue slimes may be formed which will complicate subsequent flotation steps.

In my many years of consulting, I have found that the most common problem people have when operating a flotation mill is, as you would expect, "recovery". In about 90* of these cases the problem was solved by a very close inspection and analysis of their grinding and classification circuit.

The creation of a rising current of air bubbles is accomplished by a flotation machine which produces bubbles by the mechanical agitation of the ore pulp, the direct introduction of air under pressure or both. These operations may be considered as the mechanical adjuncts of the flotation process.

To obtain the adherence of the desired mineral particles to the air bubbles and hence the formation of a mineral laden froth on the surface of the ore pulp, a hydrophobic (not capable of absorbing water) surface film (really we won't use too many big words in these articles) must be formed on the particles to be floated and a hydrophillic, or wettable, film on all other. This is done by means of collectors and modifiers, and the selection of the proper combinations for each particular ore. This constitutes the principal problem for the ore dressing metallurgist.

DIFFERENTIAL FLOTATION

All flotation concentration processes are selective or differential in that one mineral or group of minerals is floated away from accompanying gangue. Ordinarily however, the separation of unlike minerals such as sulfides from non-sulfides, is referred to as a "Bulk Flotation", and the term differential flotation is restricted to operations involving separations of similar mineral types. Differential flotation may be exemplified by the concentration and subsequent successive removal of copper, lead, zinc, and iron sulfides from a simple ore.

By proper application of modifying agents, differential separation can often be made using the same collector agent for all of the concentrates. Other ores may react differently, making it necessary to employ a different collector in each flotation stage. The variations in flotation activity between different ores containing the same minerals may be due to differences in geological history, degree of oxidation, presence of interfering soluble salts, or a number of other causes. Since it is seldom possible to control or even determine all of the variables in a flotation circuit, each ore must be considered as a separate problem. Even when this is done the character of an ore may change from hour to hour in mill operation thus creating an unending challenge to the flotation metallurgist. Here in the Western United States we have an unending problem with oxidation of our ores. Nearly all deposits will have at least some oxidation and it is these ores, when we do not know if they are sulfides or oxides, that cause real problems.

From the foregoing it is apparent that the prediction of flotation activity of an ore on the basis of examination and analysis would be difficult if not impossible. A number of correlations of course, have been noted, and on the basis of certain generals rules or procedures; however the lack of hard and fast rules should be borne in mind in reading these articles.

GETTING RID OF RATS

by Doc Isely

A pack rat is like a dog with an obsession to retrieve. It will bring home a vast assortment of objects, some worse than worthless and some prized by neighbors who will miss them. There is one difference. A pack rat works only by night. The pack rat leaves the nest with empty jaws but it soon finds some object to bring home and begins carrying it. Soon it may find another which it likes better so it drops the first object and picks up its new treasure. For this reason it is also known as a trade rat. But in switching objects it tries to trade up -your keys for a piece of burro manure is typical. It is also called a wood rat.

I once moved into an area which was suffering a plague of pack rats. Small precision parts I had made plus valuable measuring tools began disappearing even though the door was locked. The floor joist were placed directly on the surface of the earth which was mostly rocks and there was no human crawl space beneath the floor. By using a metal detector I was able to find a cache and by removing part of the floor was able to retrieve part of my equipment mixed with a large amount of filth. There were more nests in the fiber glass insulation beneath the roof so I had to remove part of the ceiling. Some of my things are still missing.

They invaded the house by gnawing through the floor everywhere there was a slight opening beside a pipe. I surrounded all pipes with hardware cloth covered with concrete but they inhabited the walls and attic. They chewed an electric wire in two stopping the refrigerator. Host of what they dragged into the house was worthless and a fire hazard. They dislike light so I began illuminating their inaccessible nest with a spot light. A nest, so illuminated is quickly evacuated. I am told that if the light is left for 48 hours that the nest will be permanently abandoned. I have never done this. There are too many bright objects which might serve as concave mirrors or burning glasses to ignite the nest from the spotlight. A 15 watt fluorescent mechanics hand light would be safer.

The rats declined all bait in traps although I occasionally caught one by placing a trap where a rat might blunder into it. Pack rats are vegetarians and an appropriate bait was hard to find. Some have claimed success by baiting traps with bright, colorful objects.

I put out poisoned grain which they consumed in considerable quantity and apparently thrived on it. Then, one day my dog was paralyzed. The veterinarian said that my dog had apparently eaten a rat which had ingested considerable arsenic from grain. He also told me that another dog in the neighborhood had the same trouble and said that owls which had migrated in to prey on the rats, would, because of the arsenic in their diet, lay eggs with thin shells which might break before incubation was complete. He suggested that putting out poisoned grain was futile and that natural nocturnal predators would, in time, decimate the rats. "These things come in cycles", he said. He was right but the rest of the cycle took about a year and during that time I learned a great deal about rats.

Ratproof buildings can be built. Use a concrete slab floor with masonry or steel walls and a steel roof. Use metal doors and window frames and cover all openings with hardware cloth. A number of large, aggressive cats will keep the problem pretty well under control. One night I set a standard rat trap with a piece of Fig Newton. Before I could go to sleep I heard the trap spring. I removed the rat and reset the trap. As soon as I had turned out the light and left the room it was sprung by a second rat. I caught five rats that night with one trap and 1/6th of a Fig Newton. After that there were no more rats in the house. I have used this bait ever since with great success. Even though the rats apparently know that this is a trap which has just killed other rats they find a Fig Newton irresistible.

In dealing with rats there are things to avoid. A story is told about some miners who discovered a pack rat nest in a rocky outcrop within their mine camp. They detonated half a stick of dynamite in the nest and rock shards flew all through the camp, doing considerable damage. Apparently the rats had been collecting powder in their nest. They have been known to start fires by gnawing "strike any where" matches in their nests. If rats are getting in, keep all valuables and all dangerous substances in ratproof containers.

I close with this poem about rats:
"I'm a dreamer. Aren't we all? Blasted dreamer. Aren't we all? In my dreams each night it seems
the pack rats have a ball. Then I waken and I find it was not all in my mind for I see that they have plundered me by gnawing through the wall."

The above is a parody upon a song called "I'm a Dreamer" which is part of a musical review with a name I don't know and written by someone or other during the 1920s, I think. You see, I'm not very good at music, but I know my rats! *****

Special Chemical Assay

By James T. Crowder

The following procedure is one that I use to assay gold in an ore that does not fire assay very well. The ore is a sandy, silty and clayey mixture. The assay uses iodine and potassium iodide and can be done in the field with equipment that can be easily carried around in the trunk of the car.

You will need iodine crystals, potassium iodide, small amount of mercury, sulfuric acid and nitric acid diluted to 25%. Also, two plastic bottles about 1 pint size with secure screw-on on lids. A small Coor's crucible and a small beaker. You can use a small beaker in the place of the Coor's crucible.

Obtain an old wine bottle or a bottle that is 1 liter or larger. Mix 40 grams of the iodine crystals together with 40 grams of the potassium iodide in i liter (or one quart) of water. Shake the bottle and the iodine will go into solution. Should some iodine crystals remain and settle out, either add an additional amount of the potassium iodide to dissolve it or Just ignore it. Next remove enough solution from the bottle so that you can add 100 ML of sulfuric acid (about 1/3 cup) and add the acid. Note: Iodine will not dissolve precious metals unless the pH is below 2. Below 1 is best.

Place your 30 to 60 gram sample of the ore that has been pulverized into the plastic bottle and then add about 300 milliliters of the solution to the on and then secure the cap and shake the bottle one minute every ten minutes for thirty minutes. On the other hand some ores should be continuously rolled or shook for up to 12 hours. Let the bottle sit for a few minutes until the ore settles and then pour off the liquid into your beaker. You may find that the ore will not settle in which case you will need to filter it through a coffee filter.

Rinse your bottle with water and return the filtered pregnant solution to the bottle. Add about 15 grams of mercury (that's about 1/2" round gob of mercury) to the pregnant solution and shake the bottle periodically for about 30 minutes. Pour off the solution and recover the mercury and put the mercury into a beaker with your 25% nitric acid. The nitric will take up the mercury and any silver and leave the gold.

Before the mercury has completely gone into solution remove the mercury and place in the Coor's crucible or small beaker with a small amount of the nitric acid solution to finish off the mercury. This will leave the gold in the crucible. This process works at ambient temperature but heat will speed up the process. You can use any suitable heat source including the radiator or block of your car. Pour off the nitric solution from the crucible and set the crucible on a heat source to dry the gold and when it is dry, the gold can be placed on a balance and weighed.

This description allows a lot of variation in the amount of the ore sample and the amount of the iodine-iodide solution to use. Sometimes not as much solution is necessary depending on the ore that you are testing. The iodine-iodide solution has an excellent shelf life and can be premixed and simply taken to the field for ready access.

Another technique that I have employed is to use a rubber-type rock polisher to run larger samples up to 300 grams of ore. Just place the ore in the polisher and add at least twice the amount of the iodine-iodide solution. To add a little extra action, I place a handful of glass marbles in the polisher. Again, either let the mixture settle when finished with your run or else filter the solution. The time to run your test can be varied but thirty minutes is a good starting point.

Good luck and I hope this will be of help to anyone that tries their ore. I would like to hear of your results. Mine have been anywhere from .03 oz/ton gold up to 1.5 oz/ton gold on assays of from 30 grams to 300 grams.- James Crowder

(Editor's Note: If you don't have a weighing scales you can add a small amount of lead to your gold and cupel it. Then you can measure the gold bead to get the weight. Issue 3 page 34 of Popular Mining gives a list of weights of gold beads. Don't throw your nitric acid with mercury out. It may contaminate. Put a piece of copper running from the surface of the nitric acid to the bottom of the jar. The mercury will precipitate on the copper and run to the bottom. When the mercury is in the bottom, pour off the nitric solution and throw some iron in it. After a few days the copper will come out of solution and you can throw the solution away (and the copper if you wish). See issue #1 on tips about how to handle mercury safely. Don't use it otherwise. - Jim)
*****

Tripod Hoist

By Mike Glenn

Here is an amazing Tripod Hoist that can be built for less than $25 if you have a welder and less than $50 if you have it welded at a welding shop. It will lift up to 5 tons, but we don't recommend that much. Of count, we say that only because we know someone who has done that, not because we did it. But we have lifted a ball mill weighing over 1.5 tons from a pickup truck with ease. This standard unit as you see it here will lift up to 1 ton from an American pickup. If you wish to lift more, be sure to add in the extra angle iron braces. We refer to a standard pickup not one of those 10 foot high 4 wheel drives.

On the other hand if you want to use one of the high trucks simply use longer pipe for the legs. Thirteen feet long instead of 10 should do the job, but then don't try 5 tons either. If you want to go to heavier loads than recommended here, try 2" pipe legs.

It's easy to use and quite portable. It can be disassembled in 5 minutes, stored in a small area, and reassembled in 5 minutes when needed. It is simple to use. A standard American pickup truck can be backed between two legs of the Tripod Hoist and any weight that such a truck can haul can be lifted by the Tripod. Be sure to connect the legs at the bottom with cable when lifting heavy objects. One might get away without connecting the legs on lighter objects weighing less than a ton, but not so on the heavier objects.

BUILDING THE TRIPOD HOIST

The drawings and pictures give most of the details on building the Tripod Hoist. If you have a welding machine, you can build the whole unit in less than a day. If not, then you can cut and get all of the metal ready and take it all to a welding shop. A list of materials follows:

Use a hack saw to cut the three pieces of 2" pipe that are to be welded to the 1/4" steel sheet. Mark the 1/4" sheet so it can be cut by a welding torch. Weld a 5/8" nut in each of the 2" pipe pieces. This allows you to clamp the legs in place with a 5/8" bolt.

At the bottom of each leg drill a hole for a 1/2" eye bolt. Use a 2-1/2" long eye bolt. This is for connecting the cable that keeps the legs from sliding outward under load. An alternative to drilling holes and installing eye bolts is to weld a piece of bent -- in place. Either way is OK, but be sure to do it one way or the other. If you plan on lifting more than a ton, be sure to weld the extra braces on the 1/4" in steel top piece.

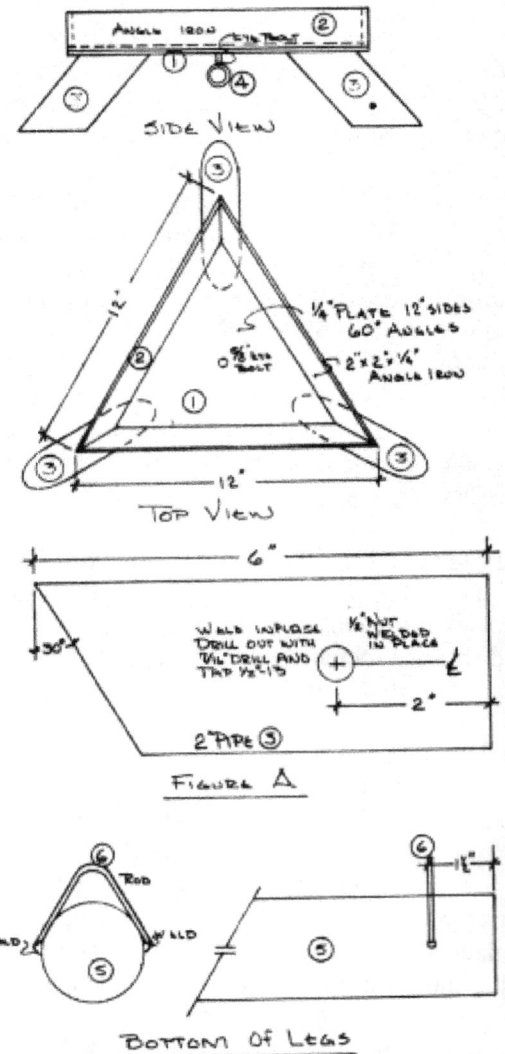

SETTING UP

(1) Lay all pieces out on the ground.
(2) Install three legs in place and clamp them down with the 5/8" bolts provided in the small 2" pipe pieces.
(3) Set the Tripod up. This can be done by one man easily. Be careful not to put too much bending pressure on the legs.

(4) Connect your lifting come-along to the eye bolt in the top of the Tripod Hoist. (We used a two ton come-a-long sold in most hardware stores for less than $20.00 until we bought a chain hoist.
(5) Be sure to connect the three legs with cable at the bottom and you are ready to go. We think that this will be a very useful addition to your operation.

(Editor's note: We cannot take any responsibility for the safety of people using our plans. This unit should be built only by people who understand the strength and operation of pipe, steel, chains, cables and come-a-longs. All responsibility for safe operation is strictly the responsibility of the builder.)

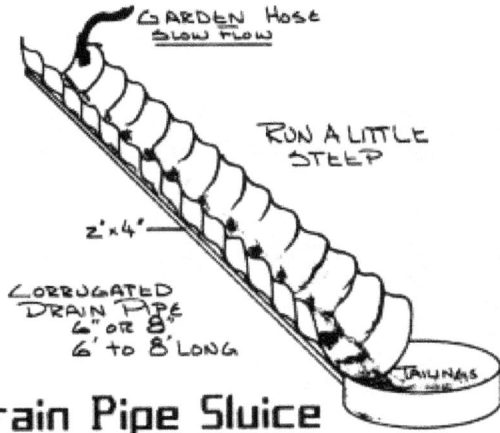

GARDEN HOSE
SLOW FLOW

RUN A LITTLE
STEEP

2'x4'

CORRUGATED
DRAIN PIPE
6" OR 8"
6' to 8' LONG

TAILINGS

Drain Pipe Sluice

A Little Drain Pipe Sluice

by Matt Bard

On the market today are all kinds of machines to separate gold from your concentrates. What follows is the best and cheapest, and also the easiest concentrator that I have found. It must be some kind of secret because I'm always showing it to people who have been mining for years and they are always amazed.

It's very simple and very effective, what it amounts to is a sluice.

(1) Use a 4" or 6" plastic corrugated drain pipe and split it down the middle. A piece at least 6 to 8 feet long is about right.

(2) The pipe cost about $4.00 and when split down the middle gives you 2 sluices. Along an eight foot length there must be About a hundred riffles.

I've used it on the fine gold beach at Nome, Alaska and it holds the fine gold there with no problem.

In the and you'll have only one small pan of cons to deal with instead of bucketsful. Spoon-feed it slowly. It takes ma 20 to 25 minutes to spoon feed a 5 gallon bucket of cons. I've seen machines or gadgets that cost a lot more and are a lot less effective than this idea. This simple device will please anyone concerned about cost and effectiveness. Yours Truly, Matt Bard.

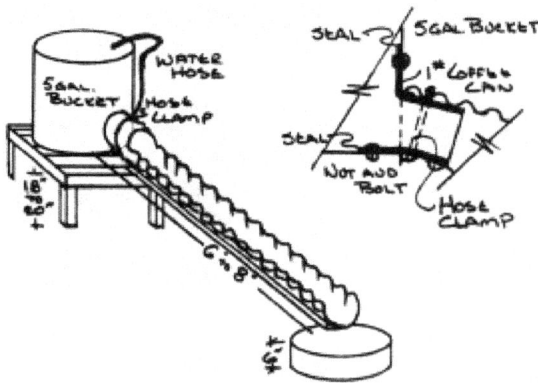

SEAL 3   5 GAL BUCKET

WATER
HOSE

5 GAL
BUCKET

HOSE
CLAMP

1" COFFEE
CAN

SEAL 2

NUT AND
BOLT

HOSE
CLAMP

18"
OR
20"

6"-8"

4"

Automatic Feed for a Pipe Sluice

By W. Pete Long

Thought I would send you a How To Do It concentrator.

MATERIALS NEEDED
1) 5 Gal. Plastic Bucket
1) 1 lb. Coffee Can
1) 4" Hose Clamp (or wire)
1) 36"x 4" Black Field Line Pipe
1) 2 oz Tube Silicon Sealer
2) 1/4" Sheet Metal Screws

(1) Take the 5 gallon bucket and cut out a 4" hole in the side just even with the bottom.

(2) Take the 1 lb coffee can and cut out bottom with a can opener. Cut slits in the can to the first rib. Bend some of the tabs thus formed so that they will fit along the outside of the 5 gallon bucket and leave some tabs unbent.

(3) Insert the unbent tabs inside the bucket at the bottom as shown in the drawing. Use the sheet metal screws to attach tabs to the bucket.

(4) Seal around the can with sealant and let it set up.

(5) Now take solid (not perforated) field line pipe and cut the pipe in half so that the bottom half serves as riffles. Cut along molded seams of pipe.

(6) Put the half pipe in coffee can and tighten in place with the hose clamp.

(7) Add 1/2 or 1 gallon of 1/4" or less material, stick hose pipe in and turn it 1/2 on. Watch for the gold in the first 4 or 5 riffles. Adjust flow and angle according to type and size of material. Works best with 1/4" to 1/8" classified material (from dredge, tailings). You can turn on the water and walk off and leave this one. The Gold will stay in riffles. Pete

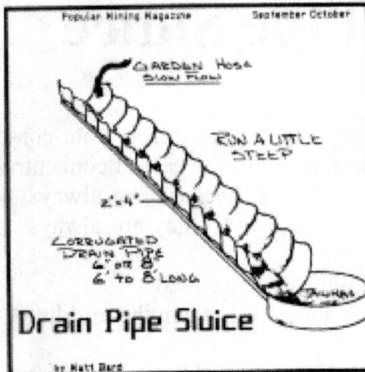

Drain Pipe Sluice

Pretreatment for a Plastic Sluice
by Ernie Wells, Elgin OR

The designer of this unit may not have known why it was so efficient. If your gold is clean and highly charged with a cat-ionic charge as you pour your cons over the rig the particles of gold will adhere to the plastic which carries an anionic charge on its surface.

Try this little trick:
1. Take some cons (about 2 cups) with fine gold you can see but is very hard to pan and put it in a gallon jar and add 1/2 jar of water with the chill taken off.
2. Clean with sandpaper or acid, a dozen nails with the heads snapped off and add to the jar.
3. Clean 6 or 8 pieces of rubber or cut 3 old rubber shoe heels in two and throw them in.
4. Put the top on the jar and shake for 4 minutes. Remove the nails and the rubber and add to the water 1 tablespoon of water soluble varnish remover.
5. Put on the top and shake gently for 1 minute.
6. Now take a garden hose and let it run slowly and carefully rinse the cons, keeping the jar upright and once in a while dipping your finger in <u>shampoo</u> (any good shampoo without hair conditioners or
oil) and letting the water run over it. Don't use soap or detergent as we are not after a flotation unit Use very little shampoo and don't rinse long.
7. Now pour the goodies out of the jar down the sluice as shown in the picture and slowly feed both the water and the cons. Be sure to clean the plastic pipe with ammonia and water using paper towels before using. <u>Right before using the pipe</u>. Windex will work.

You may be in for the surprise of your life for I have seen this set up take gold out of the sand down to a size of about 75 to 90 microns. I have never tried this with the magnetite yet in the cons so 1 don't know if it will work with the magnetite present. Wet magnetite does not allow the fine gold to separate using other methods and I am so used to taking it off first 1 just have not tried.

By the way - I cut a v shaped piece of plywood and mount the split pipe inside with screws (metal screws) near the upper edges of the pipe securing it to the plywood then I mount the whole works on a fulcrum near the middle so I can tilt it as need be.

I never thought of the pipe till I read about it in Popular Mining. I was using 2" pipe in another application to do a similar job but this is a good rig.

Mining with a Slusher

A slusher is a standard muck (broken rock) moving machine used primar[ ] extremely versatile ad flexible. Basically, it is a motor attached to 2 drum[ ] depressed, drags a scraper basket back and forth.

I consider the slusher to be a much under-rated instrument. It has a bucke[ ] (about half) and it is safer to operate - much safer to operate. When the [ ] mucker is less gentle.

I was asked some time ago how a slusher would work in a drift heading. ([ ] digging, it's the point of advancement of the tunnel.) It is used in large op[ ] moving machine mounted on tracks. I think the slusher will work very su[ ] and am anxious to try it out.

A small mucker should have 55" minimum width to operate in a drift hea[ ] much decide the width you want to operate. You decide - not the machin[ ]

This is very important because narrow width lowers costs by making les[ ] creases safety. Added width permits larger equipment but is offset by the[ ]

Be on guard against the "one-armed bandit". This is a "slusher" with 1 co[ ] offered at a bargain price. There seems to be a great abundance of these n[ ] but not to my knowledge. This machine simply moves the bucket back an[ ]

A real slusher is much more versatile - a new dimension. A little finesse i[ ] if desired. It can be raised at will and set back down in a different spot. The handles are the controls that make the drums rotate. When you rotate one drum, it pulls the bucket one direction; and when you rotate the other drum, it pulls the bucket in the opposite direction. When you press both handles at the same time, the bucket is being pulled from both the front and rear at the same time. If the slusher pulley is higher than the bucket, the bucket will raise and remain suspended 'til one or the other handle is released.

If the bucket on the "one-armed bandit" should turn over, the operator is pretty much committed to turn it back by hand. Not so with the* double-handled slusher.

Good cable is essential for efficient operation. A slight drag on the drum opposite the one in use is usually required to control back-lash. There is a drag built into the machine, but an additional light touch is usually required.

A slusher has 2 drums. There is a 3 drum model used in very large operations. The drum nearest the motor is the pull drum and pulls the loaded bucket. The pull side normally has larger cable than the other. Common cable sizes are 3/8" on the haulback and 1/2" on the pull.

The effective mucking (moving broken rock) distance is determined primarily by the amount of cable that can be wound on the haul-back drum. It will require twice the amount of cable as the pull drum. It is a mistake to crowd cable on the drum because it doesn't always wind evenly.

Good slusher blocks (slusher blocks are heavy duty pulleys) are important and the hookup is very important. Several methods are in use. The method that does the job is the eye-pin and shell. A hole is first drilled into the rock and then the shell is placed in the hole. As the eye-pin is screwed into the shell, the shell expands to make a secure anchor for a slusher block. (See illustrations). Eye-pins are easy to make, but shells have a bit of a price tag (have to be on guard here too, because there is an economy style that is a source of trouble). The good ones are 4" in length. The short, stubby ones work well when the ground is perfect (maybe in heaven.)

There is a hookup called "feather and wedge". This works up to a point and I consider its use to be a false economy. You don't use shells, thus its lower in cost, but I've found it to be undependable.

Installing The Slusher

Installing a slusher is always a marvel to me because I've seen it done wrong so many times. I've seen slushers secured with chains in all directions and stalls on top, bouncing all over the place. Operators keep adding chains and wood in an effort to hold it down - with minimal success. This is a clean case of quality does and quantity doesn't!

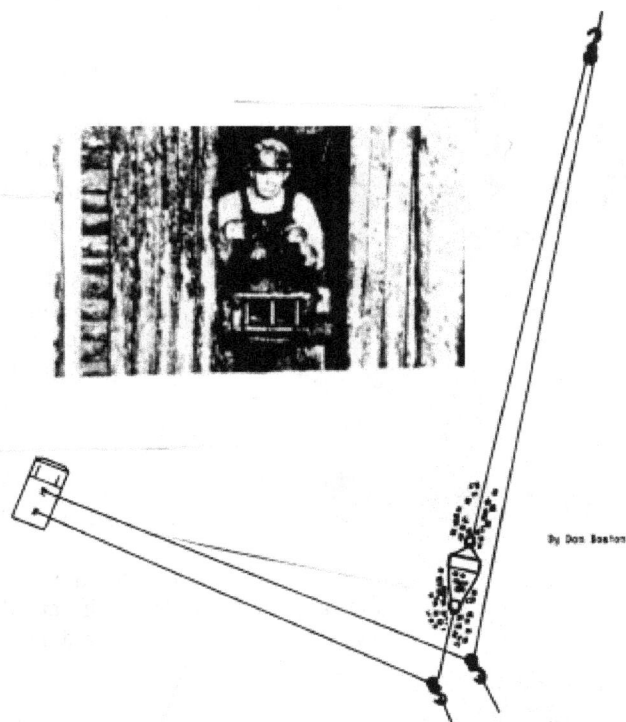

Two chains are required. Short tow chains are usually used. High tensile steel pipe chains can also be used when available. Hook 1 chain on the motor side at or near the top and hook the other chain on the other side of the haulback, also at or near the top. Hook the other end of each chain down behind you at an angle. It must be hooked up at a point below the base of the slusher. This is the key. If it is hooked even with the base, the slusher will still bounce. Hook below and it will draw up snug and firm and tamed. To get the low hookup, it is often necessary to raise the slusher - build up a base and set the slusher on the base. (See illustration A).

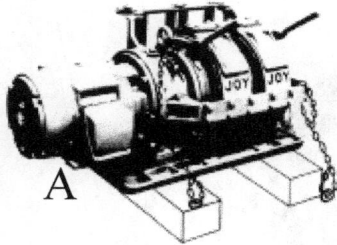

Illustration A - this is the best setup I know of. The base could be improved with a few laggings (boards) across the stalls for a tighter floor in case of side movement. The chains are connected to eye-pins drilled into solid bottom (rock floor) or low on the walls. A hook-up direct to timber is also possible in timbered areas.

Illustration B - this shows the principle we are using. The laggings are placed at the top of the slusher. This setup works, impressively. Place one or two 3* laggings, (1 is enough for a small slusher), across the slusher lane and against timber or something solid. Scoot the slusher up against the laggings as shown and it is ready to go. No chains or stulls (temporary wooden supports) are required. The laggings are an obstacle to climb over, but this is a quick setup and sometimes feasible for a fast temporary job.

Illustration C - it is hard to get stulls to stay in place on a vibrating slusher. It is easier in timber because the stulls can be spiked to timber and when they vibrate loose from the slusher, will remain close to position. The cute little slusher on the right is in a similar fix but there appears to be room to move around the stulls. Chains are always best when a hookup can be found or created by drilling, but a different tiedown is required. I'm still experimenting.

Using The Slusher

An idler block (a pulley placed to take up cable slack) placed strategically will keep things running smoothly. This keeps the haulback winding uniformly on the drum and prevents cable interference with the bucket.

Wide areas can be worked easily with well placed eye-pins and with a chain stretched across the area. The block can be connected to different locations on the chain by a movable bitch-link.

Slusher buckets usually have a hookup in the upper center of the back. This center hookup is the one most commonly used. If the bucket should turn over, simply press both handles simultaneously, raising the bucket and the bucket will right itself.

This tight-line technique will save much hand-mucking. The bucket can be floated along a wall digging every pound of broken rock.

Some buckets also have a hookup on each side in addition to the one in the center. By raising the bucket when hooked to one of the side hookups, it is possible to cover a wider area. Raise the bucket and it will hang sideways. Drop it and it flops over, the blade straddling a new area.

Slushers come in many sizes. They are all heavy. A 8.5 HP air slusher weights 325 lbs. and will carry enough 3/8" cable to muck 50' comfortably. A 15 HP air slusher will carry enough 3/8" cable to muck 150' comfortably.

Helpful hints

A large bull-hose, about 2" in diameter, connected to the exhaust is often a good move. This channels the exhaust away from the immediate area. Without this, in some mines, the fog from the slusher will cut visibility to zero.

The rollers in front are intended to be there when needed but not intended to be used regularly under load. The setup should be made to keep the cable away from the rollers as much as possible by means of well placed slusher blocks and a well placed slusher. These rollers should be greased regularly as should the blocks.

The 15 HP slusher has plenty of power. Run wide open the bucket will skip along, not digging properly and snapping cables.

It's ok to haul the bucket back at a pretty fair clip. After reaching the block, allow some slack in the haulback before starting to retrieve the bucket. Tension in the haulback will cause the bucket to float away from the face and fail to dig. Start the retrieve slowly and the bucket will dig right in, usually filling the bucket each pass.

There is an illusion to watch for that occurs regularly. When the muck is mostly against the face and all of the easy muck has been moved, it may appear that the bucket isn't digging at all. The miner will grab the pick and loosen up the muck pile and the condition will soon repeat. The solution is to make a series of short passes with slack in the haulback and a very slow retrieve and build up enough muck to fill the bucket before retrieving it all the way. If the operator can control his impatience the muck pile will disappear without any hand digging - faster, too. It may take 10 or more passes each trip to get a full bucket.

The slusher is a producing machine. It is incredibly versatile. I've used a slusher in bold raises that were a bit too flat for the muck to run (that's a tunnel going from one level to another that is not timbered and not steep enough to allow the rock to slide down), also in flat timbered raises with the slusher serving as a hoist or tugger to hoist timber and supplies.

A phosphate mine in Montana uses a slusher with results that would be hard to beat. The stopes are about 60' long and 3' wide. The vein inclines about 35 degrees and a 3-man crew makes a 5' round
across the back every day. Big time production from a narrow vein.

Purchasing A Slusher
Before purchasing a used slusher, certain precautions are advisable. A guarantee or a warranty would be the best thing. The problem is that they all look good. The drums can turn and the machine make plenty of noise and still not do the job. The last job I was on I ran into 3 slushers in a row that were on the fritz. They made noise and the drums turned, but they wouldn't reverse properly. I'd never experienced this problem and to find 3 of them in sequence was interesting. I have to suspect faulty repairs.

A slusher that is working ok will continue to work ok, seemingly forever. It is one of the most durable of machines. Proper usage means keeping dirt out of the motor by blowing hoses before they are hooked up and plugging air inlets when moving from one location to another - and, of course, lubrication.

To Make Life Simpler
As soon as the small mine operator advances past the wheel barrow stage and begins using any of this heavy, hell-for-stout mining equipment, I strongly recommend a few small inexpensive items to make life better for him. I personally consider these items to be mandatory:
1. A small cable come-along with sliding pulley.
2. A generous supply of eye-pins and shells - shells retail at about $1.00 each. The eye-pins are made from 5/8" rod and are 24" in length. (See Illustration D). The shells can be purchased in either left or right hand thread. I recommend left hand on all rock bolts, pins, and shells because the drill is left rotation and bolting can be done efficiently with the machine when required. More than one type thread is unnecessarily confusing.

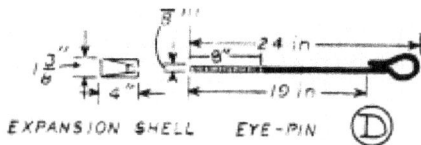

EXPANSION SHELL   EYE-PIN   (D)

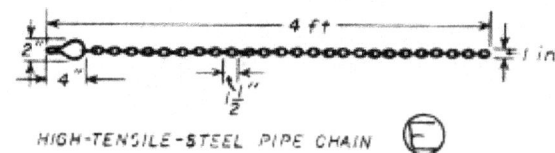

HIGH-TENSILE-STEEL PIPE CHAIN   (E)

3. A few high tensile steel pipe chains. These chains are the handiest things since applesauce. They solve so many problems, in combination with the pins and come-along that I crippled trying to work without them. They are about 4' in length and can be joined for added length. (See Illustration E).

Pinhole
This is the hole into which you put your shell and eye-pin. Bit size is a very important concern when drilling pinholes. A new 1 3/8" bit works well. If the bit is too small, the shell won't follow and if too large the pin won't tighten properly. A little smaller size bit than 1 3/8" gives the best results. A snug fit that can be gently tapped into the hole assures a good effective pinhole.

There is a right time to stop turning the pin when tightening it down. It is just a little tighter than hand tight -a touch with the pipe wrench handle inserted thru the eye. Too much twist and the shell will pop noticeably and begin turning easily. It may still do the job but has a better feel when tight and firm.

To Summarize:
1. A secure slusher setup -take plenty of time and do a solid job.
2. Good cable - it is usually necessary to replace all of the cable that comes with a used slusher. Rusty, barbed cable with knots and splices will effectively cripple the slusher performance.
3. Quality slusher blocks - be sure there is a very close tolerance around the wheel with no room for the cable to force in alongside the pulley.
4. Hookups for the block - I recommend being very, deliberate with the eye-pins in the drilling and selection of locations.
5. Small cable come-along, pipe chains, eye-pins and shells to be used.
6. The compressor must be big enough (CFM rating) to meet the CFM requirements of the slusher. This subject requires separate consideration in depth - a biggee. The Badger

Assaying Complex Ores

By Jim Humble

There are a number of tricks to assaying complex ores. We know a few and I am sure that there are many more. If you know any, why not drop us a line?

Here it an interesting trick that may allow you to prove that gold does exist in your ore or in other ores and sands that you have. In many cases if you can prove that gold does exist, you can get the financing necessary to develop a process for that particular ore. This method might just help in that case even if it is not a method that would be a practical production process.

We have found that there are a number of metals that if present in the ore will carry your gold and other precious metals right into the cupel. (Note: see Popular Mining Issue #3 for a description of a standard assay and for definitions of words not explained here.) A very small amount of nickle will do this as well as arsnic and antimony. There are other metals that will do the same thing. Some of the metals are listed in the literature on assaying and some are not Known. The theory is, or at least this theory is, that the gold does get into the lead during the assay process. But you must then be able to extract the gold by the cupelling process and that's where the problem lies.

During the cupelling process, the lead oxidizes and becomes a yellow liquid called litharge which soaks into the cupel. At the cupelling temperature all metals will oxidize except the precious metals and all the metals that do oxidize are carried into the cupel. The gold, silver and platinum that are not oxidized are left behind as a bead on the cupel. However, evidently there are a number of metals that will carry gold even when they are oxidized and these are the metals that carry the gold into the cupel, and in that case you will not be able to prove that the gold exists.

We have found a way around that problem. It isn't exactly a new idea. Other have used the idea b fore us. But it does work, at least on a limited basis. You must re-assay the cupel. That's right. If the gold has been carried into the cupel you can re-assay the cupel and get the gold sometimes, but don't expect to get the gold on the second assay. We have found that sometimes you will get a little gold on the third or fourth assaying of the cupel. In other words grind up your cupel and add assay flux to the grounds and re-assay. Then take that cupel and grind it and re-assay it. On the third or fourth time you may start getting a little gold and on the eighth or ninth time you may be still getting a little each time. Add them all together for the total gold in your ore.

The reason that this method works is probably because all other metals have a greater tendency to go into the assay flux than the precious metals. The assay flux is totally void of all the heavier metals and it tends to act as a blotter for those metals. So each time an assay is done more of the problem metal will blot up into the assay flux thus there is less problem metal to carry the gold back into the cupel.

It is our suggestion that you add calcium fluoride (about a teaspoonful) and 1/2 teaspoon of flour to your standard flux when trying this trick. We have found that the calcium fluoride (fluorspar) seems to do the best job at blotting up the problem metals. You may be surprised if you will try this simple trick. Jim

Annual Assessment Work

by Michael McKinley

In order for you to hold the possessory right to your unpatented mining claims, not less than $100 worth of labor or improvements must be done on each of your claims annually. If you own a number of contiguous claims, the total expenditure that would be necessary for you to hold all the claims may be made on any one claim. Evidence that this annual labor has been done must be filed with the proper BLM office and the County Recorder's office in the county where your claim is located. Failure of a claim owner to perform and file evidence of his annual labor or assessment work on time, may constitute abandonment of his claim and subject the claim to relocation.

When you perform your annual assessment work, remember that it should meet 2 requirements - 1) it must tend to develop the claim or group of claims, and 2) it must be done in good faith. Claims can be challenged by relocators (though usually unsuccessfully) on the grounds that the annual labor was not actually performed, or that the labor done does not directly tend to develop the claim or group of claims.

Types of Labor
A mining operation where materials are being continuously extracted satisfies the requirement for annual labor. Other types of work which are considered to be beneficial to the development of claims are roads, mill sites, and drilling or trenching for analysis. If you own a number of contiguous claims you may develop a system or device which is beneficial to all adjacent claims, such as a common road, or a common mill site, as an alternative to performing assessment work or labor on each claim.
In addition to the several types of labor, the requirement can also be satisfied by conducting geological, geochemical, or geophysical surveys on your claims. The surveys cannot be applied as labor for more than 2 consecutive years or for more than 5 years on any one mining claim, and cannot be repetitive of any previous survey. And they do not apply toward the $500 requirement for mineral patent. The surveys must be conducted by qualified experts and be verified by a detailed report. This report must include the following: 1) the location of the work performed in relation to the point of discovery and boundaries of the claim, 2) the nature, extent, and cost of the work performed, 3) the basic findings of the survey, and 4) the name, address and professional background of the person(s) conducting the work.

Example of Filing Time Limits
The General Mining Laws allow the required annual assessment work to be initiated in the assessment year following the assessment year in which the claim was located. Assessment years begin and end at noon on September 1st. Evidence of annual assessment work must be filed prior to December 31st of each calendar year following the calendar year in which the claim was located. For example, for a claim located June 1, 1984, the assessment year during which the annual labor must be performed begins Sept 1, 1984 and ends Sept 1, 1985. The evidence of this annual labor must be filed before Dec 31, 1985. In the case of a claim located Sept 2, 1984, the assessment year during which the annual labor must be performed begins Sept 1, 1985 and ends Sept 1, 1986. The evidence of this labor must be filed before Dec 31, 1986. In this second example in order to comply with the filing requirements, a notice of intention to hold the mining claim must be filed with the BLM and County Recorder's office during the 1985 calendar year.

In order to timely file evidence of annual labor or assessment work with the BLM, the letter containing the evidence must be postmarked on or before Dec 30th of the year the evidence is due, and be received no later than Jan 19th of the following year. Most state mining laws require that the annual evidence of labor be filed from 30 - 90 days after the end of the assessment year. You must file your evidence of annual labor with both the proper BLM office and the County Recorder's office in the county where your claim is located. It doesn't matter which one you file with first, as long as you file your evidence with both before the deadline. To be sure of when you must file with the county, check with the State Geological or Mineral Survey in the state where your claim is located.

Necessary Form Information
The contents of evidence of annual labor or assessment work should contain the following information: 1) the BLM serial # and name of the claim, 2) any change of your mailing address, 3) the number of days work was done, and the character and value of the improvements, 4) the date the work was performed and the number of cubic feet of earth removed, 5) at whose request the work was done, and 6) the actual amount paid for the labor and who paid it.

State regulations regarding the contents of evidence of annual labor vary from state to state. The contents given above should cover most of them, but you should check with the State Geological Survey to be sure. Fill-in forms for recording evidence of labor may be purchased from most stationary stores and mining supply stores, but check the forms out first to be sure they will supply all the necessary information.

For Mill or Tunnel Sites

Evidence of annual labor need not be filed for mill or tunnel sites, however a notice of intention to hold the site must be filed each calendar year with both the BLM and the county. A notice of intention to hold a mill or tunnel site or a group of mill or tunnel sites should be in the form of a letter or other notice signed by the owner or his agent giving the following information: 1) the BLM serial number of the site, 2) any change in the mailing address of the site owner, and 3) in the case of a mill site, that the site is in present and continuous use, or 4) in the case of a tunnel site, that you as the owner will continue to prosecute work on the tunnel with reasonable diligence for the discovery or development of a vein or lode.

Temporary Deferment of labor

Under certain circumstances that prevent labor or improvements from being done on time, a temporary deferment of that performance on a mining claim may be granted. In which case a notice of intention to hold the mining claim must be filed in place of the evidence of annual assessment work.

The deferment may be granted where any mining claim or group of mining claims it surrounded by lands over which 4 right-of-way for the performance of annual assessment work has been denied, or it in litigation, or it in the process of acquisition under state law, or where other legal impediments exist which block the right of the claim owner to gain access to the surface or boundaries of hit claim or group of claims.

In order to obtain a temporary deferment, you mutt file with the manager of the proper BLM office a petition in duplicate requesting the deferment. No particular form of petition it required, but you mutt attach to a copy of the petition a notice to the public which shows that it hat been filed or recorded with the County Recorders office. The petition and duplicate mutt be signed by you and mutt give the names of the claims, the dates of location, and the date of the one year period for which the deferment it requested. Bach petition must be accompanied by a $10 non-refundable service charge.

If the petition it bated upon the denial of right-of-way, it must state the nature and ownership of the land or claims over which it it necessary to obtain a right-of-way in order to reach the surrounded claims. The land mutt be described by legal subdivisions if the land it surveyed. Give full details at to why present use of the right-of-way it denied or prevented, and as to the steps which have been taken to acquire the right to use it. Your petition should state whether any other right-of-way it available and if so, give reasons why it it not feasible or desirable for you to use that right-of-way.

If the petition it bated upon other legal impediments, they should be given and their effect described in detail. Upon receiving a decision on the petition, you mutt file or record in the BLM office in which you filed your notice of petition, a copy of the order or decision disposing of the petition.

If the showing made it satisfactory, the authorized officer of the BLM will grant a deferment for an initial period of one year. The period will begin on the date requested in the petition unless the approval sets a different date. Upon petition, the one year period may be renewed for another year if justifiable conditions exist. If the conditions justifying deferment are removed before the expiration date of the deferment period* the deferment will automatically be ended at that time.

You may begin all deferred assessment work at any time after the expiration of the deferment, but it mutt be completed no later than the end of the assessment year beginning after the removal of the causes for the deferment or the expiration of the deferment. All deferred assessment work will be done in addition to the annual assessment work required by law for the present year.

Notice of Intention to Hold a Mining Claim

A notice of intention to hold a mining claim or group of mining claims may be filed at the election of the claim owner* regardless of whether the assessment work has been suspended* deferred* or not yet done. Upon filing petition for deferment of annual assessment work, a notice of intention to hold a mining claim it filed in place of evidence of annual labor or assessment work. A notice of intention to hold a mining claim will be effective only to satisfy the filing requirements for the year in which the notice it filed. The filing of a notice with the BLM does not relieve the claim owner of complying with Federal and State laws pertaining to the performance of annual assessment work. (The work will eventually have to be done).

A notice of intention to hold a mining claim or group of claims filed with the BLM should be an exact duplicate of the instrument, signed by the owner of the claims, which was or will be filed for record with the County Recorders office where the claim is located giving the following information: 1) the BLM serial number assigned to each claim 2) any change in the mailing address of the owner of the claim, 3) a reference to the decision on file in the BLM office by date and serial number which granted a deferment of the annual assessment work, or a reference to a pending petition for deferment of the annual assessment work by date of filing and serial number and proper BLM office.

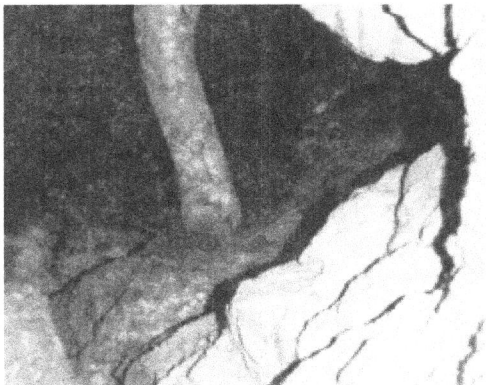

Exploring Old Mines

by Robert L. Desmarais

It doesn't take any special skills to explore an old mine. It takes brains and logic. Throw in a little common sense, and you have it all. There are many unforeseen dangers and hazards in a new or old mine. Take for instance, some wild animal living inside of a tunnel and you unsuspectingly stumble upon it, or you lean against some old rotting or boost shoring timbers and the whole thing caves in on you. You never know what you're up against. I'm sure you have heard or read in a newspaper about how someone had fallen in a mine shaft or was killed in a cave-in. You will never know what to expect from an old mine, so be prepared.

Let's start off with some basic definitions:

1. Mine: a man-made cave, can be very deep and/or long.
2. Tunnel: an opening going in a horizontal direction.
3. Shaft: a hole going in a vertical direction.
4. Shoring: large timbers in a tunnel or shaft that are usually placed to hold up loose rocks or unstable earth formations caused by blasting or earth removal.
5. Air shaft: an opening from a tunnel going upwards to open air to help ventilate a long tunnel.
6. Open pit mining: usually done without a tunnel or shaft generally starting out with a small hole and enlarging it to use large equipment.
7. Hard rock mining: usually internal mining following the drift of a vein with tunnels and/or shafts and offshoots of other tunnels.
8. Placer mining: the tabling of gravels and sands from stream beds or former dirt flows usually using sluice methods or dredging.

Old mines have been known to hold a few surprises. A hard hat can save you a lump on the head from an unseen low hanging rock. A little awareness can save you from falling down a sudden shaft right in the middle of a tunnel. Many mines have sudden shafts just out of sight of the light from the opening, or anywhere along its length. Some shafts may have been covered over with lumber and dirt hiding a real danger if the lumber slips or rots away and you have stepped on it at the wrong time. Some mines have many tunnels which could be miles in length and have so many offshoots that one could get lost in it. It might be advisable to make readable markings on the walls so you can find your way back out of a mine. An old deep mine can develop some bad air in it. The air could be either toxic or explosive. Can you imagine hitting a pocket of explosive gas while exploring with a candle or lantern? I'm sure it will blow your flame out. I can recommend for your safety, using a flashlight or battery-type light.

A pile of rocks can hide a rattlesnake or two or three, as I have stumbled upon, so be careful with bare hands when picking up rocks or moving old boards and tin. In certain areas one can find scorpions. A favorite hiding place of theirs is in a cool pile of rocks or under a piece of tin or wood. Sitting on a pile of rocks can cause enough movement to stir up a sleeping scorpion or other unwanted creature. Old timbers or wood lying around can have nails sticking out of them, which if you are not aware of can cause a bad scratch or puncture wound.

It would also be good to know what can happen if you pull on a certain loose rock inside of a tunnel. Some rocks are Just like a wedge in holding other rocks in place. I have pried on an interesting ore sample and brought down more than I bargained for.

One last item, please remember that a lot of existing mines are claimed and being held for future operations, or are being worked on a part time basis. Treat the mines like you would want your own mines treated. Be prepared and be careful. End

Editor's note: the last paragraph is often more true than not. Our own mines have been "vandalized" by those who have broken up our ore cart, thrown trash around, and stolen items. Those individuals may have thought our mine had been abandoned or not. Rarely do abandoned mines have pieces of equipment lying around. Another good rule for exploring old mines is never touch the overhead for any reason. Remember areas with really dead air spaces will always have poisonous gases if nothing but carbon dioxide.

A Homemade Trommel

by B. Hoffman

Trommel: a sieve, usually a revolving cylindrical one, used in screening ore, coal, etc.

Over 3 years ago when this project was started we were hand screening our placer material then concentrating it over a table. We soon found out that some gold was being carried out with the water at the screening process. Not having the funds to buy a trommel, we built one from mostly scrap iron and items found lying around. By the way, this trommel works very well and we still have it for a museum piece (someday, we hope). I am sure this will help someone out there that is trying to make a go of it. This trommel consists of several things that are hard to pick out from the pictures

#1 THE BARRELS

Three old 55 gallon barrels, the one that has a ribbon of iron is an old acid barrel. These barrels are gas welded together - we could not afford an arc welder at the time.

The first and second barrels have small blades inside them, about 1 1/2" high and about 12" long offset half-way between each set. They can be seated with rubber, or silicon (which is easier). The third barrel has hole! drilled all through it (as many as we could get in it). It also has blades in it which are covered with 1/8" screen to only let the fines through. The screen is bolted at both ends and overlapped, then wired together. On the outside of the third barrel, we used 16 mesh screen, which is regular window screen. This was fine for our use. Any size can be used. The 1/8" screen inside protects the outer screen. The outer screen is held with thin stripping and mounted to the barrel with small sheet metal screws, approximately every 1" apart.

On the last barrel, do not cut the hole end out of the barrel. Leave about a 1" or 1 1/2" rim around the outer edge. This will hold any excess water and make sure that all fines will get worked into the screen. Under the last barrel there is a half barrel that is cut lengthwise. It will catch the fines you want to retain. We cut a 4" hole in one end to mount a 4" PVC pipe into it, so that the fines would be carried to our concentrating tables. This is all accomplished with gravity. The half barrel will not catch all the droppings, so a thick piece of metal or aluminum can be used, mounted to the top pipe of the frame. This allows all droppings to wash into the barrel.

AT THE END OF THE TROMMEL

At the end of the trommel, we used an old piece of sheeting for the oversized material to roll down the hill. The sheeting is wired from side to side to give it a covered roll; thus acting as a chute.

We extended the 1/8" screen about 6 inches out the end of the trommel allowing us to save the 1/8" material for further testing. All the oversized material would work itself out. You can see the 1/8" material piled up at the end of the trommel.

#2 THE WATER LINE

The water line you see at the front of the trommel has 4 lawn full-spray heads mounted, one to each side and one at the end. This helps to break up clays, muds, and grasses while the trommel is turning. The water line is 3/4" PVC and hose. A piece of thin metal about 1/8" to 1/4" in diameter is bent and bolted to the top frame. This is to hold the water line from hitting the trommel. On the side that you can't see in the picture, is a set of 3 lawn head sprayers which only spray on 1 side. They are set at a distance to spray 12" of screen. This helps the final washing and also helps Keep the screen clean for new material on the next batch.

#3 THE FRAME

The frame is made of old 1 1/2" iron pipe* gas welded together. The first barrel (with the ribbing) is sitting on an old conveyor belt with rollers about 12" wide. These have bearings on the ends and mount to the frame. This type roller also has adjusting capabilities. On the second barrel, you can see the black grease. These rollers are the same type except that they have no adjusting ability, just what we have added to them. They are supported by 1" old steel pipe welded at the bottom of the frame. The pipe is split in half and heated to bend it into a "T" shape. The rollers are mounted on 1/2" bolts with two nuts each for adjusting capabilities from the T pipe.

On the side of the top frame is a curved piece of metal about 1/2" thick that is drilled and bolted to the top frame. This metal also has two 1/2" bolts from it to the roller and it also has the same type of adjustment capabilities. These rollers have no bearings, so we used a 2" thick rod, and the roller works on the rod. (The rod doesn't move). Once or twice a day a little oil is squirted in the bearings -graphite mixed with oil.

Under the end (or third barrel), we made our own rollers using heavy rod again and finding some steel pipe to fit over the rod. The heavy rod is mounted to bolts like the second set, but more underneath the last barrel to hold up the weight. The pipe rolls on the ribbed edge of the last barrel.

THE DRIVE

The drive is a 2 horse Briggs and Stratton gas motor with a small pulley, which drives the pulley on the jack shaft which is attached to a reduction gear (about 24 to 1). The other side of the reduction gear also has a pulley V-belt that goes all the way around the trommel. Pulley sizes don't really make that much difference - whatever you find to match together with a belt. With the gas motor you have some control on the speed desired.

Trying to use a big belt is hard to accommodate to your needs, so any belt can be used, the slack taken up with an old water pump and pulley from most any junk yard. The old pump already has a bearing in it.

If you look closely at the picture, you can see the throttle control to the motor on the right side of the middle area of the trommel.

Sluicebox Troubleshooting Guide

by Tom Bryant - Alberta, CANADA

To read this table find the problem you are having and look to the right of it to see possible solutions, ie. "BLACK SANDS IN TAILINGS". The chart says to - a) maybe decrease the volume of water, b) decrease the angle of the sluice, c) increase the width of the sluice, d) maybe increase the height of the riffles, e) increase length of the sluice, f) maybe decrease the amount of material being fed to the sluice, g) maybe clean out the sluicebox, and h) maybe classify material to run finer stuff separately.

This is just a basic table, a more complete one will come out later.

PROBLEM	Volume of Water	Angle of Sluice	Size of Solids Being Run in Sluice	Width of Sluice	Height of Riffles	Length of Sluice	Amount of Material Being Fed Per Unit of Time	Clean Out Sluice	Classify Material to Run Finer Material Separately
RIFFLES BURIED UNDER LIGHT COLOURED SAND WITH MODERATE WATERFLOW	+	+				?−			
RIFFLES BURIED UNDER BLACK SAND THAT DOES NOT CLEAR OUT OF THE SLUICE WHEN GRAVEL FEED STOPPED MODERATE WATER VOLUME	?+	?+		?+				!	
FINE GOLD BEING BLOWN OUT INTO TAILINGS	?−	−	−	?+	+	?−	?		
RUNNING FULL VOLUME AND RIFFLES LOADING WITH LIGHT COLOURED SAND		+		−	?−	?−			
RUNNING FULL VOLUME AND RIFFLES LOADING WITH BLACK SAND THAT DOES NOT CLEAR OUT WHEN THE GRAVEL FEED STOPS		?+		?−				!	
MODERATE WATER VOLUME PRONOUNCED BUMP OR WAVE OVER EACH RIFFLE	+	?−		?−	?−				
FULL WATER VOLUME PRONOUNCED BUMP OR WAVE OVER EACH RIFFLE		?−		−	?−				
ONLY FINE GOLD IN YOUR AREA		?−	−	+	−	+			
ONLY COURSER GOLD FOUND IN YOUR AREA	+	+	?+		+	−	+		
BOTH FINE AND COURSE GOLD FOUND IN YOUR AREA									
INCREASED GRAVEL FEED TO SLUICE	?+		?+	?+	?+				
DECREASED GRAVEL FEED TO SLUICE	?−		?−	?−					
BLACK SANDS IN TAILINGS	?−	−		+	?+	+	?−	?	?

Notes embedded in the chart:

- (RIFFLES BURIED UNDER BLACK SAND…) DEPENDS ON THE COURSNESS OF THE GOLD TO BE SAVED SOME BLACK SAND LOSS IS ACCEPTABLE IF NO FINE GOLD IN THE AREA
- (BOTH FINE AND COURSE GOLD FOUND IN YOUR AREA) CLASSIFY THE MATERIAL AND ADJUST THE SLUICE TO SAVE BOTH FINE AND COURSE GOLD BY RUNNING THEM IN SEPARATE SLUICES
- (BLACK SANDS IN TAILINGS) CHECK TO SEE IF ITS HEAVY BLACK SAND OR ONE OF THE LIGHT WEIGHT LOOK ALIKES. PAN SOME TAILINGS. IF YOU SEE LOTS OF HEAVY BLACK SAND AND PROBABLY SOME FINE GOLD DO AS ILLUSTRATED

CODES USED:
INCREASE +
DECREASE −
DO THIS BEFORE ANYTHING ELSE !
MAYBE INCREASE ?+
MAYBE DO ?
MAYBE DECREASE ?−

Air Hose Adapter

by Stephen B. Barnett

After dredging with a snorkel AND a sore back, I decided this year to go with AIR. I was told the air would make dredging easier, just lay in the water or on the bottom and work with the nozzle, and also NO MORE SORE BACK at the end of the day.

So I started to read about diving and breathing underwater, and the safety involved. The local public library was my first source of information. Then a friend told me I could borrow some of his books, three all told. I could see from reading that diving was going to be a new experience. SAFETY was foremost in my mind.

I purchased the normal equipment - T-80 compressor, (which would fit my motor without trouble, as my motor was made to accept the T-80), a . small air tank, regulator and mouthpiece with harness, short hose, long hose and weight belt with 40 lbs. of weights. Boy! Was my wallet losing weight! Next stop was at an Air Plumbing shop, for air fittings (air disconnects). These were purchased with a lot of planning, my wallet was losing weight fast.

On the compressor, I installed a bayonet (male fitting). On the short hose, a receptacle (female fitting) was threaded to one end. The other end was threaded into the air tank. At the other end of the tank I installed a bayonet. The long hose was threaded into the regulator. On the other end of the long hose, a receptacle was installed for hookup to the air tank.

Disconnect the air hose , from the compressor and air will blow out of the bayonet and you can run without air and not overload the relief valve,

To carry the tank, just plug the two connectors together, put the hose over your head, and carry like a sling.

If at any time a fitting gets accidentally disconnected, you have a small air supply in the hose (from you to the disconnect). Maybe enough to allow you to drop the weight belt and swim to the surface.
Remember SAFETY and PLANNING when starting a project or trip.

EDITOR'S NOTE: Diving and use of diving equipment is a field in which one should respect all the safety rules. Read the books, contact a Dive Shop, take it slowly, and play by its rules. Always dive with a buddy, think that's the first rule learned. Nancy

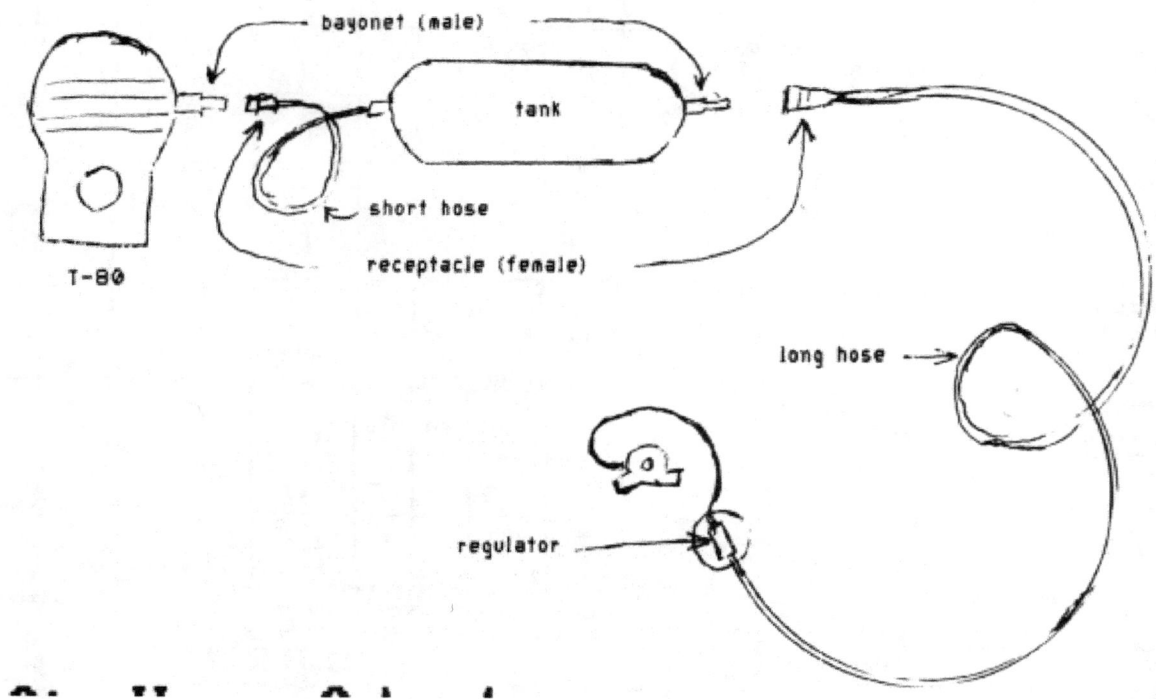

Build Your Own Triple Sluice Dredge

I have in front of me a booklet of plans to build your own triple sluice, written by C. Dennis Boring of Oroville, California. We were originally going to try to publish the whole thing but realized that room prevented that, but I want to give you the basic idea behind these plans. Should you decide that you would like to buy the complete set of plans (consisting of parts list, 2 summary pages, and 14 pages of drawings) you can get them directly from Dennis Boring, P. 0. Box 1619, Oroville, California 95965 for $20.00. Included as an additional bonus is a drawing of a grate attachment for a suction or suction-nozzle tube. As you can see from the pictures, Dennis has been using this very sluice this year. He wants to sell it now so that he can build a bigger one for next year.

Here are excerpts from the booklet.....
The sheet metal used in this project is all 20 gauge galvanize sheet metal from an air conditioning/heating & sheet metal shop and the man who used these drawings to fabricate the various pieces did a nice job and only charged about $100.00 for it all. I also checked with the various hardware stores and found a reasonable price for the PVC pipe and adapters to make it all fit together. The hardware came from the same store and the frame was sandblasted and painted to prevent any premature rusting or adverse contamination of the gold settling areas. It is also a good idea to use a silicon sealer like RTV OP other clear sealer to insure a good tight sea) at all sheet metal joints after they've been soldered or riveted as the case may be. I installed the pump and motor at the center area to insure proper balance and ease of access. Be sure to position the motor so the exhaust blows away from the diverter box. It makes for a cleaner unit and you won't get a hot spot that might prove to be pretty uncomfortable since the water you're working in should be quite cold and you certainly won't notice a slight burn as quickly, the inner tubes were fastened to the framework with "rubber snubbers" and they were positioned in a triangle and secured in the center where they join, put some RTV silicon around the seams to seal against any loss of concentrates. 1 steam cleaned all metal areas after I was through drilling and messing around so as to clean the gold settling areas of any grease or oil (it'll carry your gold right over the riffles if you let it!) seam clean the nozzles too, and don't forget the screens and riffles. Every little bit really helps, remember...the odds are against you anyway, don't make them greater than they are.

The framework was created to fit in the bed of a 1/2 ton pickup truck and if you maintain the 4', 2" max. width figure you'll have no problems. The overall length isn't as crucial since I run around with the tailgate down most of the time anyway.

I used my own welding machine which is a Hoart Beta-Mig 200 but if you have the use of a gas welding setup it would work just as well. I soldered the sheet metal in the water-flow areas and pop riveted the balance of the pieces or bolted as necessary. Also enclosed is a cost breakdown of areas that you'll spend your hard earned money in. They are listed in groups of general importance. Use some imagination in placing a particular piece of material in a particular category, it really doesn't matter much.

Permit to dredge (Calif; $ 5.00
Motor/pump (used $300.00
Square tubing 15.00
Sheet metal work/ material 100.00
Muffler tube & horseshoes 18.00
Carpet strips 10.00
Latches 15.00
Misc. items/labor 75.00
Inner tubes 10.00
Total $548.00

The above items listed are approximate and your costs may vary depending on your location. I had pretty good luck finding the things I needed by approaching friends and others interested in the gold dredging idea. I also looked high and low for a pump and motor of sufficient size to run the 3" (triple sluice) requirements, which by the way run approx. 200-300 gallons per minute and that takes a 5 HP or 6 HP Briggs & Stratton engine; although any engine would work so long as the pump and motor are matched in power curve/ delivery rate.....

The screen through which the 3" PVC pipe is passing will classify the incoming material and only allow approximately 3/4 minus to pass through to the "gentle" washing areas of the side sluices.....

It took me about 2 weeks to build this piece of dredging equipment and I didn't work continuously. A lot of the time I spent looking for parts and sort of drawing/fabricating at the time of construction....

I placed a piece of waxed paper under my screen and then some masking tape strips on either side of the area I wanted to seal off (which coincides with the riffle location) and spread some silicon into the diamond pattern. I used a business card to insure it layered flat and when it dried I just carefully peeled away the waxed paper. Make sure you use some boards or something to make the screen lay flat while inserting the silicon. Whenever you insert the screen after a cleanup or inspection, just locate the screen on the carpet so the riffles will lay flat across the screen area containing the silicon. I saw this method in Popular Mining Issue #2, so many of you will already by aware of its' use. Believe what it says, it's the gospel truth.

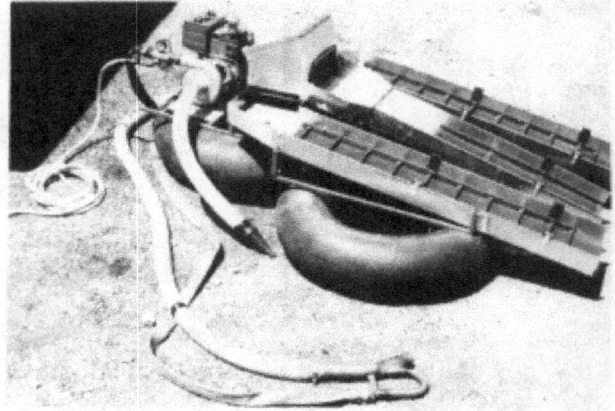

Triple sluice - 2.5" suction line 6 HP B & S engine
P300 pump by Keene

Idea Corner

Boil black sands with salt
by W. H. McKennan

Several years ago I was dredging on the Stanislaus River about 1/2 mile downstream from the old Melones Mine. This was the biggest mill in the Mother Lode. It had 100 stamps and they dumped all of their tailings in the river.

The black sand concentrates were very fine and dense. It was impossible to make the fine gold settle to the bottom of the pan. There was quite a lot of fine gold also a silvery-gray trail at the end of the black sand with the gold.

I had heard about boiling the black sand in salt water, so I took a small amount of the black sand and rolled it in the pan to get an idea of how much fine gold there was, so I could see if there was more afterward. There was no liquid mercury in this sample.

After boiling it in salt water for 30 minutes, stirring it quite often, I put it back in the pan and rolled it again. There was no fine gold on top as before, but when 1 got to the bottom there were 3 large beads of mercury (liquid). And all of the fine gold was in the mercury.

My only explanation is that there was quite a lot of oxidized mercury in the black sand that had been lost from the mill upstream and the salt water returned the mercury to its liquid state and it picked up the cleaned fine gold. The silvery-gray trail was gone also. It must have been fine gold coated with mercury partly oxidized.

So if you are dredging downstream from an old mine and mill, always boil some of your black sand concentrates in salt water. It may be very interesting.

This area is gone for mining now since the new Melones Dam has been built and it's under about 250 feet of water now.

The salt I used was just table salt that you can buy in the supermarket. .
soluble gold

Water soluble gold
by James Newsome -   Hi Hat, Kentucky

I would make this suggestion concerning water soluble gold. When a person is looking in the field for gold, it would be a good idea to look for water holes where the water is not clear. Carry along with you some pH paper. If the pH is lower than normal for water or it is not very clear, you may just be onto something.

Use Anti-Static Solution on Vials

Miners who decide to use the body of hypodermic needles (the glass part) to fill 2 pennyweight bottles with gold -and it is hard to find funnels small enough to fit in the vials - are now faced with plastic barrels and plungers. These carry an anionic charge upon their surfaces and the fine particles of gold carry a cat-ionic charge which causes the gold to stick to the inside of the tube. Solution - first treat the plastic tube with anti-static optical fluid available from opticians. A Q-tip works quite well to apply the juice on many. Friend, Ernie Wells

Economic Furnace Hood

 For the small miner and prospector who does a limited amount of lab work or the prospector or miner who has a small assay furnace and who has shied away from a lab-type hood and blower for adequate ventilation to take away the toxic fumes, you can go to a second hand store and purchase a hood that was used over an electric or gas range in the kitchen. Then purchase an old blower for a gun-type oil furnace - when the oil pressure pump or nozzle gives out and yet the blower is still good, you can usually pick these up very cheap.

Now get from the hardware store 3 sections of galvanized stove pipe. Sometimes you will want 4 so you can have one to come through the wall of the building where you do your assaying or chemical work. Also get a galvanized elbow.

With a top cap, a few metal screws, a piece of tin or aluminum and a pair of tin snips you can readily fashion a very effective and safe exhaust system. The oil furnace blower for the gun-type burner has adequate capacity and the hood will catch all the fumes you don't want in the room. Discharge the blower into the pipe and the fumes will enter the air intake of the unit. Ernie Wells

Crucibles
by Jim Suchor, Montrose, Co.
If you melt borax glass in a new crucible, that will seal it up and it will last a lot longer.

Cover that Metal!
I personally get rid off any and all metal baffles, paddles or obstructions inside any trommel and replace them with rubber. Rubber lasts longer, does a better job and does not contaminate the concentrates. Just cut old tire casings into u-shaped strips and bolt to the inside with carriage bolts. You will be surprised at the results. Ernie Wells, Elgin, Or.

Rule of Thumb for Hydraulic Motor Applications
Where Gasoline Engines or Electric Motors Are Now Used
1 Electric Motor H.P. Equals 1 1/2 Hydraulic Motor H.P.
1 Gasoline Engine H.P. Equals 2/3 Hydraulic Motor H.P.
1 Hydraulic Motor H.P. Equals 1 2/3 Gasoline Engine H.P.
1 Hydraulic Motor H.P. Equals 2/3 Electric Motor H.P.
1 Electric Motor H.P. Equals 2 1/2 Gasoline Engine H.P.

Catch Your Fine Gold
by Ernie Wells
A careful light sanding with some finishing sandpaper (very fine grit) and then cleaning as I mentioned in another idea improves efficiency of new plastic drain pipe. I also attach with waterproof cement a sheet of magnetic material (like used in making magnetic signs) to the final 18 inches of the sluice, wrapped around underneath. As very fine gold has an affinity for magnetite and as you have it charged and have magnetite standing on end, this also improves efficiency.

ANOTHER AREA FOR PROSPECTING
You might want to advise the readers of PM to look at the shavings around oil or gas wells and any types of drilling. Then the information on a well log is very helpful. The Kentucky Geology Department keeps these well logs and they will make you a copy for $.35. James Newsome, Hi Hat, Ky.

FREE SCREENS
Old clothes dryers have a round screen that fits perfectly over a standard gold pan. The screen has 1/4" holes, so it makes an ideal classifier, and won't float away like 2 of my plastic ones did.

CREVICE BLASTING
I use PVC fittings to come off of my dredge pump and use 1 1/2" fire hose and nozzle for crevice blasting. Not my own idea but it sure works good! ABS fittings are common. Some swimming pool suppliers have a good assortment of PVC valves and plastic king nipples.

TAPED TOGETHER
Use teflon tape on threaded fittings where they contact iron or steel. Rust occurs and makes it a job to get apart. (For big pumps only, size your hose and nozzle down appropriately.) Teflon tape also makes a better seal.

"Y" DO IT
A friend of mine uses a double "Y". One is for a crevice flusher, the other to power a small 2" dredge for clean up.

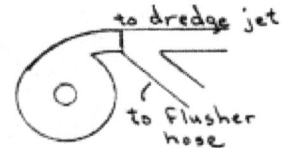

BURLAP BAGGING IT
A burlap bag with a few large rocks in it makes a decent anchor plus it's biodegradable if it needs to be left behind because it got covered with tailings.

TOOL BOX
I bolt or rivet a small tool box (such as socket set boxes with latches) to my dredge frame. In it goes a pipe wrench, pliers, a few small combination wrenches, sockets, small bolts, nuts, gaskets for hose ends, teflon tape, etc......

Moving a boulder

Using a Cable Hoist to Remove Large Boulders

by Steve Barnett, San Carlos, Calif.

Sometimes when you wish to remove a large rock from your dredge hole, try the method and tools shown in the pictures. It works with time, muscle, sweat, and hard work, tempered with patience. We removed 5 or 6 rocks like the one in the picture. It works, you can tell by the SMILE on Dan's face. We are dredging in a pool that is right at the top of a waterfall, all of our tailings from the dredge go over the waterfall (no tailings pile).

We were down about 4 feet deep when we found rocks too large to get out without some danger to one of us. We were stopped. The creek is too far from the road to use the winch on the jeep. We removed all of the rocks we could SAFELY, a man on each side pushing, prying and cussing the rocks and each other. During a rest break, we ran all kinds of ideas back and forth; the question was "how to get the big &!*@* rocks out safely?"

With a hammer (3 lb) and a 5/8" star drill, we drilled a hole in a HUGE boulder, installed a 5/8" x 5" red head expansion bolt with an oversized washer and nut op top of the cable EYE. At last we had a solid anchor. With some random lengths of steel wire rope, we spliced an EYE in one end. Two or three U bolts will work, or cable clamps, to make an EYE in the cable. To rig or hitch the rock to pull, we used old tire chains bolted together. You can splice chain with a bolt, 2 washers and a nut. If the bolt will just fit through the chain it will work fine. If you study the pictures, I think it is almost self-explanatory. You can see everything but the anchor.

Ii your cable is long enough, you can wrap it around a stationary rock or tree. We used the expansion bolt because there was not a tree or rock in the direction in which we wished to move the rocks. Everything went over the water-fall out of our way.

The cable ratchet hoist we purchased at a hardware store, the cable grip or puller at a local power company tool auction ($20, about (60 list), the cable we had. You can find cable at a steel scrap yard or surplus store. The cable grips come in different sizes to fit the cable diameter. Ours will fit 1/4" to 3/8" cable.

The cable grip, once it is on the cable will slide towards the anchor very easily, but pull on it and it grips the cable. The harder you pull, the tighter it grips the cable.

After you are hooked up, you slide the grip towards the anchor to remove the slack. Then start using the ratchet handle and PULL. Just before the hoist was used up we would stop, slack off the hoist, and slide the grip towards the anchor, and pull some

The cable hoist

more to get the rock closer to where we wanted it. Several times we had to block the rock with smaller rocks to keep it from sliding back into our hole while we were sliding the cable grip forward to get a fresh start on a new pull. We had to "winch and slide the grip towards the anchor" several times to get the rock out of our way.

We used this method and we are happy with the MOBILITY of the cable grip. To use a cable with an EYE in each end is almost impossible because of the unknown length needed. The cable we used was about 60 ft. long. We only used about 20 ft. of the cable, but since the cable grip doesn't damage the cable, we still have 60 ft. of usable cable.

Remember SAFETY when working with cable. If a cable snaps in two, it will WHIP. DO NOT STRADDLE the cable while pulling.

At one time during our pull, we thought we had too much tension on the cable so we stopped pulling. We sat down and looked at it for 5 minutes or so, when Dan suggested one man work the hoist and the other work in the water with a pry bar to HELP the rock along. It worked fine.

The rock in the picture we thought must weigh about 10 TONS but after we got it out and took a long look at it we knew we had over estimated its weight by about 9 tons! The rock measured about 70" x 40" x 24-30". We now think it weighed only about 1500 lbs, but even that would have been tough without the cable grip hoist.

Quartz Cracker

by Ernie Anderson

This device was built to serve as a primary rock breaker in preparation of sample pulverizing with impact mill #140.

Construction is simple and low-cost, using a few pieces of old iron found around the shop.
Here are the items used for this one (of course dimensions can be adjusted to fit your materials):
1. 40" of 30 lb. mine rail
2. 14" of 8" I beam
3. 20" of 5/8" threaded bolt stock
4. 6 5/8" nuts
5. 12" of 1 1/2" pipe
6. 12" of 1" angle iron
7. about 2 1/2 square feet of sheet iron

I used a circle saw with an abrasive disk to size the parts and a home-run buzz box for the welding. Total cost was less than $10.00

STEPS OF CONSTRUCTION
1. Cut (3) 13" lengths of mine rail 3/8" shorter than the I beam. Leave a 1/4" step on the wheel surface to center the grid ends.
2. Cut (6) spacers 2" long from 1 1/2" pipe and oval with a hammer.
3. Cut (2) 7" lengths and (1) 6" length of 5/8" bolt stock.
4. Drill (3) 5/8" holes through the web of each mine rail length; one in the center and one through each end. These holes need to match.
5. Insert the 6" bolt through the center hole and a 7" bolt through each end hole.
6. Place a pipe spacer over each bolt end. Add another rail, then more spacers and then the third rail.
7. Screw on the nuts and torque really tight. The end bolt ends should nearly touch the I beam sides. There should be 1/4" space between the ends and sides of the I beam compartment to allow broken rock to fall through.
8. Tack one sheet iron end and the sides to I beam.
9. The sliding door should be cut to fit closely within the open end of the I beam and sides.
10. Tack 1" angle iron door guides to the I beam open end. Allow about 1/16" space between the door guides and door faces to allow easy sliding.

Use a single jack and short bar to contact rock; or a long bar with a 2" square striker surface.

See the two additional pictures showing the grid removed and the grid in place with the sliding door open.

Two impact tools are shown here; one with a flat striker face and one with a rounded striker face. Each one has a 4' handle to allow the user to stand upright while breaking rocks. The broken rock fragments fall down through the 1/4" slits between the mine rail.

The grid is self-centering at a distance of 1/4" from the ends and sides. This allows rock fragments to fall below through this space also. Now, since rock fragments fall through the slits, there is a minimum of powdered rock between the mine rail grid and impact tool. The space around and between the mine rails is a reservoir for several pounds of rock fragments.

The high sides minimize flying rock fragments. Long handles on the striker tools avoid battered hands. The I beam web, being elevated, helps in placing the broken sample on a tarp by tipping the rock breaker on its end.

It's a handy item to carry in the pick-up on prospecting trips and is very sturdy and ready for long and vigorous service.

grid in place

grid removed

Wet Suit Heater

Attaching hoses to heater core

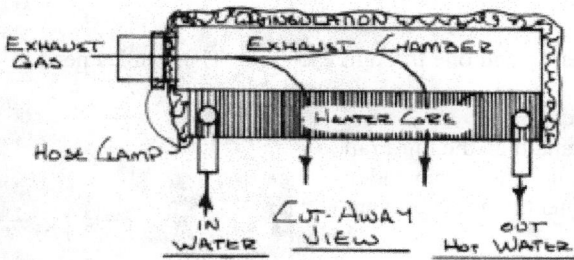

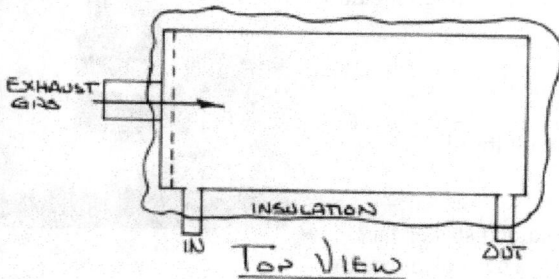

by Gary Walker

This is my version of a cheap do-it-yourself wet suit heater using an automobile heater core. It works very well. The water temperature in my creek generally runs about 48°F and is usually a "dry suit creek" for prolonged dredging. I run the hose in the chest area of my suit, then once in a while pull it out and put it in my gloves for a few minutes. I glued 1/4" insulation on outside of heater and still lose a lot of heat. More insulation could make water scalding, if it were used without a cold water mixer.

I didn't have time to fabricate a box out of metal so my "wet suit heater" is a cut down metal cased heater from an old pickup. I have the exhaust coming out the bottom so as to Keep heat trapped in.

Also it is essential that water is running through the heater core while the engine is running otherwise the solder would probably melt out of the core making it leak. I used a garden hose water valve for regulating water flow. I regulate heat with water flow. About 1 quart every 20 seconds gives water that won't start to burn the skin.

Good luck in your operation. Gary walker.

attaching heater to exhaust on engine

Poor Miner's Vacuum System

Pump hooked to filter flask

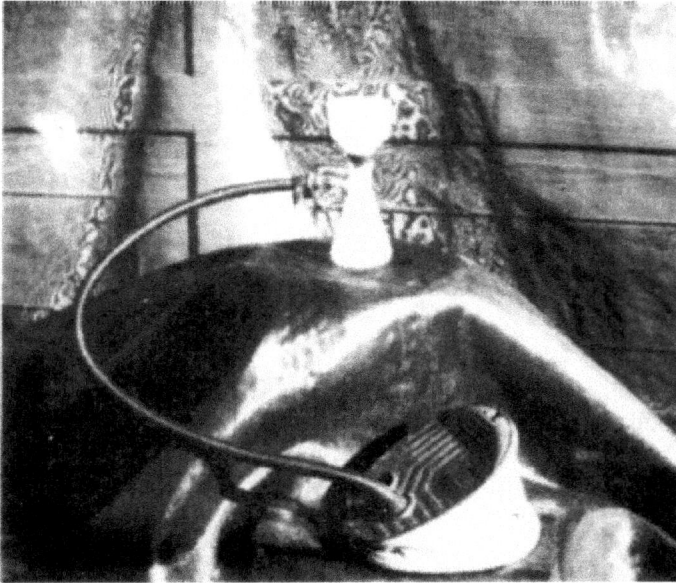

by Bill Thomas

Tired of waiting for your test filtrate to drip through the filter paper? Wish you had a nice vacuum system like they had at the seminar you attended at Crow Canyon? Can't afford $349.50 for the system? Be of good cheer....there's still hope.

While casting about for personal answers to this dilemma, my eyes fell on something in the backyard. It was a foot-operated pump my daughter had bought to inflate a. rubber raft. There are 2 holes on top of the pump, one marked "inflate" and the other "deflate". I reasoned that the deflate connection might provide a small amount of suction to a filter flask. It works!

The photo shows the setup. One end of the plastic hose is inserted in the "deflate" hole, and the other end attached to the vacuum flask. It just so happens that this unit, made by Sevylor, fits the flask connection nicely. All the parts are plastic, so a little acid vapor won't hurt it. It is slower than a real vacuum pump, and requires your foot action to pump it, but it beats heck out of gravity.

TWO PRECAUTIONS:
1) clamp the hose to the edge of your work table so you don't accidentally knock the flask off the table and break it;
2) step gently on the foot pump. Stepping on the pump expels the air through the "inflate" port.
but it also pressurizes the tube slightly. That's okay if there's plenty of liquid in the fitter funnel, but if not, too vigorous pumping will puff up the filter paper and move it around. The suction occurs on release of the pump any way, so step gently on it.

Happy inexpensive filtration!

Mine Safety - Lack of Oxygen

by State of California Mine Safety Training Dept. Donald! Steven, Associate Safety Engineer

Editor's note: The following is the beginning of a series of articles written for us by the Dept. of Mine Safety, State of California. Although some of their information refers to California rules, most of the data is valuable anywhere in the world.

In the state of California there are so many abandoned mines that I don't believe any one person could quote you an exact number, but that number would run over 500,000. That's a lot! That's a load of potential problems for many people who are unaware of what happens underground.

With more and more people moving into the Mother Lode areas and as the price of gold and silver goes up as the economists have predicted, people who are looking for the fast, easy dollars (boy, is that a false idea!), or just out of curiosity, or the honest to goodness prospector, are going into abandoned mines and should be aware of what could be some of the hazards they might encounter.

One of the 2 main hazards is irrespirable (unbreathable) air. The other is the possibility of ground collapse, In this and the next few, articles we will deal with underground air, its hazards, gases which could be encountered, where they come from, how they can be detrimental to one's health, how to detect them (without a canary). Further, if you must still, enter into irrespirable air, how to do it safely.

On the surface one hardly ever considers the air one breathes. As one enters a foreign environment, (underground is a foreign environment as much as underwater), we need to take a look at this gas called air.

Air is a combination of gases. It is made up of 78% nitrogen, 21% oxygen with remaining 1% made up of argon, carbon dioxide and small traces of other gases.

Underground the amount of oxygen in the air can be easily depleted by a number of causes. First, let's see what happens to the body if the oxygen content of the air becomes less than 21%. If it drops to about 17%, you'll breathe faster and deeper because your body is trying to compensate for the lack of oxygen. When the oxygen content drops to 15%, you become dizzy and will get headaches. If oxygen drops as low as 9%, you lose consciousness. A 6% content of oxygen in the air will almost always be fatal.

The State of California Mine Safety Orders require at least 19.5% oxygen in the mine at all times. (Mine Safety Orders 7098-e).

CAUSES OF OXYGEN DEPLETION IN A MINE
There are 4 main causes of oxygen deficiency underground:
1) insufficient or improper ventilation which fails to bring in enough oxygen.
2) displacement of the air's oxygen by other gases
3) a fire or explosion that consumes oxygen
4) consumption of oxygen by workers, explorers, prospectors, or animals.

Looking at these in more detail....
1) insufficient or improper ventilation - when these mines were abandoned the ventilation systems could have been removed by whomever excavated these holes. Or the oxygen could have , very easily through the years, been used up because of the decaying process of the mine timber supports, the oxidation of the iron left behind in tracks, air and water lines* or oxidation of some types of minerals. Also without ventilation or air movement* it could be stale air or oxygen deficient air left in the mine.

2) different types of gases can displace oxygen and cause deficiency. Each gas will be discussed later.

3) fire and explosions consume oxygen. It does not matter if it was an ancient or recent fire or explosion. Once the oxygen is depleted* and without ventilation to bring in more oxygen* the deficiency could be extreme.

4) In small holes the amount of oxygen used by a person depends on how hard he/she is working and how much oxygen was there to start with. In an area that is borderline* it is quite possible to deplete the oxygen content without knowing it. After the first symptoms of oxygen deficiency around 17% oxygen, show up and the brain starts to get starved for oxygen* you might not know the real condition of your body and not realize how serious the trouble is until too late.

FACTS
Oxygen is colorless, odorless, and tasteless. It has a specific gravity of 1.1054. If you are going to check for oxygen deficiency you need to know that it is slightly heavier than air. Air has a specific gravity of 1. So in checking for oxygen deficiency, you would have to check Just below or at waist level.

CHECKING FOR OXYGEN DEFICIENCY
There are several ways to check for deficiency. One which hasn't been used in a long time is with a canary. (I think the SPCA would come down very hard on us if we tried it today!)

Another way is with a Flame Safety Lamp. It will tell if there is an oxygen deficiency or if there is an explosive atmosphere present. But they also have their drawbacks, for without a lot of knowledge and practice, it is possible if improperly used, to cause an explosion. (We'll try to get an article on how to use one safely.)

The best way of detecting oxygen deficiency is with a meter and there are several on the market today that are very good and easy to use, small enough to carry in your pocket or on a belt, and they sound an alarm in an oxygen deficient atmosphere.

Railroad Series Laying Rail

by M. D. Isely (Doc)

HISTORY

Fifty years ago America had thousands of small, family sized gold mines. The price of gold had risen from $20.50 per ounce to $35.00 per ounce. As long as the price equaled or exceeded a week's wages the mines were active. Gold price is in the range of a week's wages today, but there are few small, active mines. The advancements in mining technology have been principally in the large, open pit variety. There have been notable advances in small mills and in "do it yourself" assaying but getting the ore out of small high value pockets hasn't changed much.

One thing that hasn't changed at all is the small mine railroad. Why use a railroad?
a) the car is steered by the track. The track can be laid so that it clears the walls of the tunnel by a small fraction of an inch if necessary. One can avoid moving a lot of worthless rock by boring a small tunnel.
b) A steel wheel on a steel rail raises almost no dust.
c) Power requirements for a railroad are extremely low. This is most important in an unventilated tunnel where our usual source of high horsepower, internal combustion engines are impractical.

25 years ago there was a lot of mine railroad equipment and at scrap prices in salvage yards. Now most of it is in museums. However, the technology is simple and can be re-created. One thing which seems to be in considerable supply is light rail although most of it is in inactive mines.

Railroads have many advantages in mines. One person can push a quarter ton load, on a level track, without great effort. Because of the low power requirements, small electric or compressed air motive power may be used. There is no need for high horsepower, internal combustion engines with their related health hazard and ventilation expense. Animal and human power are even used in some small mines today. The principal disadvantage is that a railroad follows a fixed route and moving the track is an added expense.

There are certain advancements which can be applied to this old art.

Railroads were the high tech of a century ago and were brought to a high level of perfection. But people today, including some otherwise very knowledgeable miners, are generally unfamiliar with them. For this reason railroads in small mines often are a source of unnecessary trouble. This article and others to succeed it will deal with the basics of getting the most out of a small mine railroad.

RAIL

The weight of rail is always given in pounds per running yard. Thus 12# rail weighs 4# per running foot. However, the lengths are given in feet but measured in yards. A 60' rail is 20 yards long, a 39' rail 13 yards long, a 21' rail 7 yards long and so on. Mine supply houses carry new rail down to 20# weight but this is much heavier than is needed in a typical small mine. 16, 12 and 3 pound rail can be purchased used although new rail of this weight is rolled, to American standards, in West Germany. Delivered in the Mountain west this costs about $.38 per pound in 30 ton lots. So you can buy 15,000 feet of 12# rail for $1.52 a foot. This is enough to lay 7.5 miles of track. This price involves about 3 marks to each dollar. New track hardware of all kinds is available from American suppliers but you won't find it at your local hardware store. A 200# keg of spikes contains about 1600 spikes and costs $320 or about $.20 a spike. You need 4 spikes per tie.

If there is already a railroad in your mine, your problems will be with maintenance rather than with construction. If you contemplate new construction or an extension, the use of used material will enable you to "poor boy" it. I recently bought 220' of track (440' of rail) complete with ties and all hardware for $125. This was mostly 8# rail with some 12* and about 1/10 of it was covered by a landslide. If ties are old, dry and rotten, it is possible to pull rail, by means of a winch, from under a cave-in or landslide. In this case the spikes held and it was necessary to dig through the landslide to unspike each tie. The mine had been abandoned for 10 years and the access road was obliterated. There were 3 miles of rocks and gullies plus some steep grades on gravel. If you have more time than money, this is one way to get a railroad. Prices of used rail vary with condition and location. $.15 per pound for rail in usable condition at a material yard seems current here in the Mojave Desert. However, it is getting scarce and the yard may not have any.

rail car to haul rail into the mine

GAUGE

The wheels of each car are machined at an angle of 20 to 1 and tapered toward the outboard side. If everything is in good repair the flanges on the inboard side of the wheels won't touch the rails on straight track or on moderate curves. On sharp curves the flanges keep the car on the track by contacting the inboard side of the rails. It takes more energy and causes more wear to negotiate such a curve so, if possible, keep the track straight or with moderate curves. It will be desirable to widen the gauge a little on each curve, depending on the radius and if you have uneven rail it may be well to increase the gauge of straight track slightly. Be sure that your car rolls freely over any new track and be sure that the car's bearings turn freely when making this test.

TIES

Wooden ties maintain the gauge of the railroad and provide support for the track. The rail is a form of I beam which provides rigidity between the periodic support of the ties. Mine ties are usually 4x4's and are twice as long as the gauge of the rail. Thus an 13" gauge track will have ties 3' long. There is nothing absolutely necessary about this length, and with other methods of maintaining the gauge and a smooth rock tunnel floor, conventional ties can be dispensed with as will be described later.

SPIKES

Rails are usually secured to the ties by means of spikes. A spike should be driven straight into a tie - not at an angle. This will be difficult to do with a doublejack because the head of the rail is so close to the spike. A spike maul has a narrow but heavy head. If none is available, use a large punch or other bar to clear the rail head or a specially made tool like a rivet set. A light jack hammer may be used but be careful not to overdrive the spike. The head is offset and can be bent if you keep driving after it has contacted the base of the rail. The spike is made with a tapered underside of its head to fit the tapered base of the rail. If it is driven too hard the head tends to spring back and if it is driven at an angle the underside of the head never contacts the rail in the first place. A spike driven at an angle will, if it works loose, fail to maintain the gauge since it will pull out at an angle away from the rail. It will be helpful in driving spikes straight to stay on the same side of the rail as the spike. A railroad can be operated safely with a number of loose spikes if they are driven straight because they will still maintain the gauge. The weight of the car holds the rail against the tie.

Below: Note this man is standing on same side of rail on which he is driving spike. This is regarded as the

When spikes are unavailable, nails are sometimes used. These are not very satisfactory but if you must use them be careful not to break off the heads by driving them too hard. Stop when they contact the rail! You can do the best job if you don't work too hard driving nails or spikes. Lag screws are sometimes used. They are time-consuming and expensive, but if they are as thick as the spikes they replace, they do a good job. Set them straight. Large area washers are a good feature when using nails or lag screws.

JOINTS

Joints between the rails are the weakest part of the track and cause the most trouble. They have been such a large maintenance expense that mainline railroads use rail in several hundred foot lengths just to eliminate joints. Long sections of track are now pre-fabricated and moved into place on special trains. In a small mine, especially where rail must be moved frequently, short sections are best and the welded tie, prefabricated sections (see PREFAB TRACK), have some of the same advantages.

When a railroader refers to a low joint, he doesn't mean a disreputable establishment. When rail ends sag at a joint, they form a low joint. All joints should be supported on ties. The angle bars, bolted to the rail web, with their angled edges engaging the angles on the underside of the head and the top of the base can always give a little to compensate for expansion and contraction of the rails and support from below is most essential. (See illustration of a potential low joint). Avoid excessive gaps at rail ends. If the track is outdoors in a climate with great swings in temperature, it is well to leave a very small gap no more than 1/3" with a typical short section of mine rail, to permit expansion in hot weather so that the rail will not buckle. In a tunnel with uniform year-round temperatures, ignore expansion gaps. If the rail ends are close enough together they place less strain on the angle plates and provide extra rigidity in that way.

If two different weights of rail are used, splice bars can be used where they are joined. Be sure to put shims on the tie under the lighter rail. If splice bars are unavailable, angle bars made to fit the smaller rail can be welded to the heavier rail at that joint. This isn't the best method, but with care, it can be made to work. Before welding the bars, align the rails vertically and horizontally so that there is a continuous smooth surface at the upper inboard side of the rail. Then bend the angle bars, which have been bolted to the smaller rail so that they conform to the joint and weld them to the larger rail. The best joint, one which wears well and offers no appreciable resistance to the passage of the car, is a rigid joint which is very near to offering a continuous surface on the upper and inboard sides of the rail.

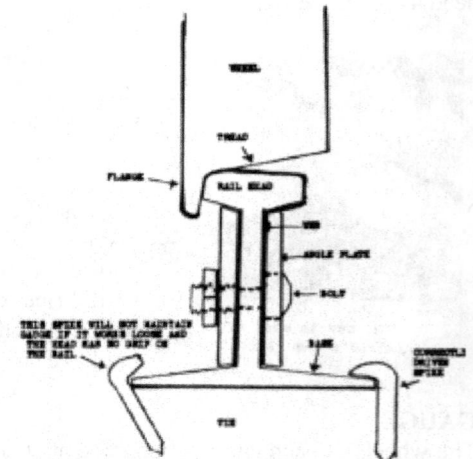

PREFAB TRACK

Phelps Dodge Morenci mine is an open pit operation with about 30 miles of bench track which is moved frequently. They use prefabricated track sections 39' long. This is standard 4' 3.5" gauge track with 90# rail. A crane advances over each new track pulling a car loaded with prefabricated sections and rotates 180 to set each section in place. A 21' section of 8# rail with low profile steel ties welded in place weighs about 130# and can be manhandled by 2 men. Several sections can be carried, stacked on a little push car. Shorter sections can be used to advance the track when boring a tunnel.

Ties in mine railroads are sometimes set 3' apart. This means that the rail must be heavy enough to support the cars between the ties. Ties and spikes are expensive but so is heavy rail and heavy rail is hard to move. Most mine tunnels have fairly flat floors and the rock is solid enough to give continuous support to the base of the rail. Gauge can be maintained by using through bolts in tubes located between the rails. Be sure that the tubes are low enough to clear the wheel flanges.

Another method is to prefabricate sections of track by welding the rails to angle, channel or flat plate ties. At one end of each section a broadplate is welded extending beyond the end of the rails to support the joint with the next section. Grooves can be made in the tunnel floor to accept the angle or channel protrusions. On a soft fill, stringers in the form of planks laid beneath the rail and between the ties will provide flotation. In making prefabricated sections of track, it is most important that each rail of a given section be exactly as long as the other and that the ends of the section be square. A simple jig can be made to assure alignment. Pieces of discarded rail can be used for ties. Curves must be made to order. It is a good idea to have a rail saw, a drill and a bender to do this work. The joints, of course, must always be adjacent rather than staggered as is usually the case on a high speed railroad.

COMPATIBLE COMPONENTS

Rail comes in many weights and the related hardware in many forms. A variety of angle bars with a variety of hole locations may have to be modified if a good job is ddne and considerable time must be expended. If you can obtain standard components, the laying and moving of track will be greatly simplified and expedited.

Investment Money

Over the past 5 years I have talked to at least 2 persons a day on the telephone about mining. Including weekends and holidays that makes over 3600 people that have called here. I have heard every story known several times over. The story that I hear most often, I would say, is the story of how a miner has failed to secure investment capital and thus has failed in his mining venture and in most cases continues to fail.

There are many reasons why miners fail to secure capital for their mining ventures. But I want to say a few words about the major reason that I have found. That reason is simply that most miners are willing to give to little in return for the money.

Let me say that it has been my experience that a miner that has had a mine for years, and has worked that mine diligently, and has spent money on that mine, and has really "bent his back to it" will seldom give too little in return for investment money. It's only the newcomer, the guy that has done little work, the guy that has kept his claims for years even lying about the assessment work, or the discoverer of new claims that will try to get away with unreasonable small percentages for investment money.

I know of miners holding very good claims, but who have never done anything more than turn in assessment work papers and who have no real investment in their good claims who expect the larger mining companies to come in and invest millions, setting up a milling and mining operation. They then expect to wind up owning over 70% and to have control. If you are an experienced miner you will think to yourself, "how could anybody be so dumb". But that is not as unusual as you might think. There are a lot of newcomers around and some not so new who expect that. And there are some miners who have developed processes who expect the same thing.

So many deals are not made. And many miners walk around with 100% of nothing rather than taking 10% or 20% or 30% of a fortune. It would be wise to understand that even if one has the best mine ever known, one that will make any mining company the richest company in the world, they still will not invest millions and leave the mine owner in control with over 50% interest.

Most miners and in fact people of all businesses say, "why should money be king. Why should we give him control just because he has money?" Let me say that probably the most important point to keep in mind is because that is the fact of life. Money in the bank is a reality. Money in the ground is still a dream. Money in the bank is proof of success. An idea, even a fantastically good idea, is no proof of success. Even evidence of lots of hard work is no proof of success. The money in the bank is the proof. And obviously the guy who is successful is the guy who should have the control. Thus money is almost always the control and when it isn't there is usually a disaster.

A miner who has had a mine for years, who has worked at it and spent money on it and has earned a mining education costing between a quarter of a million and a million dollars can ask for a lot more - if he has shored the mine up and has built a mill and needs a few bucks for operating capital, and if he has definite proof of ore bodies then he can get it without a great deal of trouble. Such capital is available at reasonable prices in many cases even as a loan. If not a loan for such a person, investment capital can be found with the miner retaining 90% or 80%.

But a person who wants a mining company to take over his claims and do all the work had better start talking at 10% royalties and in many cases accept 5%. Few mining companies, especially the bigger successful mining companies, will pay over 7-1/2%. This depends on the claim and the amount that must be invested before a profit it obtained. Persons looking for investment capital to help then develop a mine should expect to retain no more than 49% and in most cases no more than 20% or 30%. Investors are not interested in the possibility of making money they want sure fire results.

I hope this will help in some way. Jim Humble

On the Lighter Side...
Gold Nuggets

by Robert L. Desmarais

In that everlasting search for gold, I have been to many gold bearing areas but none had nuggets as big as this one. There they were for the taking, stumbled upon by accident just like most gold finds were.

The area was up in Holcomb Valley in Van Dusen Canyon which is behind Big Bear Lake in Southern California. The old mine that the gold nuggets were found in was what I believe to be part of the Lucky Baldwin.

After working all week, it's quite relaxing to go prospecting on the weekends. With gold pan, pick, and shovel in the truck, an early morning start will get you there ahead of the crowds. This Holcomb Valley area is quite popular since gold has been taken from there for years. Some gold panners still remove gold from there today, which makes this area a good place for further prospecting.

Many old mines dot this area in which mining began around 1860. Earthquakes in the past have damaged some of the old mines, putting an end to some of the workings before they were really worked out.

Upon doing some searching for a likely spot to do some panning, like around certain tree roots, a large depression in the soft earth near a large pine tree revealed a deep crack going down into the earth. Who brings a flashlight when going panning? Well, now I carry one on my belt when I go prospecting. After discussing and studying the opening for awhile, a gleam of gold was thought to be seen. A light was quickly sent for and upon shining it down the hole, the light shined back with the color of gold. Further playing of the light revealed many large thumb sized gold nuggets. Some nuggets seemed to be in small bunches like grapes. They were scattered all over the floor below. It sure looked like this was going to be a large strike. Our first thought was that the recent rains had just created this opening since it hadn't been found before.

The hole looked to be about 15 feet deep, and with no rope on hand, part of the group was sent down to Big Bear City to rent a ladder and get a couple coils of rope. A couple of pails were to also be brought back to haul the nuggets up from the hole. The wait for the hardware to retrieve the gold seemed endless and the gold was being spent before it was even hauled out of the mine. The excitement was as intense as any I have known, having such riches just outside of your reach and knowing that the equipment was soon to arrive to get to the gold. Finally the equipment arrived and with a 10 foot extension ladder let down into the hole, it was an easy climb down to the gold.

The hole was quite damp and muddy inside. It could be seen that previous rainfall had washed through the opening partially filling in what seemed to be an old mine tunnel. Upon playing the light on a large nugget, the color seemed to be different than from above. The nugget now appeared to be a transparent amber. A quick grab of the nugget put an end to the problem of the different color appearance. Upon touch, the nugget was sticky like sap, and squashed with the slightest pressure.

Oh, how the laughter erupted as everyone found out that the gold was actually pine sap from the large tree above the caved in mine tunnel.

It was quite a thrill for awhile, but sometimes the color of gold can really fool you and get your hopes up for quite a big letdown. But don't let that make you quit trying. Well, let's get back to prospecting....   I

Patenting Your Mining Claim

By Mike McKinley

The patent process is the process of purchasing unpatented mining claims from the U.S. government. Patents can be issued on lode claims, placer claims, and mill sites. Once you have been issued a patent on your mining claim, you will own the land as well as the mine, and you will no longer be required to perform and file annual assessment work.

The most essential requirement for obtaining a patent is that there be within the boundaries of each of your claims that you wish to patent, a discovery of a valuable mineral deposit which can be mined and marketed at a profit. Although the mining statutes do not specifically define a valuable mineral deposit, the courts have established the "prudent man and marketability" test to determine what is a discovery of a valuable mineral deposit. Requirements of the test have been met where minerals have been found and the evidence is of such character that a person of ordinary prudence would be justified in further expenditure of his labor and means, with a reasonable prospect of success in developing a valuable mine. The evidence must also show that the minerals can be extracted and marketed at a profit.

Another essential requirement is that not less than $500 must have been expended in labor and/or improvements in the development of each claim. Or, if the application is to include several contiguous claims held in common, that an amount equal to $500 for each claim has been expended for the benefit of the entire group. Geological, geochemical, and geophysical surveys conducted to satisfy the annual assessment requirement do not apply toward the $500 requirement for patent.

The first step to obtaining a mineral patent in the case of a lode claim, a placer claim on unsurveyed land, or a placer claim which does not conform with the legal subdivisions of the federal surveys is to have your claim surveyed. Where you wish to patent a mill site in connection with your claim, the survey requirement for your claim would also apply to your mill site. Placer claims and mill sites described by legal subdivisions on surveyed land do not require this special mineral survey. In cases where the survey is required, it must be made under the authority of the chief cadastral engineer of the local BLM office.

An application form for survey may be obtained from the BLM office. At this time they will give you a list of U.S. mineral surveyors. You must choose your surveyor from this list and enter into a private agreement with them as to payment for the survey. If the original location of your claim was made by the survey of a qualified mineral surveyor, the original location survey cannot be substituted for the survey required for patent.

The job of the mineral surveyor is to perform the survey and return the preliminary plat (map or plan) and field notes to the BLM. The surveyor will also make a report of expenditures and mining improvements made on your claim, and return it to the BLM.

The copies of the plat, field notes, and certificate of expenditures you will need to post on your claim and include with your patent application will be made and supplied to you by the BLM office. You will be required at this point to make a cash deposit to the BLM to cover their cost of making the necessary copies.

Where a survey has been required, you must now post a copy of the plat (map) of the survey and a notice of intention to apply for patent, in a conspicuous place on your claim, or if you are patenting a group of contiguous claims, post on one of the group of claims. In the case of a placer claim described by legal subdivisions you need only post your notice of intention to apply for patent. This notice of intention should give the following information: 1) date of posting, 2) name of the claim owner, 3) name of the claim, 4) number of the survey (if a survey was required), 5) mining district and county, 6) names of adjoining claims and any conflicting claims as shown by the plat of the survey. As proof of posting you will need the written statements of 2 credible disinterested witnesses, stating that the plat and notice of intention to apply for patent are posted conspicuously on the claim, giving the date and the place of posting. A copy of the notice of intention to apply for patent must be filed with the county recorder's office, in the county where the claim is located.

PATENT APPLICATION

Once your claims are posted you are ready to complete and file your patent application. There is no particular form established for the patent Application. The required information should be written down in a narrative fashion. The application must be filed in duplicate, in the proper BLM office, and be accompanied by a $25 filing fee. The application and all supporting statements must be signed within the land district (usually within the state) where the claim is located. If you are absent from the district, the application may be signed by your attorney-in-fact within the land district. An attorney-in-fact must have an original or a certified copy of power of attorney. Your application should include the following information:

1) 2 copies of the plat and field notes, and the certificate of expenditures if a survey was required. When your patent is issued, one copy of the plat and field notes will be returned to you as part of your patent.

2) 2 copies of the statements of proof of posting the notice of application and survey plat on your claim. One copy of the notice of intention to apply for patent should be attached to each set of statements.

3) Evidence of title. Unless you can prove full possession of title the patent process will be suspended until full possession is proven. Giving evidence of title in the form of a certificate of title or a certified abstract of title is one of 2 ways of satisfying the evidence of title requirement. The alternative described here is much easier and less time consuming.

EVIDENCE OF OWNERSHIP

As evidence of your ownership of the claims being patented, you must furnish a certified copy of the statute of limitations applicable to mining claims in the state where the claim is located, along with a written statement giving the following information:

1) The origin and maintenance of your title. State whether you purchased, inherited or staked your claim, and that you have performed the required annual assessment work on the claim.

2) State the area of the claim or group of claims.

3) State the amount and extent of improvements made and labor performed.

4) State whether your title has been disputed in court proceedings or otherwise. If so, explain the dispute in detail,

5) State any other matter you know of which bears upon your right of possession.

You must support your statement relating to the 5 points above, with a written statement from any disinterested person of credibility who has first-hand knowledge of the facts relating to the statement.

You must also file a certificate , under seal of the court having jurisdiction, that no suit or action involving right of possession to the claim is pending. This certificate should also state that there has been no litigation in the court, affecting title to the claim, for the time fixed by the statute of limitation other than those which have been decided in your favor.

As proof of U.S. citizenship, you must include a written statement giving when and where you were born and your current address.

If your claims were located after August 1, 1946, you must state whether you have had any direct or indirect part in the development of the atomic bomb project. If so, you must describe your participation in detail.

Your application must include sufficient details for the mineral specialist to determine whether a valuable mineral deposit has been found. A field examination will eventually be conducted by a government mineral examiner to confirm the statements contained in your application.

EVIDENCE SHOWING VALUABLE MINERAL

This next section pertaining to showing evidence of a valuable mineral deposit can be broken down into 4 parts.

1) Describe the character of the mineral deposit and the natural surface features of the claim. Give the yield per cubic yard of precious metals as demonstrated by prospecting and development work. State the thickness of the gold bearing (if that's the valuable mineral) gravel, the thickness of the overburden, the distance to bedrock, and the formation and extent of the valuable mineral deposit. Also describe the natural features of the claim. Describe the streams by course and volume of water carried. Describe the kind and amount of timber and vegetation on the claim and their adaptability to mining and other uses. In the case of a placer claim, state whether or not your claim is all placer. If your placer claim also has lode formations on it, describe all of the known lodes or veins. You must give the location of each lode that you intend to include in your patent. All known lodes on placer claims must be surveyed and plotted. All other known lodes not described or surveyed are by the silence of the applicant, excluded from any claim by the applicant. Where there is no known lode on a placer claim, written statements attesting to this fact must be given by 2 disinterested witnesses.

2) Describe your workings and improvements. That is, describe in detail shafts, cuts, tunnels, discovery points, and drill holes claimed as improvements, giving their dimensions, value, course, and distance to the nearest survey corner. In the case of a placer claim conforming to legal subdivisions where no survey was required, this statement of description and value of improvements must be supported by the written and signed statements of 2 disinterested witnesses.

3) Describe the method of extraction of the raw material from the ground. Also describe the methods by which you process the raw material once removed, and the method of transportation of your material from mine to mill or processing plant, and if appropriate, to market.

4) Give an economic analysis of your mining operation. This is, the actual or estimated mining, processing and miscellaneous costs of the operation,. including reclamation, the value or price of the valuable mineral, and the estimated or actual profit from the operation.

BLM CHECKING

When received, the BLM will check your application for sufficiency and conformity with statute and regulations. The status of your claims will be investigated and the Office of the Solicitor will offer a title opinion.

PUBLISHING NOTICE

When your application is checked and found to be acceptable, the BLM officer manager will publish a notice of application for 60 days in a newspaper published nearest to the claims, and will furnish the applicant with the name of that newspaper. You must make an agreement with the publisher to pay the total cost of the publication of the notice.

If the notice is published in a daily paper, it will be put in Wednesday's paper for 9 consecutive weeks. If published in a weekly paper, it will be put in 9 consecutive issues. If published in a semi-weekly or tri-weekly paper, it will be put in the issue on the same day of each week for 9 consecutive weeks.

A copy of the first publication must be given to the BLM where it will be checked for accuracy. One copy of the notice of application must be posted on your claim, and one copy must be posted in the state BLM office for the entire 60 day period of newspaper publication.

Any adverse claims must be filed by the adverse claimant within the 60 day publication period. After the 60 day period of newspaper publication, you must obtain from the publisher a sworn statement that the notice was published for the full required period, giving the first and last date of publication. You must also write your own statement giving evidence that the plat and notice of intention to apply for patent were posted conspicuously on your claim during the 60 day period of publication, giving the date of posting.

PAYMENT OF LAND

Upon the filing of the proof of posting and proof of publication, if all the papers are in order you will then pay for the land. Payment is made to the BLM at a rate of $5.00/acre and each fractional part of an acre for a lode claim and mill site in connection with a lode claim. Or $2.50/acre and each fractional part of an acre for a placer claim and mill site in connection with a placer claim.

FINAL PAPERS

You must now file with the BLM a written statement of all charges and fees paid by you for the survey, publication, BLM deposits, and filing fees. At this time your Final Certificate of Mineral Entry is issued.

The issuance of Final Certificate does not mean that the patent process is complete. At this point you are no longer required to perform annual assessment work. Now the validity of the valuable mineral deposit must be verified by a government mineral examiner before the patent is issued.

The government mineral examiner will advise you of the date set for the property examination and invite you to accompany him. You must make the points of discovery available to the examiner. The examiner then takes samples of your material for testing. You may request that the precious metals obtained from the testing be returned to you.

When, upon testing of the samples taken, your claim is shown to have a valuable mineral deposit, you will be notified that a patent has been issued on your claim. Within a week or two you will receive your patent papers. Copies of the final papers should be filed in the County Recorder's office in the county where your patented claim is located.

EDITOR'S NOTE: The Rocky Mountain Mineral Law Foundation, Fleming Law Building B405, University of Colorado, Boulder Co. 80309 publishes the "Digest of Mining Claim Laws", 2nd edition, which covers the Federal Regulations as well as the laws for each state. The digest costs $24.00.

Simple Amalgamator

This amalgamator works very well for detecting gold in any rock that you suspect carries gold.
Have a machine shop turn out a ball race for the size bearings you have. They can be most any size. 1 used 1/2" bearings in mine.

Next weld it in the end of a piece of tubing the size you choose. It can vary.

For the spinner that drives the balls, you can take a disk and weld a stem on it and have it trued on a lathe so it will run even.

Ore should be ground to 100 mesh to start, or finer. The balls will grind it down to 800-1000 mesh. Use about 1/2 teaspoon ore. Add 3 to 4 drops mercury and then add water to cover ball bearings. Add i drop dish soap to break surface tension. Chuck spindle in drill press. Set on slow speed and let it turn for at least 1/2 hour. If there is a trace of gold in your ore it will pick it up.

Next dump everything in a gold pan, ball bearings and all material, and take a magnet and pull out the bearings. Pan out the mercury.

Put in parting and drying dish and add nitric acid, a little at a time. (NOTE: do this outside or under a hood as mercury fumes are poisonous) until all the mercury is evaporated.

Then all that is left is the gold. This is a great little machine for testing.

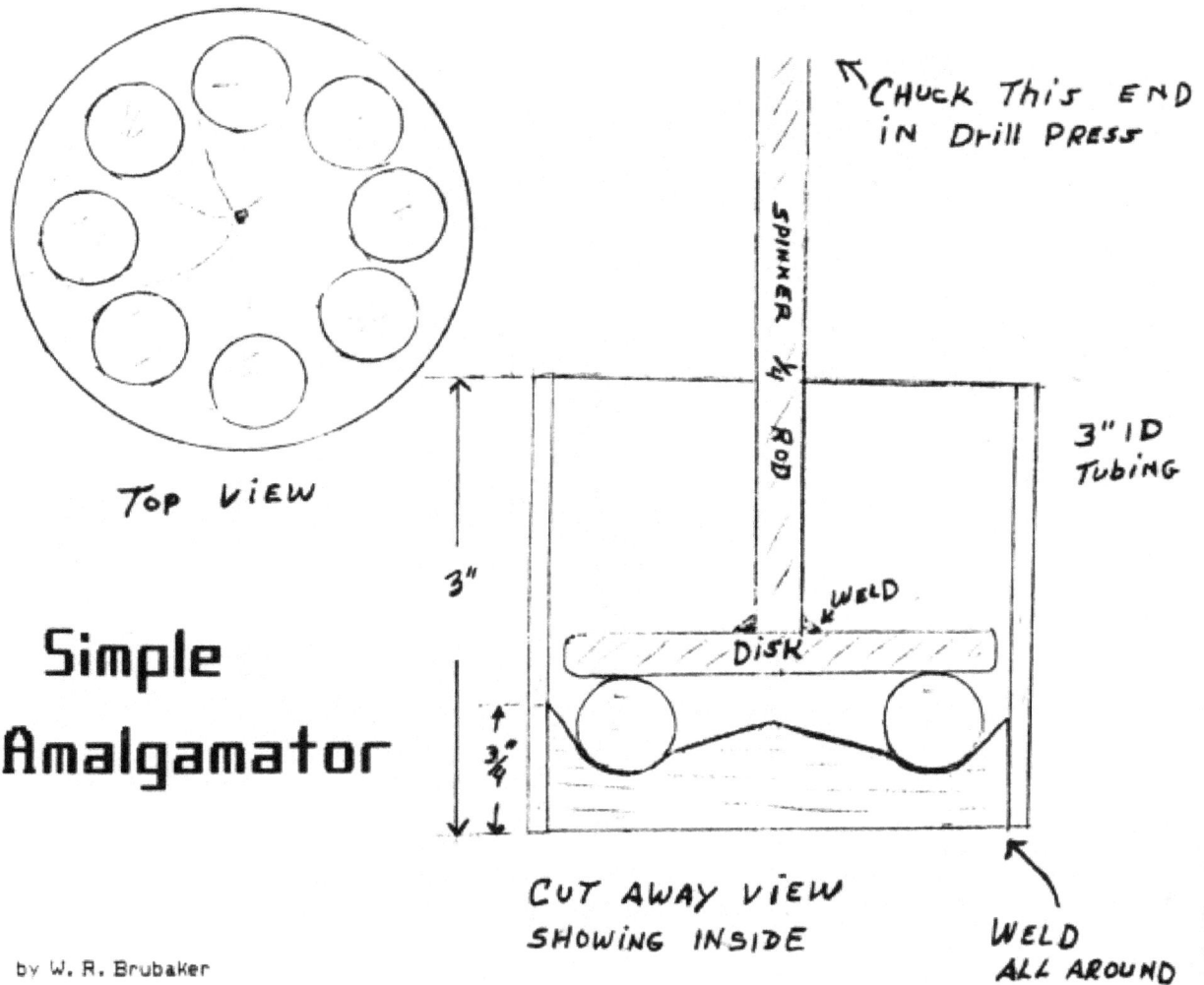

Top View

Simple
Amalgamator

by W. R. Brubaker

CHUCK This END iN Drill PRESS

SPINNER ¼ ROD

3" ID Tubing

3"

WELD
DISK

¾"

CUT AWAY VIEW SHOWING INSIDE

WELD ALL AROUND

A Simple Ball Mill

by Dick Day, Boise, Idaho

I had the pleasure of an acquaintance letting me read his copies of PM. In #3 issue, I saw your plans for building a small ball mill. I have built 3 of them which are still in use; the first about 12 years ago. I use it every summer.

I use wooden bearings for the mill to roll on and have had very little wear (if any) on the wood or the pipe from the ball mill ends.

This is a more compact way than to use wheels and tires and less expensive as I get the hardwood (locust, maple or walnut) from an old tree or someone's stove wood and cut it out completely with just a chain saw. Wipe a little grease on the pipes between washers 3 or 4 times a day and Keep it covered so dirt and rocks don't get down in there.

NOTE: this ball mill can be driven by belt or chain from a gearhead motor mounted in the lower part of the frame. I cut 10 or 12 teeth off of an old chain sprocket and welded them in a straight line around the mill for the chain to run on and turn the mill.

The mill itself is made from piper whatever size you choose. The bearings are carved out to fit the end pipes of the mill.

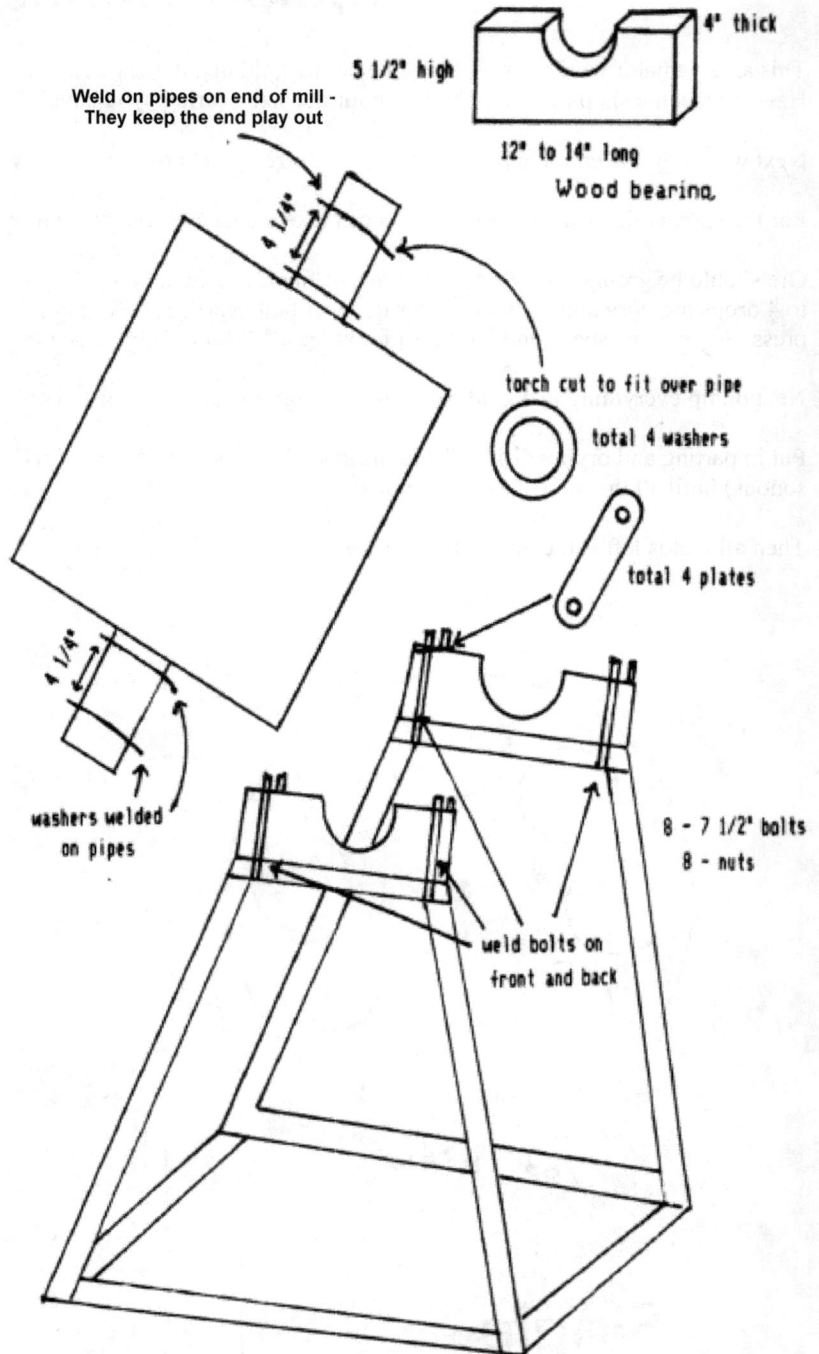

Weld on pipes on end of mill - They keep the end play out

5 1/2" high

4" thick

12" to 14" long

Wood bearing.

4 1/4"

torch cut to fit over pipe

total 4 washers

total 4 plates

4 1/4"

washers welded on pipes

8 - 7 1/2" bolts

8 - nuts

weld bolts on front and back

Ball mill frame made completely from 2" to 4" channel iron, (takes 16' to 20' for complete frame).

Use the plate to clamp the wood bearings down (see picture).

Precious Metal Extraction

By Douglas Stolk, Spectrum Laboratories, Dallas TX

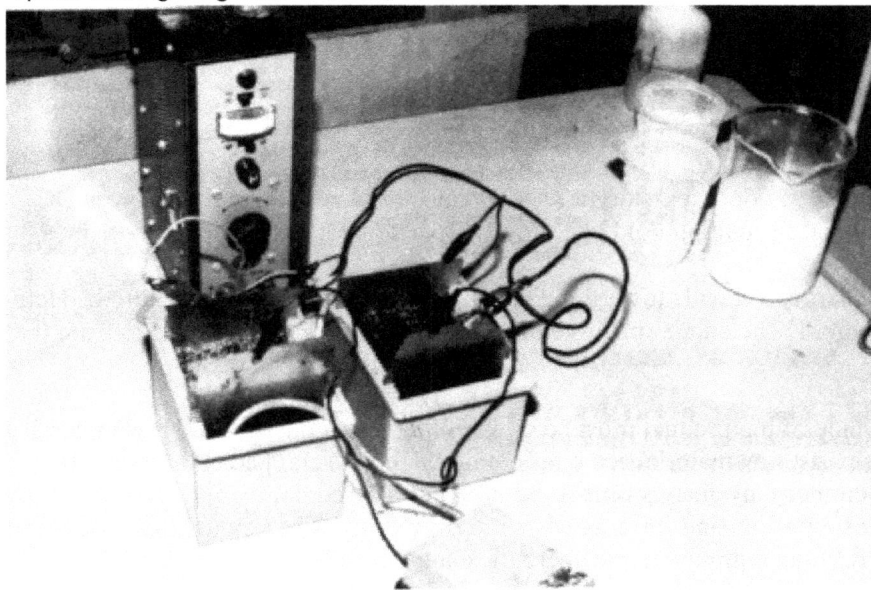

The cell & pump

INTRODUCTION

Each year many pounds of our nation's precious metal resources are lost through various uses. Some of these applications are space vehicles, dentistry and medicine, coinage, jewelry, photography, commercial buildings, and especially the electronics industry. The latter industry uses large quantities of Au, Ag, and Pt in electronic parts. Some of these are familiar to most people, including small parts for computers, video games, missiles, electronics banking, and telecommunication products. My project will outline the current state-of-the-art in precious metals recovery and add a new and novel method of gold recovery.

RESOURCE RECOVERY

One source of gold comes as a by-product of other industrial processes such as copper mining. Often the gold is contained as a fine powder or sludge from copper electroplating tanks. This gold by-product is called a "concentrate". Typically the gold content ranges from 2% to 20%. Concentrates are produced by mining, and salvage companies mostly in the western United States.

Another source of gold is the electronics industry which produces millions of scrap electronic parts every year. Some of these parts are printed circuit boards, micro chips, wire, connectors, and reject parts. Through proper chemical and metal treatment, much of this scrap will yield valuable quantities of gold, silver, and platinum.

Other sources of precious metals are photographic and x-ray film, scrap from dental and jewelry castings, gold coated window glass (like many of Dallas' new buildings), silvered mirrors, tableware, decorations, and machine parts.

All of the above sources of gold, silver, platinum are in addition to their primary source - metal mines around the world. One can see that the precious metals recovery industry is a vital part of conserving our country's limited and strategic resources.

A novel extractive metal-collecting process makes use of a combined, 2-part, leaching and electro-winning system. In the process, chemical catalysts and metal ion formers additionally enhance the extraction rate of desired metals, such as gold and silver.

When running, a minimum amount of energy is consumed, no hazardous or toxic byproducts are released to the environment, and the process produces exceptionally fast metal recovery rates by an unique arrangement of physical components, industrial procedures, and equipment.

At the heart of the gold extraction process is a chemical reaction of cyanide and gold in solution. In the presence of oxygen (which is a necessary catalyst to make the chemical reaction go), gold is dissolved into solution from scrap.....like sugar dissolves in water. Now the gold and cyanide ion (CN) are stuck together as a "complex". By chemical symbols, the reaction looks like this $4\,Au + 8\,CN + O_2 = 4(Au(CN_2)) + 4\,OH$.

This reaction must take place in an alkaline solution; that is, a detergent-like water solution. The OH ion, added by sodium hydroxide, keeps the above reaction working as long as it's not too strong or too weak.

The process of dissolving gold by chemicals is called "stripping". During the process, the solution (1% NaCN by weight to gallon of liquid) must be always moving to get good circulation of all the ingredients. This movement is done by air bubbled through a basket full of parts. This air also contains oxygen (21%) which is vital to drive the reaction forward. Once the gold is stripped and in solution as a cyanide complex, it is ready to be plated out.

Electro-plating, sometimes called "electro-winning", is done by stainless steel plates hooked up to a low-power DC power supply. Reference books say 0-6 volts is ideal for plating gold or 1 amp DC per square foot.

Since the gold complex has an overall positive charge, it travels toward the negative steel plate, called the cathode. Here the complex gains an electron and deposits pure gold. The anode (positive plate) loses an electron and gets one from the dissolving gold.

For any chemical process to be profitable, the value of the product must be greater than all production and raw material costs. In reclaiming gold from scrap electronics parts, raw materials cost little. For example, scrapped computers often sell for 2 to 3 cents per pound. Thus, a 1 ton computer may cost as little as $20.00, but may contain 3 ounces or $924.00 (at $308/oz) worth of gold. However, labor, transportation, and refining costs rapidly add up. The key to profits lies in the processing stage. Therefore, the cost of the refining equipment and energy use must be low.

In Victor, Colorado, 2 young brothers recently developed a profitable gold recovery operation. Because of proper plant design and clean feed materials, they have been processing low grade gold ore for $160 per ounce of gold. So a low cost extraction process like mine is not unusual. Compared to the above costs, the cost of chemical supplies (which consist mainly of water and plating chemicals) is very small - less than 10% of total costs.

The main drawback in running this extraction process is twofold. First, design and construction of the plant must ensure that spilled chemicals or rinse water is not allowed to escape into streams, rivers, or lakes. Such would kill fish and wildlife and even endanger man. To prevent this, a safety catch pond with neutralizing chemicals is built (uses chlorine - household chlorox; or pool chlorine in granules -sodium hypochlorite). Then if a tank breaks, the toxic chemicals are harmless. The second disposal problem consists of cyanide disposal. Also, if the cyanide extraction chemicals become acidic, poisonous and deadly hydrogen cyanogen is released. This gas has been used in gas chambers. To prevent worker injury or harm, the cyanide may be converted into harmless N2 and C02 and safely released into the environment.

Many projects have met their goals of making a process to extract precious metals at low cost. As long as gold and silver sources are available, this process can make a profit . A small industrial gold and silver production plant could be built around the process on a much larger scale. When running, a very small amount of energy is used and metal is quickly recovered by special chemicals, equipment and procedures.

EDITOR:
LEACHING SCRAP
1. Set up a tank so that a solution of water, 1% cyanide and enough sodium hydroxide to keep the pH at 10 can be bubbled through the electronic scrap.
2. Note in the picture that a fish tank air pump is used to pump bubbles up through the scrap.
3. Then pump the liquid into the second tank for electro-plating or electro-precipitation.
4. If the metal plates you will see it as gold. It can be peeled off the stainless steel plate.
5. If it does not plate it may produce a black powder or sludge that can be smelted into gold.
6. Be careful with cyanide. Don't dump it without adding chlorine to kill the cyanide. A bottle of Chlorox will do for ten gallons or less. You should check it with a cyanide kit to prove all cyanide is destroyed.

An Easy to Build Classifier

by E. T. Cole, West Covina, Calif.

Made from a Bucket to Work in a Bucket.

Most miners know that to save fine gold in any mechanical concentrating system the ore must be crushed or classified first. This holds true whether the system is panning, dry-washing, sluicing, tabling or any of the other many methods. Classifying takes time though, and this is the phase of mining that most of us cut corners on.

Here is a classifier, easy to build, easy to operate, and best of all, easy on the budget, which will do the job quickly and help to save the gold that we now lose.

To build this useful device all you need do is shorten a 5 gallon bucket to 1/2 its present depth and replace the bottom with a screen. This is a project which you can complete in 2 hours time using tools and parts that you probably already have.

I make these units for resale and therefore have designed jigs to simplify the various steps but you can do it easily yourself and save a couple dollars.

STEPS
1. Draw a line around the bucket 4" below the top, draw a second line around the bucket 4" above the bottom. Saw the bucket on these lines, making 3 pieces.
2. Turn the bottom piece upside down on a flat surface and cut an 8" hole out of the center.
3. Using sandpaper, smooth all cut surfaces.
4. From whatever size screen you want your classifier to be made of, cut a 9" disc.
5. Using an electric soldering iron with a 1/4" tip, carefully "weld" the screen right into the plastic of the inside of the bottom of the bucket. Weld the screen all around the circle so that no rough edges of the screen remain exposed.
6. To assemble the completed unit, put the center piece into the top piece and put the bottom into this unit.

Your classifier is now ready for use, either wet or dry, with another 5 gallon bucket.

I make these classifiers using 4 different size screens from fine to 1/2". They save much time in respect to the commercially available handheld classifiers, and I save much more fine gold.

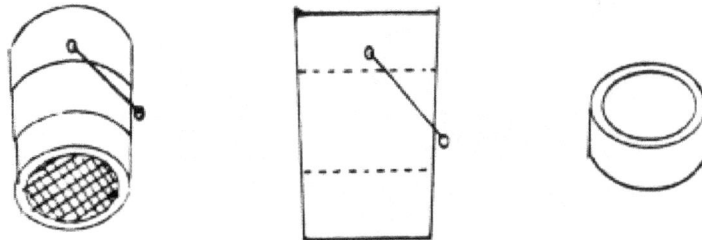

Using a Feather on your Concentrating Table

Feathers ⅜" x ⅜" wood

by D. D. Peterson, Forks, Washington

It is my hope to pass onto others some helpful hints in the use of an ore separation table. The subject and use of a table of any kind or make is a very exhaustive one which would require several thousand words and drawings to just outline the table's use. However, a few helpful hints would , we hope, be helpful!

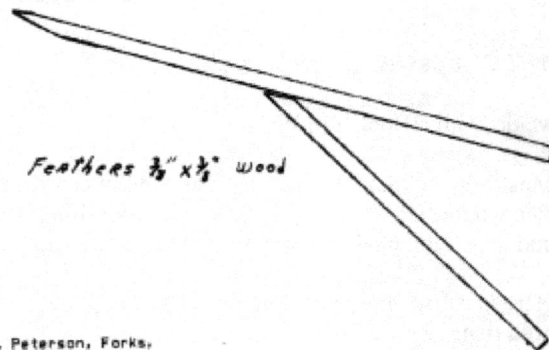

When working with gold ores, I have found it helpful to use a feather which is explained by the drawing. What is being done by the use of the feather is to pull or separate the higher specific gravity minerals such as gold, platinum, palladium, rhodium, iridium, etc. which will (or should be) the top band of the ore or pulp on the table. This is normally referred to as "pulling a high con."

To pull any concentrate on your table, the first thing to do is to tune the table to the ore being worked - using whatever action your table has. The water with whatever depressant you are using is turned on, as well as the table. A dime is thrown onto the table just below the pulp feed box and its action and movement is observed. It should move in a uniform jerking motion toward the con-end of the table. The approximate point the dime exits the table is the approximate point the feather is placed, at least to start with. I have tuned tables where when the dime was placed on the table it moved up to the head-end of the table, instead of towards the con-end of the table. Often the dime will simply not move at all. These problems can be due to improper head motion or any number of reasons. I have found the simplest way to overcome these problems, on a non-manufactured table, is to elevate the head-end of the table 1/4" at a time. This will usually cause the dime to move in the proper way.

Once this is done, the pulp or ore is fed into the pulp box and the movement of the ore is observed. The feather is so placed, usually with a screw or nail in the center to allow for adjustments, so the feather cuts the high side of the concentrate, as shown by the drawing.

The ore feed to the pulp box must be very uniform. If this is not done the con band at the top will widen and narrow to such a degree that you will either lose values in the middlelins or the concentrate will not be as clean as would be wanted.

The second feather is installed to widen the separation between the cons and the middlelins. Of course, the catch basin or trough that is used to catch the concentrates must be divided in such a way as to save the various cons, middlelins, etc.

The use of a feather is most helpful in "re-coning cons". I have found it helpful if only 1 table is available, to first concentrate your ore without the feather. This is done so as to pull a fairly wide band of ore in the concentrates. These concentrates are then re-tabled with the use of the feather. In this way I have upgraded cons from 2 oz gold per ton to over 50 oz of gold per ton, with a tail loss of .015.

Because there are so many variations in ores, and the noble minerals are combined in so many different ways it is impossible to make any hard fast rules as to their tabling. The feather can be used in many different ways to accomplish many different things.

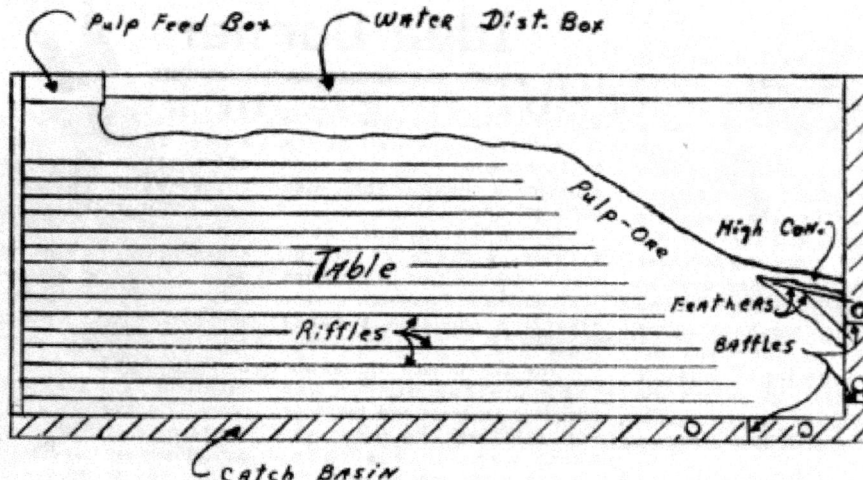

Pulp Feed Box   Water Dist. Box

Pulp-Ore

High Con.

Table

Feathers

Riffles

Baffles

Catch Basin

Bottom side of the whilpool concentrator

Whirlpool Concentrator, another version

by Robert W. Terry, Anderson, California 9600?

I think I have a much improved design of the whirlpool concentrator. Probably not better in performance to the original design, but I think it is equal. (See Issues #2 & #5 for other ideas. The original article was in Issue #1)

Basically it is much more sturdy - not as easily damaged by a little bump, no hard-to-fit reducers needed used to build up pressure in the jets....

1. Shallow radius to fit bucket - grind, file, whittle or whatever. 3/4" PVC elbows and pipe caps. Don't have to be perfect but fairly closely fitted.

2. Glue nipples then drill. It's easiest to first drill small holes then enlarge them.

3. Slip in the drilled pipe caps and nipples in the elbows but don't glue. Proceed with the rest of the setup and finish strapping and gluing.

4. Pull out pipe caps and glue. Make sure holes are opposite direction so water spins. Easiest if you fit everything before gluing

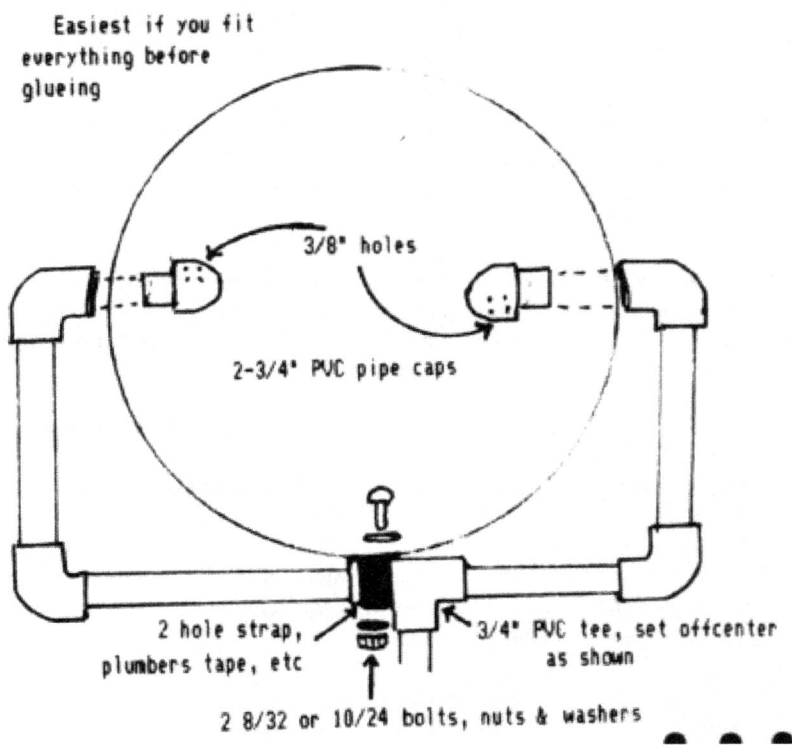

Easiest if you fit everything before glueing

3/8" holes

2-3/4" PVC pipe caps

2 hole strap, plumbers tape, etc

3/4" PVC tee, set offcenter as shown

2 8/32 or 10/24 bolts, nuts & washers

1930s Forest Service/Miners Working Together

Copy of a letter to Mr. Ronald McCormick, Supervisor Siskiyou National Forest, Grants Pass, Oregon from Mr. Ben Baker dated February, 1984

Dear Mr. McCormick
The Siskiyou Nat. Forest is a very special area for me and many others like me that have their past welfare tied up into the production of natural resources of this forest area.

During the first years of its existence, prevention of forest fires was a full time job for a few and during fire season was the job for many. Trails were built to the right places and lookout towers were built. Some elderly man and his wife would take the job as lookout and stay there all summer. Communication was by messenger. Telephone wires were eventually strung so better reporting could be accomplished.

The problems were met day to day. The 1920s passed. The lumber industry did not have a chance to expand because there were no roads. The mines were serviced mostly by pack horse and trail. The beef raisers drove their cattle to the surrounding mountains for the summer. Briggs Creek and the upper Illinois took care of hundreds of cattle. The streams running into the ocean were prime spawning beds for salmon.

Then the 1930s and depression arrived along with thousands of Dust Bowlers ready and willing to work hard to make a living. The Forest Service was confronted with a problem - to try 'to keep some semblance of a regulated and lawful society in the mountains. Back in Washington the Civilian Conservation Core was organized.

Thousands of young men were recruited and sent to the western states; several hundred to the Siskiyou Natl. Forest to build roads, bridges or anything else that would be necessary in a National Forest.
These young men never had anything going for them except that which is the most valuable - "a willingness to learn and to do the best job they possibly could".

Mr. Hershall Obie was Siskiyou Natl. Forest Supervisor and Mr. Don Cameron was the Siskiyou Natl. Forest Engineer. Mr. Cameron and Mr. Obie soon brought order out of confusion. Good roads were being built. Young men learned how to run construction machinery as well as how to maintain their equipment. Later these same men (CCC's) were accepted by the Army Engineers and Navy Sea Bees for military service as qualified workmen and given rating as such.

This task of organizing gave Mr. Cameron and Mr. Obie the experience that would be needed so much after Pearl Harbor day, Dec. 7, 1941. All at once the people in Washington found out we were almost a have-not in regards to several strategic materials. What the Siskiyou National Forest could produce was all strategic. We had to produce all of these articles as fast as possible. Chrome was Number One on the list.

We had very little development and practically no roads to the deposits in the mountains. Chrome only occurs in the peridotites along with nickel and cobalt, which are in the laterites. Chrome goes to depth. Cyclone Gap is over 400 ft. deep and the Oregon Chrome at Oak Flat is over 700 ft. deep. No one knows where more can be found.

GOV'T HELPED THE MINERS
The USGS under Dr. Francis Wells came here with a crew of geologists to help the prospectors find more chrome. The Bureau of Mines helped wherever possible. The Forest Service helped on roads and in any way possible that the regulations permitted.

There had to be authority from Washington to do a lot that was absolutely necessary. First Senator Holman of Oregon along with a few Bureaucrats called a meeting in Grants Pass. The Forest Service, if my memory is correct, chaired this meeting. It was for information as to what it would take to get as much of these resources to where they were needed. All 3 resources were discussed - chrome production, lumber production and beef cattle. Senator Guy Cordon of Oregon pushed the Congress and the different departments to make it easier to cooperate and to cut the Red Tape. Senator Clair Engle of Calif, held meetings at Yreka and took the same steps to assure that a production of as
much as possible could be had from these resources as fast as possible.

The Army and Navy had requisitioned all construction machinery so what was left was in very poor condition but we had no choice. The Siskiyou Natl. Forest had a couple of almost junkers to do the job with. There was no private equipment available. The Forest Service had to be very careful and use the tractors for road maintenance only. But somehow Mr. Cameron and Mr. Obie would go the limit to help get out that extra ton of chrome. On these occasions the miner would pay the driver and for the fuel.

The Sanger Peak road was built in the spring of 1942, by mineral access money. August 26, 1942 there were 12 trucks in line at the Cyclone Gap Chrome Mine to be loaded with chrome ore as fast as they could. The ore bin had been built previously by using pack horses to bring the poll timbers from Youngs Valley 2.5 miles away. 30 men with wheelbarrows, pick and shovel ready to load those trucks, 15000 lbs. each and running 24 hours a day making 3 trips each. We continued this pace until the snow came on the first of November and forced a shutdown for the winter.

This road was a Number 1 priority for Mr. Cameron and he and sometimes Mr. Obie would be out to the end of the road to see that nothing would stop its progress. The trucks loaded directly into a gondola car and the ore went to some steel mill in the US to be used to make material that was needed in the war effort - several thousand tons of metallurgical grade chrome. Not only did we mine the big deposit of chrome at Cyclone Gap, but had 6 pack horse strings packing chrome from surrounding deposits 5 miles away to the road to be picked up by truck and taken to Grants Pass.

The FS did not have the appropriation to do all the necessary construction. Mineral access money was asked for. The Wimer Road, the High Plateau branch and Sourdough branch was granted mineral access money. The bridge across the West Fork of the Illinois, Boulder Creek and the Cement bridge across the Illinois River is there because Mr. Cameron and Mr. Obie cooperated with all the people having an interest in developing and producing natural resources in these areas.

MULTIPLE USE CONCEPT
Mr. Cameron and Mr. Obie conducted the business of the Siskiyou Natl. Forest in a fair and impartial way. By doing so they served their country in a far greater way than any single person could say; I am just one of perhaps a hundred. I presume there are a very few living yet that could add their experiences with these men. They used the multiple-use concept years before Congress made it a law. They used it because it was fair to all the different resources and I believe they thought that it also was most fair to all the people of our nation. How much more products our nation had makes very little difference because it is the idea. The very act for them to do their best so that our country could survive, to me makes them heroes. I believe there is a quote that says, "to do the greatest good for the greatest number of people."

The effort to produce from the Siskiyou National Forest lands in 1942 was a success because the miners, the loggers and lumber operations, and the beef cattle raisers had the full support of Mr. Cameron and Mr. Obie* who won "all the people of our nation".

SUPPORT & COOP OF THE F.S.
But had we not had the support and cooperation of the Forest Service, we could not have produced a fraction of what we did. What has happened can happen again.

It is criminal legislation that places public lands in a position that the resources cannot be developed in case of need. It takes lots of time and lots of money to develop any natural resource. Example, nickel and cobalt from the laterites. Over 60 years of research and still answers to be solved...a better way to recover chrome, to make pure chromium.

We here in America depend upon our own initiative to gain a living. There has to be retained that opportunity for each of us to get a better life for hard work and money expended.

MINING & RECREATION
I want to say something about recreation. How thousands are enjoying a day or a few days in the mountains at a lake or meadow because there are circumstances like the Sanger Peak Road and Youngs Valley. Besides being a road to haul out a natural resource product, it also has been possible since the spring of 1943 for anyone with a car to go to a beautiful lake and meadow. With the facilities the Forest Service has installed, thousands of people now enjoy this spot. The CCC built the Molan Lake Road; through the years thousands of people of all ages, whether crippled or not can enjoy that beautiful lake.

I have been in these mountains all my life. I have spent a lot of time in the Kalmoapsie Wilderness and Trinity Alps wilderness. The only people that go there are the wealthy that can hire pack
horses and guides or the young and strong that can carry a load on their backs all day. Recreation is for all the people to have an equal chance. As a resource* recreation can be served better by Multiple-use. As for timber* minerals and fisheries, it has to be Multiple-use.

In closing I want to say "thank you Don and Hersh for all your understanding and help that made it possible for many like me to produce more for our country in a time of need".

Very sincerely yours, Ben Baker Ft. Jones Calif.

Using Stock 2-Wheel-Drive
Vehicles in
Off Road Situations

by M .D. Isely , Trona, California

The Four Wheel Drive Automobile Company of Clintonville Wisconsin built its first FWD in 1910. It was a touring car which also had 4 wheel brakes. Yet, during the era of extensive exploration by automobile which followed, I never heard of one being used. The Citreen half track and the stock Dodge were the standard explorers' cars. FWD and Jeffries/Nash-Quad trucks were built in some quantity and FWD is still in business but, long ago they turned exclusively to large, commercial vehicles. Their automobiles were heavy and flotation tires were a development of the 1930s. Their trucks were unexcelled in plowing snow from a frozen road, but in an era of unpaved, poorly drained rural roads, when the road thawed they could get stuck in places where a farmer could skim over in his Model T.

I once owned a 1920 Dodge touring car. It was stock, just like the ones which explored Death Valley and the Gobi Desert. It seated 5 and 8 more could ride on the running boards and front fenders. It weighed less than 1500 lbs. and had a long stroke engine which developed high torque at low speeds. It was geared low and had torque tube drive. There was 1 universal joint behind the transmission and the drive shaft ran through the tube which was riveted to the rear axle housing. This stabilized the rear axle much as radius rods do and made it a good car for dunes or the beach. The torque tube kept the drive shaft from picking up chicken wire or barbed wire (in seconds creating a mess which might take an hour to cut off.) The wheels projected beyond the ends of the car giving approximately 90 degree approach and departure" angles. It stood so high off the road that one could work on the underside of it without jacking it up. The weight was concentrated over the front axle so if the wheels began to slip, I would ask some of those riding outside to stand on the plate above the gas tank which was behind the rear axle. The car itself was so light that it didn't need much shift in the passenger weight to get most of the weight over the driving wheels. It could climb grades it could not safely descend going forward because it had only rear wheel brakes. With the top down it had excellent visibility so I backed down steep grades to keep the weight on wheels with brakes.

My 1920 Dodge is long gone and there is no comparable car on the market today. Since then I have owned 3 FWDs and a 2 wheel drive trail bike but my present boondocker is a 1954 Chevrolet pickup. Like the Dodge it is simple, durable and is easily maintained. Like the Dodge it has torque tube drive and a long stroke engine and it has a 4 speed transmission. I have removed the rear bumper so that I can go through gullies.

IN SAND
Recently I stopped on some soft sand and when I tried to go again the wheels started to spin. I couldn't back up because of the grade so I lowered the tailgate and shoveled about 500 lbs. of sand onto the tail gate and into the bed behind the rear axle. And went on my way. I was alone at the time. If I had had passengers they could have climbed into the back of the bed with the same effect as the sand. Weight is weight, whether it's your girl friend or a bale of hay. Your car doesn't know the difference! Speaking of hay...in areas where there is much ice and snow the farmers often fill the bed of a pickup with baled hay for rear wheel ballast. The family car stays in the barn until the bad weather is over.

ADHESION & FLOTATION
The reason for loading the drive axle is to improve <u>adhesion</u>. The added weight causes the wheels to adhere to the slick road. Another way to improve adhesion is to install chains. Of course, if the mud is so soft that the wheel sink in, you have another problem. You have inadequate **flotation**. When I stopped my pickup on the sand I had both problems but the main one was adhesion. If, after loading the sand the wheels had slipped again I would have improved flotation by letting some air out of the tires. I carry 32 lbs. psi and if I had reduced that to 16 psi I would have roughly doubled my flotation. In such a situation don't spin your wheels if the car won't move. They will just dig holes from which the car cannot climb. To improve flotation I use the largest tires that will fit under my fenders. This reduces the turning radius and fuel mileage a little and the tires won't fit the tire carrier but these are necessary tradeoffs to get where I need to go.

The one advantage of 4 wheel drive is that all the weight is on driving wheels. However the extra driving axle increases the total weight so if adhesion is adequate and flotation all important an ultra-light 2 wheeler such as a dune buggy maybe most appropriate. I have a miner friend who has both a converted VW and a Bronco and he uses them about equally. The Bronco for rocky trails and the VW for sand. The body of a dune buggy is removed and very large, super flotation tires are installed on the rear wheels. The resulting car is so light that the increase in gear ratio caused by the tires can be handled by the engine but this is a short stroke, high speed engine. It will lug out in low unless speed can be maintained. Also there is little weight on the front axle and especially in climbing steep grades, front axle adhesion is poor and steering not as precise as with a four wheeler. You need to practice with one of these things for a while to get the hang of it.

DON'T PANIC

So in spite of knowing all this you got stuck! Don't be embarrassed. It happens to just about everyone sometime. This is the reason for taking at least 2 vehicles on an excursion and if one has a winch it is most helpful. But if you are alone and have a long way to go for help don't panic. Pretend you are advising another motorist, who has done what you just did, over the CB radio. As a former tow truck driver, I have been impressed by the fact that vehicles I recovered from the wilderness could have, in most cases, been driven away if the driver had known what to do and had carried the right equipment. In one case I was able to back the abandoned car to my tow truck without doing anything except to straighten the front wheels which were turned at a sharp angle and digging into the sand. The driver didn't know that he couldn't make a sharp turn on sand. He wasn't stupid, in fact he was an M.D. He was just ignorant of off-road driving. He did know enough not to spin his wheels and get really stuck.

Look your situation over carefully. If you have 4wd make sure that both front hubs are engaged. You may be able to move a rock with your pry bar or chop out part of a stump with your hatchet. If the wheels are sunk much below the surface of the ground you will probably not be able to go again by removing soil in front of or behind your wheels. You will probably need to jack the car up and build a roadway under it. Use your bumper jack to raise the body of the car enough so that you can get your low starting point jack under the axle. Bach jack must have a rigid flotation plate, of course, or they will sink just like your car did. Now build a substantial roadway from whatever material is available to get back to solid ground. I carry old belting for this purpose. I use standard railroad spikes (these have rounded heads) to hold it down so that it won't slip on a low adhesion surface. You can use rods with the ends bent over to avoid puncturing a tire, too.

AIR JACKS

Air jacks, which have been recently introduced from Europe are a good idea. These are bags, like dunnage bags, which can be put into a limited space and inflated by the exhaust of your automobile. They are bulky but light and provide their own flotation. I say they are a good idea but the only ones I have seen are of very imperfect design, hard to use and unreliable. If you get air jacks, test them out before you depend on them and still carry another jack.

I suggest that you test all of your equipment before you start. Manufacturers practice quality control. This means that they test one item every so often and so defective merchandise shows up on dealers shelves. The dealer will gladly replace it of course, but if he is 75 miles away, well that's a long walk.... I have bought a jack and a tire pump both of which proved defective when I bought them. Don't be like the aviator who never had his parachute inspected. He said, "Why should I? The manufacturer will give me a new one if it ever fails to open."

• • •

Letters to the Editor

Using Quicksilver

Dear Sir, I have some fine ledge gold here that has a lot of fine lead and iron and other metal mixed with it and I'm not sure just how to clean it up. I believe that quicksilver would pick up the lead, but I'm not so sure about using nitric acid and water either. Could you tell me how this could be cleaned? Thanks very much, W. Clyburn, Yreka,Ca.

ANSWER:

Dear Woodrow, Nitric acid and water would indeed be a bad way to go. See the story of amalgamation. The best way to clean gold at a reasonable price would be to grind the ore. Grinding cleans the gold. Some of the other metals may go into the quicksilver but this is not a problem as they separate out easily in the refining stage. This would also be true of the lead. Quicksilver is the cheapest, fastest way to go. Sincerely, Jim.

Use of Magnets

Dear Sir. I read in your Idea Corner about using magnets to create a magnetic field on bottom of a sluice box. How would this apply to your whirlpool concentrator? Also to different types of sluice material, steel, plastic, aluminum? Would it be (magnetic) across entire width? Fastened directly to the material? Or would you space them away a short distance? Would these magnets create a loading problem, in riffles, if you have a lot of magnetic black sand along with the gold and silver and other black nonmagnetic sand? I believe my magnetic sand probably contains some precious metals if it can be leached or fired with flux.

Do you have ideas on this? Would appreciate any information you could supply, as many of us small miners probably would like to know.

ANSWER:

Dear Miner, I would say that using magnets as suggested would be best when there is no magnetic sands in the ore. That would mean removing the magnetic sands first. Magnets should be easily removable and fixed so that they do not contact the ore. The advantage of a magnetic field is that any metal, including gold, that passes through that magnetic field in a conductive liquid will be slowed down. The water must be conductive, so generally one would have to add sodium hydroxide or some electrolyte to make the water conductive. Some magnetic black sands contain gold and others do not. I would suggest roasting, grinding very fine, and amalgamating rather than leaching. Jim.

Coat your retort cap with chalk

Editor, I was reading in Issue #7 about the mercury retort written by DG Piper. In this article the 2" cap which holds the amalgam and takes the heat is fine, but I would like to add a suggestion to this.

Before putting the amalgam into the cap, give it a good coating of chalk or possibly bone ash and this will stop the gold from adhering to the cap. Maybe not stop the problem entirely but help. I have also read that if you put the amalgam on paper this will work. I have never tried this and do not know what kind of paper.   D. Doehring, Mill Sup., Golden Arrow Mill, Groveland, Ca.

Furnace torch

On the torch for your furnace, which I am going to build soon, would it be possible to use an ordinary propane torch; such as we use for burning weeds, thawing pipes and so forth. That way it already has hose, needle valve, regulator and air chamber. It wouldn't need auxiliary air supply; so could be used out in the field where electricity is non-existent. I realize the pipe in furnace would have to be bigger to accept the larger torch. Would be a good way to go (if it would work) as I use a torch quite often thawing pipes, ice and so forth. The possibility would be not enough force maybe to heat a barrel furnace to desire temp. Also would be more expensive than yours, but has numerous other uses as well. What do you think? Don Feck, St. Anthony, Id.

ANSWER: Should work good. Jim

Lighting my pot furnace

My brother and I have completed the 5 gallon pot furnace. It sure does the job. I happened upon a handy way to light the furnace. I had some old half-burned charcoal brickettes so I doused one with starter fluid and put it in the furnace, lit it, turned on the blower, then the propane. Ideal!

 L.M, Wiley, Orem, UT

Proper combustion in the pot furnace

I just read the "Letter to the Editor" in the July/August 1985 issue. I would like to add a comment in answer to Don Peck, St. Anthony, Id. in regard to using a weed burner for a furnace torch. Without an auxiliary air supply I would question complete enough combustion to raise the, temperature high enough to do any melting except for lead or zinc. These are designed to work in an unenclosed area.

I am using an up-scaled model of a propane blow torch head and get good results without an air blower. The shape and size of the furnace opening is critical for good combustion. I also use a flue pipe to help create the necessary draft. My furnace temperature is well over 2000 F.

Sincerely, Joe Reschke, Montrose, Co.

Separating sand from mercury

I've gotten almost all of your magazines, "they're great". But I need a little help. When you grind your black sand gold and put mercury to recover it, I was wondering if there is any other way to get the sand away from the mercury without taking any mercury with it. I don't have a mercury screw. Could you siphon it off with a small clear hose? And there are 6 mines I found mostly in quartz. I know they only high-graded out here but the sulfides are ungodly abundant. What should I do? I need a little advice and maybe a partner with equipment (small scale). I want to keep a low profile. Drew Galiano

ANSWER:

One way to get the mercury wax from your black sands it to pan it. Another way is to dump the mercury from one bucket to another back and forth each time leaving a bit of black sand behind. In that case you have to pan the smaller particles of mercury out, because the smaller beads of mercury usually have more gold in the body of mercury. You will find several sluices that will recover mercury from your black sands in earlier articles of Popular Mining.. Jim

What's micron gold?

Thoroughly enjoy the mag. This letter might skip hither and yon. Some things I think you should cover for us'ns greenhorns here in the east and maybe those out yonder (west).

1) Examples of micron sizes. They don't mean much if you never used them.

2) When does gold quit becoming grains or flakes and becomes flour gold.

3) What does platinum look like in concentrates? Can you tell by appearance or must an assay be run. According to the GPAA mag, Coker Creek Tenn. streams have quite a bit of it. Thanks again,   Stan Brookshier, Sr., TN

ANSWER:

Dear Stan, Everyone has their own idea about what size micron gold really is. One micron is one ten thousandths of an inch. That is much smaller than anyone would ever see with his eyes. In fact, it takes the very best of microscopes or an electron microscope to see a particle that size. But what most miners call micron gold is any gold smaller than about 200 mesh (that's .005 inch or about 75 microns). That is also about the size of particles in talcum powder for babies. 625 mesh is 20 microns and 1250 mesh is 10 microns. Gold becomes flour gold at about 200 mesh. (See above)

Platinum sometimes looks very dull like lead and sometimes is very bright like tin when in sands. A particle of platinum will not shatter when pressed on with a hard object like a knife. If the particle is lead or silver it will dissolve in hot nitric acid, but platinum will not. Jim

NOTE ON WILLAMITE SCREEN

In reference to the article "How to Check for Mercury in your Ore", the mercury detector screens that fold up use powdered willamite coated on the screen. You can find out more about them from Ultra-violet Products, Inc. 5100 Walnut Grove Ave, San Gabriel, CA 91778 (818) 285-3123 or any shop that carries ultra-violet detecting products.

Water Distribution Box on a Table

by D. D. Peterson, Forks Washington

In my last article I spoke of the use of a feather on an ore table to pull a high grade concentrate. In this article I would like to pass on some other valuable tips in the use of a table.

The water distribution box and one's ability to absolutely control the flow of water over the length of the entire table is a necessity.

All parts of an ore table must work in unison as with any other piece of equipment. The action, the feed, the counter-flow of water, the kind and quantities of depressants used must all work together to pull any kind of a decent concentrate.

Of course, the ideal situation is to purchase one of the better brands of manufactured tables, with the engineering done for you by the manufacturer of the table* using your ore. This is however, quite expensive and not always needed until a mine has sufficient amounts and grades of ore to warrant such expenses.

A SIMPLE WATER DISTRIBUTION BOX

On most manufactured tables some form of diamond-shaped gates are used to control the water flow on the table. I have drawn a diagram of a simple water distribution box that works quite well, and is simple to build. The box itself can be made from a metal or plastic rain gutter. Spaced holes are drilled at intervals to allow the water to feed at a given rate. Wooden pegs are made and when too much water is being fed over the table in any given area, one or more pegs are inserted in the holes drilled. In this way the water feed can be controlled. If you use too much water, values are washed into the middlelins or tails. If not enough water is used, one does not pull a good concentrate.

RIFFLES

Riffles on a roughing table are used primarily to pull a coarse concentrate on a volume basis. On a finishing table, if you art working with the noble metals and they are in a fairly uniform mesh or size, very small amounts of these metals should go over the second or third riffle. I, and others I have worked with over the years, actually prefer to use no riffles on a clean up table.

The concentrate is pulled as one would in a gold pan by the action of the table and a control of the water feed to both the pulp feed box and over the balance of the table. As has been stated previously there are so many variables in ore sizes, shapes, and combinations of ores that there is no way to make any hard fast rules. Generally speaking, you should have a fairly good separation of the heavy metals you wish to save in the first 2 or 3 feet of your table. (I am speaking about a clean up table.)

The balance of the table is used to do the final lighter separations as is needed. You must of course, have a balance of water used in the pulp feed box and the distribution box. On a production basis, a good table man is adjusting these ratios quite often.

Valuable metals are seldom uniform in size or quantity in a given lead. There will be hot spots and lean areas. These changes in the ore must be constantly watched for. When they occur, they must be adjusted for if they are not to be lost.

FINE GOLD

In many gold ores there is very fine gold. Often this is so fine that it can not be seen with eye alone. The use of a strong white light held at an angle to one's view of the ore will often show you this fine gold. You will see it not be actually seeing the particles of gold, as they can be too small for that. What you can set often is the light as it reflects on the gold as it moves and turns. There is an ore of iridium that is very white and when light is directed on it at an angle, there will be flashes of light like hundreds or even thousands of exceedingly small mirrors. When you see something like that that rides high on your table, or hangs back in your gold pan, it is advisable to check for the platinum group.

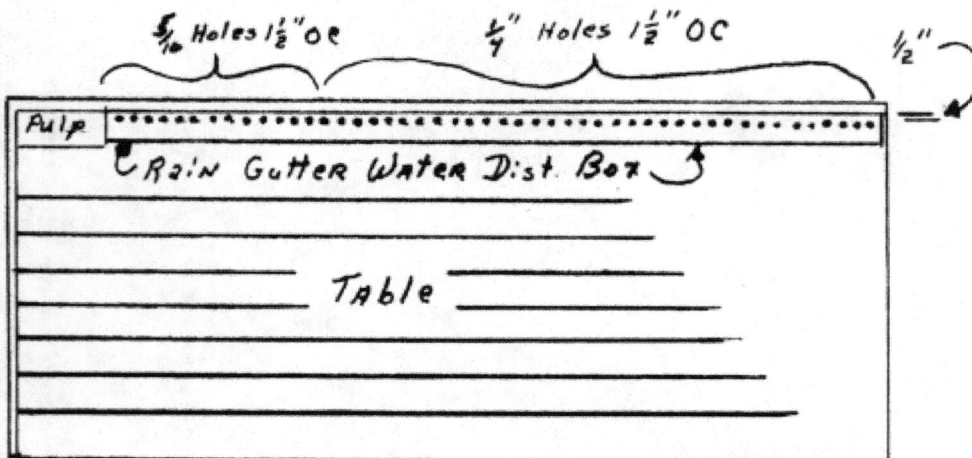

The hole sizes and spacing can vary as per one's ore and table

A Drive for Trommels or Ball Mills

by Ted Neeley Eagle Idaho

Please find enclosed a method to Make a more positive drive on trommels and small ball mills.
I got to see your PM #5 and in that was a story of a man who had built a homemade trommel and he drove it with a belt. That belt drive never works out very well. When you load the trommel with gravel and the belt gets wet and slips, a man can learn to swear pretty quick.

My Method works much better and the drive never slips and is much cheaper to install.

This sprocket had 13 - 4 tooth sections. By using only 12 makes it much simpler to space around a much bigger diameter. The speed can be as much as 125 RPM without any problem.

This drum could be a small ball mill or a trommel screen. This uses #60 roller chain.

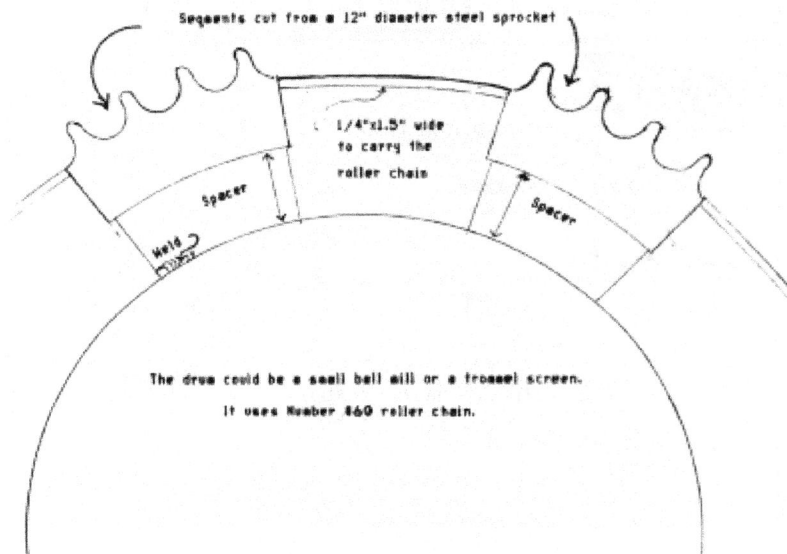

Segments cut from a 12" diameter steel sprocket

1/4"x1.5" wide to carry the roller chain

Spacer

Weld

Spacer

The drum could be a small ball mill or a trommel screen.

It uses Number #60 roller chain.

Poor Man's Hammer Mill

by F. Fletcher, Gadsden

Hammer mill made out of old lawn mower

Funnel added to top as ore inlet. Rubber tubing is outlet for ground ore into bucket

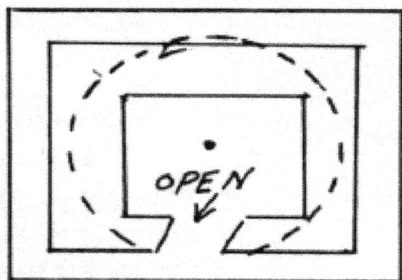

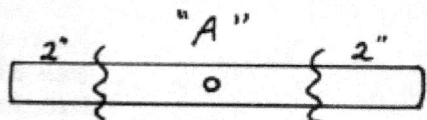

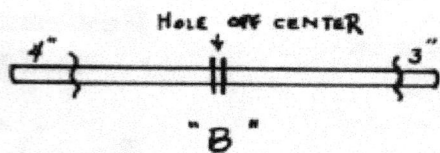

I call this the Poor Man's Hammer Hill and is rather simple. Source - a used or flea market lawn mower, preferably 31/2 HP. Remove wheels, place on a piece of sheet metal. Place sheet metal on at least 1/2" plywood. Mark the outside of the mower.

Place 4 rough 1"x 6" planks. Now mark the inside floor of the mill for the circle of the blade.

Cut the radius of the blade with 1/2" clearance between blade and wood. Leave an opening at the discharge port on the lawn mower.

Cut 2" off Blade 1 at each end as shown in Diagram A. Cut approximately 7" off of Blade 2. Drill the hole in the exact center of Blade 1. On Blade 2, drill a hole between the 3" and 4" mark so the heavy end can •wing out or loose.

Now secure the blades to the shaft. Use a high tension shoulder bolt to hold your hammers in place. Set the motor on the base and drill 4 holes through the mower carriage and into the base. Secure it down with silicon caulking under the carriage to prevent leaks.

Now cut a hole in the top of the carriage and secure an old outdoor light fixture or a funnel-shaped speaker horn. Flare out the bottom of the funnel and bolt it to the base. Secure it to the old handle knobs of the mower.

Put 4 metal screws in the exhaust port on the side of the mill. Leave 2 heads sticking up. Hang an old cutoff innertube on the ends of the screw heads. The other end of the tube goes into a 5 gallon bucket.

Never try to push the ore through while running the engine if the funnel stops up. Or use a small piece of wood. Turn it off and clean out.

Stand clear of the top of the funnel and use a shovel to load the mill.

You will be surprised at how fine it will pulverize and how much dry material it will work up. It is much slower on wet material and will sometimes gum up. If this happens stop the engine. Slip off the tube and clean it out, making sure your hammer blades can rotate and swing free.

It actually does a better job than we thought it would just to be made out of junk and very economical. Just be careful. I will not be responsible for your mistakes.

Some of the nation's first gold miners were located just south of here and there is still some gold found in the streams. Our state doesn't have an assay office and we have it difficult finding one economically. We are so far away from you that we are the last to know anything. Ha! Sincerely, Frank

One point I missed - you should not use a cast aluminum frame. They are too easy to burst.

OPEN

"A"
2"   2"
o

HOLE OFF CENTER
4"   3"
"B"

A Simple Scraper

by Adriane Riehm and Audrey Bradbury, Pahrump, NV

Here are photos of things we thought may be useful to the readers of your magazine. Photo #1 is a SLIP SCRAPER we built. This is used for moving material, etc. hooked up to a 4 wheel drive. We've found it very useful in mining, where the cost of heavy equipment is too much. It scrapes the ground and the teeth help loosen it up and dig in. The ore is thrown into the bucket as you drag it.

Home Built Furnace Torch

Photo #2 is a home built furnace and torch (see earlier articles). Using a valve from a gas stove, we've found it works great. Also on the end of the pipe we brazed in a ring which helps to "roll" the gas and air. This helps to mix them better. We used a regulator which produces enough heat to smelt our lead/silver ore, shown in the photo. We also used a blow drier for the air in this furnace.

We enjoy your magazine very much and also think there are so many useful ideas in them

A Simple Amalgamator

by Steve Barnett San Carlos CA

As a recreational weekend miner my mining and prospecting is strictly a hobby, but some problems I run into seem impossible, yet in reality the problems are not impossible. One problem I had was separating gold from the black sand concentrates acquired after a weekend of dredging. After trying several methods suggested by friends, or read about in books and magazines, I came up with an idea for a simple amalgamation system. I built a simple amalgamator for about $6.00 worth of parts and $4.00 for welding.

Following is a list of parts and where I bought them:
2 pieces of cold rolled steel 6" x 7" x 1/4"
1 piece of steel angle 1/8"x 1 1/2" x 1 1/2"
x 11 3/4". This length was governed by
the typewriter rollers length.
2 used typewriter rollers
1 barbecue rotisserie motor

The steel I purchased at a local junk yard, the rollers at a typewriter repair shop, and barbecue motor at the Goodwill Store. The welding I had to shop around for price. I tried muffler shops, lawn mower repair shops and finally got the welding done at a trailer hitch installation shop.

The 2 pieces of 6" x 7" x 1/4" I clamped together and drilled 2 holes. Then I used a hacksaw to carefully saw the slots. Then by use of the rotisserie motor I measured and marked the position of the 2 mounting screws for the motor.

With a file, I filed the one roller end square to fit the square drive of the motor. Care must be taken at this time as the rollers are hard steel. A post grinder tool would have been really handy at this tine.

Then I checked my mounting position marks and found the 2 holes slightly off, made a small correction and then drilled the 2 holes.

Now I carried the 3 pieces of steel to the welder, and when I got back home, I used a little paint, a little time for assembly, and I was ready to try my amalgamator.

A clean mayo jar (1 qt), some black sands, mercury, water, and a few drops of dish soap, and it was working. About 2 hours later I removed the black sands from the quart jar to a gold pan and separated the mercury from the black sands. The mercury was not mirror shiny. It was speckled which showed me it was carrying GOLD.

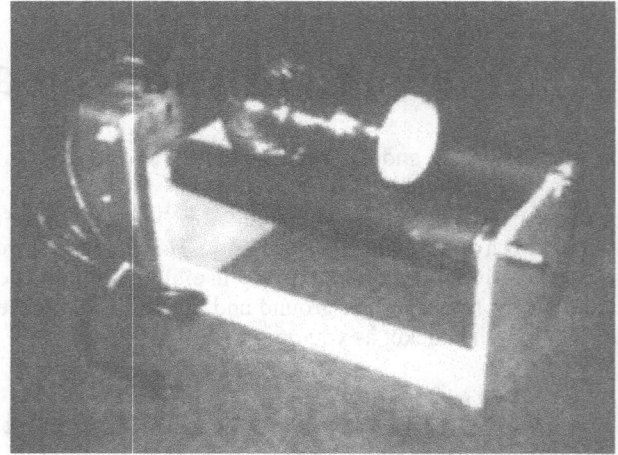

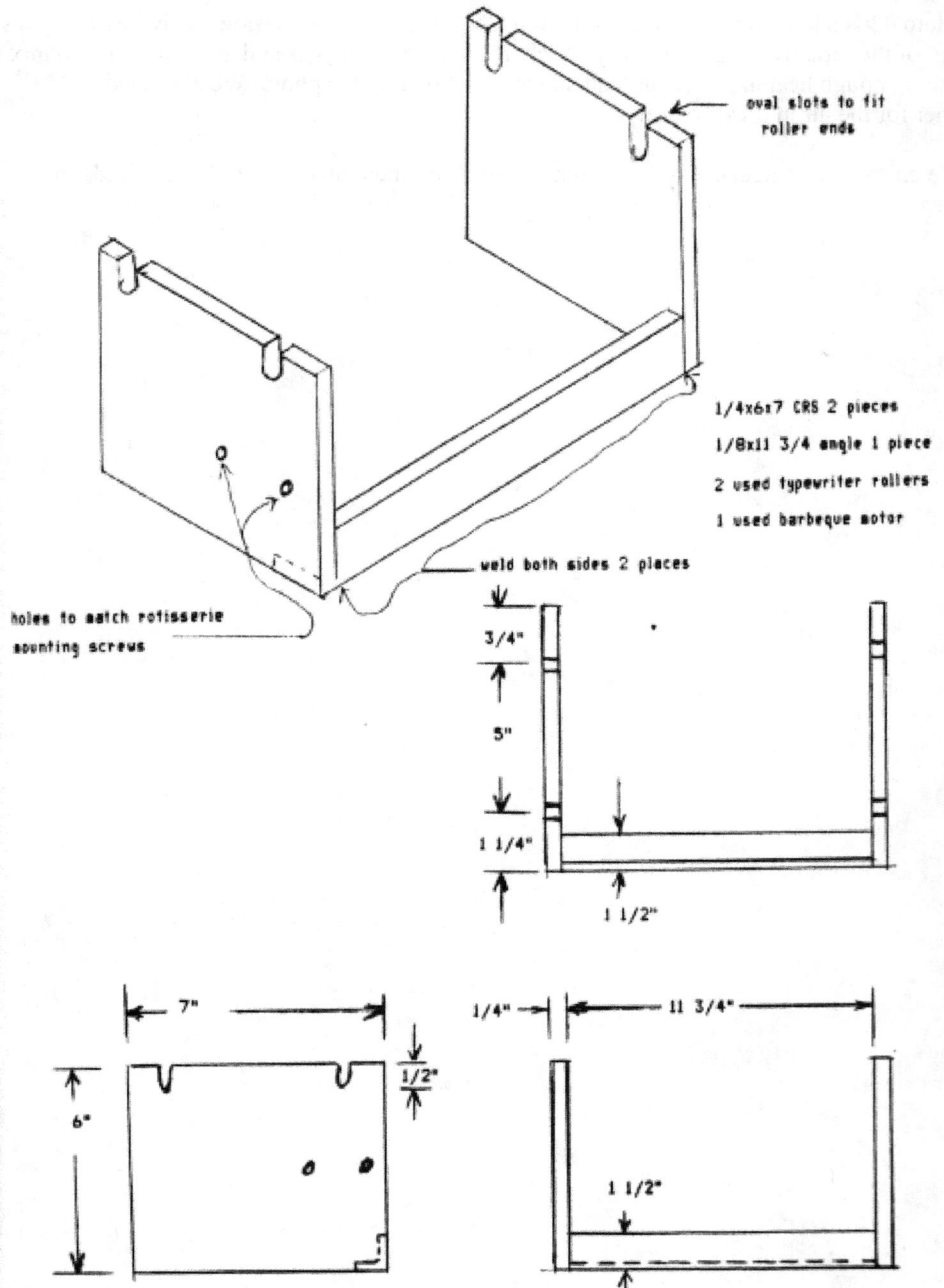

oval slots to fit
roller ends

1/4x6x7 CRS 2 pieces
1/8x11 3/4 angle 1 piece
2 used typewriter rollers
1 used barbeque motor

weld both sides 2 places

holes to match rotisserie
mounting screws

3/4"

5"

1 1/4"

1 1/2"

7"

6"

1/2"

1/4"

11 3/4"

1 1/2"

TO RETRIEVE THE MERCURY

To retrieve the amalgam of gold and mercury, I use a small wash tub and garden hose. With the garden hose I high pressure the water into the mayo jar to wash out the black sands and leave the mercury in the jar. Experiment with low water pressure and slowly increase the water flow to get the sand out and the mercury to stay inside. (Always work over the wash tub to save the sands). Then I wash the sand 2 or 3 times again to be sure I have not washed the mercury out into the tub. Save the black sands as it can be crushed later for even finer gold that is trapped in the sands.

When working with gold, black sands, and mercury always work over a dish pan, gold pan, or wash tub for safety. If you make a mistake the container you are working over will save a weekend's labor and also not contaminate the work area with beads of mercury. Remember, mercury will evaporate at room temperature (it's the fumes that are so dangerous) so always keep the mercury in water, work in or over water.

Since I have made this amalgamator, I have added more slots for the free roller, so as to be able to use smaller and larger jars. 1 have used a 1 gallon mayo jar.

I live in an apartment so space is limited. If you have a garage or work shop you are in a far better position to build the amalgamator. Good luck.

P.S. In the June '84 issue of GPAA a picture of this amalgamator was in the Miners Mail column. The Editor mistakenly credited a Mr. Don C. of Reno with the project, yet it is my picture and writing they printed.

Conversion Charts

Conversion of Troy Weights into Metric Weights

TROY	METRIC
1 grain	= 64.8 milligrams = 0.0648 grams
1 pennyweight = 24 grains	= 1555.2 milligrams = 1.5552 grams
1 ounce = 20 pennyweights	= 31.1042 grams
1 pound = 12 ounces	= 373.248 grams

Example: To convert grams into troy ounces, divide the gram figure by 31.1042. If you have 572 grams, this equals $572 \div 31.1042 = 18.3898$ troy ounces.

Common Gram Weights Converted to Troy Ounces

5 grams =	0.1607 ounce
10 grams =	0.3215 ounce
20 grams =	0.6430 ounce
50 grams =	1.6075 ounces
100 grams =	3.2150 ounces
1,000 grams = (1 kilogram)	32.150 ounces

Conversion of Millimeters to Inches

One inch equals 25.4 millimeters, or equivalently, 1 millimeter equals 0.03937 inches (which makes 39.37 inches = 1 meter). To convert millimeters to inches, multiply by 0.03937.

Average Weight of Various Materials Blasted					
		SOLID		BROKEN	
MATERIAL	Specific Gravity	Pounds per Cu. Ft.	Cu. Ft. per Ton	Tons per Cu. Yd.	Pounds per Cu. Ft.	Cu. Ft. per Ton
Basalt	2.8-3.0	190	10.5	2.57	125	16.0
Coal-Anthracite	1.3-1.8	100	20.0	1.35	65	30.8
Coal-Bituminous	1.2-1.5	85	23.6	1.15	55	36.4
Diabase	2.6-3.0	175	11.4	2.36	115	17.4
Diorite	2.8-3.0	185	10.8	2.50	120	16.7
Dolomite	2.8-2.9	180	11.2	2.43	115	17.4
Gneiss	2.6-2.9	180	11.2	2.43	115	17.4
Granite	2.6-2.9	170	11.8	2.30	110	18.2
Gypsum	2.3-3.3	175	11.4	2.36	115	17.4
Hematite	4.5-5.3	305	6.6	4.12	200	10.0
Limestone	2.4-2.9	165	12.2	2.23	105	19.1
Limonite	3.6-4.0	235	8.5	3.17	155	12.9
Magnesite	3.0-3.2	200	10.0	2.70	125	16.0
Magnetite	4.9-5.2	315	6.4	4.25	205	9.8
Marble	2.1-2.9	155	12.8	2.09	100	20.0
Mica-Schist	2.5-2.9	170	11.8	2.30	110	18.2
Porphyry	2.5-2.6	160	12.5	2.16	105	19.1
Quartzite	2.0-2.8	160	12.5	2.16	105	19.1
Salt-Rock	2.1-2.6	145	13.8	1.96	95	21.1
Sandstone	2.0-2.8	150	13.3	2.03	95	21.1
Shale	2.4-2.8	160	12.5	2.16	105	19.1
Silica Sand	2.2-2.8	160	12.5	2.16	105	19.1
Slate	2.5-2.8	170	11.8	2.30	110	18.2
Talc	2.6-2.8	165	12.2	2.23	110	18.2
Trap Rock	2.6-3.0	175	11.4	2.36	115	17.4

A Simple Dewatering System

By Mike Glenn

1. An approximate 50# Dewatering System is made out of 1 12" PVC pipe with 2 end caps with a 5" hole in one of them and a 9" hole in the other;   with 2 rings to hold the ore and 4 paddles to start rotation of the ore.
2. 6 wheel frame to turn drum in.
3. Variable belt drive from 10 rpm for clean-up and 180 - 300 rpm for dewatering,

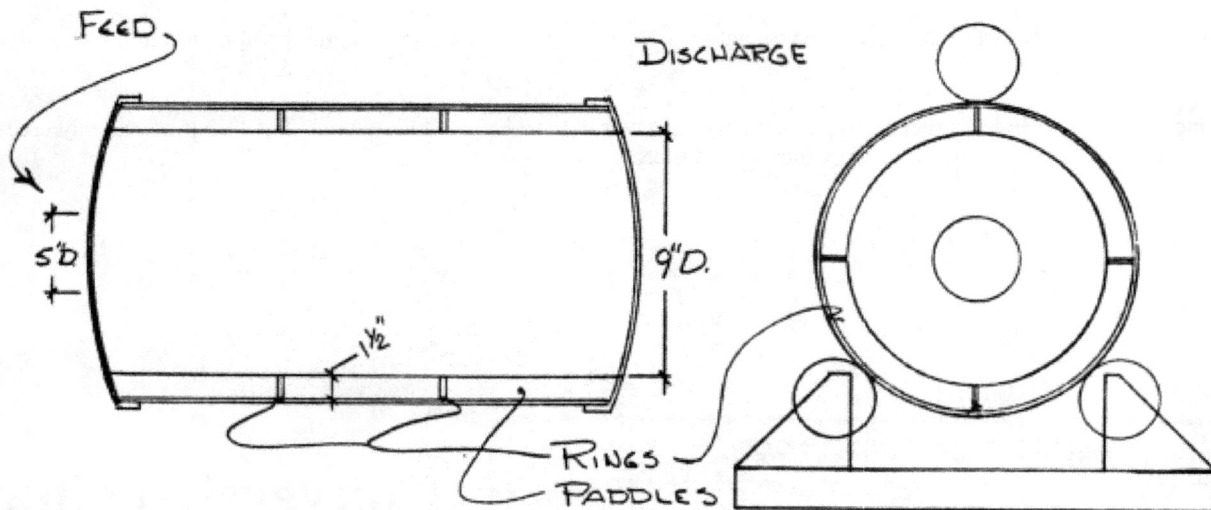

Acquiring a Mining Claim

by Michael McKinley Dannemoba New York

The 2 most common ways of acquiring a mining claim are by purchasing an existing claim, and staking a claim on open land. Choosing a good claim site and making sure it is a legal claim in the case of a purchase, or open unclaimed land in the case of staking, is imperative to the success of the prospector. A claim site is best chosen by first researching gold-producing districts in the general area where you would like to set up mining operations.

Mining districts can be researched by studying US Geological Survey reports, state geological survey reports, and perhaps the corporate reports issued by small mining companies. What one should look for in their research are the production reports of the district. How much gold, silver or platinum has the area produced? Also of interest is the geology and mineralization of the area. What type of ore would you be dealing with? Are the gold values coarse nuggets or microfine particles? Are the gold values associated with sulphides? From this type of research you can get an idea of what type of equipment you need to process the ore from the area. Corporate reports often state ore grade and reserves of current mining operations which will give you a good idea of what values are in the area. One book I've found very useful for researching mining districts is "U.S. Geological Survey Professional Paper 610 - Principal Gold Producing Districts of the United States".

It gives past production figures and geological reports for all the major gold districts in the country. Publications describing gold producing districts are available from state geological or mineral surveys. Many corporate reports are printed in mining and business publications. These reports usually include the mining companies address, so you can write to them for further information.

If you plan to stake your own claims, in addition to inspecting the proposed claim site and taking samples for assay, you must check the public land records in the regional BLM office to make sure the land is open to location. These offices keep up-to-date land status plats that are available to the public for inspection. The BLM also publishes a series of surface and mineral ownership maps that describe the ownership patterns of public lands. BLM and County claim records will help show you which areas are unclaimed by others. (Their records are not always complete or easy to hunt through.)

ONCE YOU'VE DECIDED ON AN AREA
Once you've decided on a district in which you would like to set up your mining operation, the next step is to find a specific claim site. If you wish to purchase a claim this can be done by checking the classified ads in the several mining publications, until you find one you are interested in. Or you can place your own ad in the publications, stating your interest in buying a mining claim in a particular district.

After you have lined up some prospects, you should inspect the claims to decide which one is right for you. The right one being the claim you can work for an acceptable profit. At this point you should forget about the prices of the claims and concentrate on the degree of profit offered by each claim. It is senseless to buy the bargain claim and end up losing money on your mining operations when the higher priced claim would have more than paid the difference with several years of healthy profit margins.

DETERMINING FACTORS
In determining which claim offers the highest profit potential there are several factors to consider. The assays are most important. Are the values there? Is it possible for the claim to pay? If you already own concentrating and processing equipment can it be used on this ore, or will you need to purchase compatible equipment? The availability of water, accessibility, the amount of overburden and vegetation which must be removed before encountering the gold are all cost factors which must be considered. How about the climate? In the north you will have to shut down for the winter months, thereby shortening your mining season. The distance to the mill or refinery is another factor. The greater the distance you must ship your ore or concentrate the more you will have to pay for transportation. Perhaps roads need to be built. Or you'll need a mill, leach pad, settling pond, or baghouse. The values on the property have to be great enough to offset the costs of starting up and operating the mine.

It is strongly recommended that you examine the claim before you make any deal. You will want to take samples for assay, note the degree of development work or actual mining which has taken place, and generally confirm that it is a legal claim with the potential to be worked for a profit.

BUYING A PATENTED CLAIM
Purchasing a patented mining claim is a much easier affair. You needn't be as concerned about the presence of the values as they have already been confirmed by a US mineral examiner. All you really need to confirm is the grade of the ore. One thing you should be prepared for when purchasing a patented claim is the price. When you purchase a patented claim you usually receive full title to the land. Expect to pay about the same price as you would for similar real estate in the area and sometimes more if the mine has significant value.

GET SOLID ANSWERS

Examining the claims you are interested in and taking samples for assay will greatly help eliminate the possibility of fraud. Last year I encountered a situation which could possibly have involved fraud. I answered a classified ad in which a man offered to stake claims in a "vast new placer area" for a very low price. The man sent me an impressive package describing the area and stating average assays of 1.3 oz. gold and 4.6 oz. silver/per cubic yard. It sounded almost too good to be true. All he wanted for staking, filing and assaying the claims was $10 per acre. There was a minimum purchase of several claims though, which drove the total price way up.

I wrote back requesting more information about the deposit, accessibility! availability of water, location and so forth. I also checked US Geological Survey reports and state reports for the general area he spoke of. There was no production history for the area, only the discovery of some very low grade deposits to the north. To my knowledge, there were no mining companies operating in the general vicinity.

In tine I got my reply. He stated that if I still wanted some of the placer property I had better send him my money in a hurry. He said he had a large mining company come in to test the land and they were very interested in buying it. In the next paragraph he wrote that he wouldn't tell me the location of the deposit because I might stake a few claims for myself. And in answer to my questions, of course, there was plenty of water available, the claim area was 2-wheel drive accessible, and the mining season was almost all year round. And wouldn't you know it. He had just discovered that there was a "high percentage" of platinum in the same ground!

I wrote a letter back to him the next day saying that since I could not examine the property first, I would like to pay him to stake one claim. In the event this claim proved out I would be happy to pay him to stake several more at the very least. I never heard from him again. I'll never know for sure if his deal was on the level or not. I sure wasn't going to part with thousands of dollars to find out! One thing to remember when purchasing a mining claim is that if it sounds too good to be true - it probably is.

In contrast, during the mid-70's I paid to have 2 claims staked in Nevada. I researched the area and found land open to location where there had been a producing silver operation. The claims had been abandoned in the early 60's by a Canadian firm when the price of silver was too low and the cost of processing the ore too high to return an acceptable profit.

With some research I was able to obtain the companies reports of ore grade and reserves, as well as their processing methods. I inspected the claim sites and took samples for assay. The assays were good. They came close to those in the companies report. I checked the county records to see if anyone had since relocated the claims. No one had, so I went ahead and paid a professional staking service to stake 2 twenty acre claims which contain appreciable amounts of silver and traces of gold.

PURCHASING A CLAIM

When purchasing a claim you should deal directly with the miner or prospector who is selling their own claim, rather than a broker. If you deal with a broker you will usually pay more than if you buy from the original owner.

IS IT LEGAL?

Once you have decided to purchase a mining claim, you must check to be sure that it is a legal claim. You can do this by checking the records at the County Recorder's office in the county where the claim is located, and the BLM regional office. From the owner of the claim, you should insist on seeing the claim slip from the BLM with the assigned BLM number. Upon the purchase of the claim demand that the claim slip, location notice and proof of previous year's assessment work be given to you. If taxes must be paid on the claim, they should be paid to date.

It has been said that "to buy an unpatented mining claim is to buy a lawsuit". The transfer of title of an unpatented mining claim should be made by a quit-claim deed to help avoid this. The seller of the claim should have his spouse join them in conveying title.

Whenever the owner of a claim sells or conveys his interest in the claim, he must file in the BLM office within 60 days after the completion of the transfer, the serial number assigned to the claim, and the name and address of the persons to whom the claim has been sold. This information must also be filed at the County Recorder's office in the county where the claim is located.

Owning a gold claim can be a rewarding experience or a costly burden. Some careful research and inspection of the claim or proposed claim site before you purchase or stake can save you from the headache and expense of owning the mineral rights to several acres of "cow pasture".

Safety - Lead

Although this safety data sheet was written for lead wool (used in cupelling; it applies to any type of lead, including lead oxide. You'll notice under section IV, it refers to dust, vapor and/or fume.

Inhalation of fumes is the biggest problem in assaying. You must make sure that your furnace is under a ventilated hood or out in the open, fie sure to be downwind when opening the door of a furnace. The same holds true for any use of lead oxide in "torch assays".

When handling lead oxide in a powder form, try to wear rubber gloves to Keep it off of your skin. Even wearing a dust mask will help.

Although doing these things seems like a real nuisance, your health is worth it. Lead accumulates in the system and it's tough to get rid of, although you can with the right treatments. A lot of people can go for a long time without showing any symptoms of lead poisoning (or any type of poisoning for that matter), but read what this Safety Data Sheet says about overexposure.

Safety Data Sheet

TRADE NAMES	Lead Wool
SYNONYMS	Pure Lead Wool
INTENDED USE	Industrial, Commercial and Domestic

II HAZARDOUS INGREDIENTS

MATERIAL OR COMPONENT (CAS#)	WEIGHT %	HAZARD DATA
Lead (CAS # 7439-92-1)	99.9	30 ug/m3*

*Ref: Occupational Safety & Health Standards, General Industry Standards Part 1910

III PHYSICAL DATA

BOILING POINT @ 760 MM Hg	3164°F (approx.)	MELTING POINT	621°F (approx.)
SPECIFIC GRAVITY (H2O = 1)	11.3 (approx.)	VAPOR PRESSURE	Not Applicable
VAPOR DENSITY (AIR = 1)	Not Applicable	SOLUBILITY IN H2O (% BY WT)	Negligible
% VOLATILES BY VOL	Not Applicable	EVAPORATION RATE (BUTYL ACETATE = 1)	Not Applicable
APPEARANCE AND ODOR	Gray fibrous metal; no apparent odor		

IV HEALTH HAZARD INFORMATION

Routes of Exposure When Processing or Handling

Inhalation	Dust, vapor and/or fume may be irritating to the respiratory system, and can result in both acute and chronic overexposure.
Skin Contact	Dust, vapor and/or fume may cause irritation.
Skin Absorption	Dust, vapor and/or fume are not readily absorbed through the skin.
Eye Contact	Dust, vapor and/or fume may cause irritation.
Ingestion	Dust, vapor and/or fume may be absorbed by the digestive system, and can result in both acute and chronic overexposure.

Effects of Overexposure

| Acute Overexposure | If left untreated: weakness, vomiting, loss of appetite, uncoordinated body movements, convulsions, stupor, and possibly coma. |
| Chronic Overexposure | If left untreated: weakness, insomnia, hypertension, slight irritation to skin and eyes, metallic taste in mouth, anemia, constipation, headache, muscle and joint pains, neuromuscular dysfunction, possible paralysis and encephalopathy. |

Emergency and First Aid Procedures

Eyes	Flush with copious quantities of water. Get immediate medical attention.
Skin	Wash thoroughly with soap and water.
Inhalation	Remove from exposure. Get medical attention if experiencing effects of overexposure.
Ingestion	Get immediate medical attention.

Notes to Physician
Lead and its inorganic compounds are neurotoxins which may produce peripheral neuropathy. For an overview of the effects of lead exposure, consult Occupational Safety and Health Administration Appendix A of Occupational Exposure to Lead (29CFR1910.1025).

V FIRE AND EXPLOSION DATA

Flash Point (Test Method)	Not Applicable	Autoignition Temperature	Not Applicable
Flammable Limits in Air (% By Vol)		Lower	Not Applicable	Upper	Not Applicable
Extinguishing Media	Dry chemical or carbon dioxide should be used on surrounding fire. Do not use water on fires where molten metal is present.
Special Fire Fighting Procedures	Use full body protective clothing and full-facepiece, self-contained breathing apparatus operated in a positive-pressure mode.
Unusual Fire and Explosion Hazard	Molten metals produce fume, vapor and/or dust that may be toxic and/or respiratory irritants. The product, or its dust, can react vigorously with strong oxidizing agents.

VI REACTIVITY DATA

Conditions Contributing To Instability	Not Applicable
Incompatibility	Strong oxidizers and this product may liberate hydrogen gas.
Hazardous Decomposition Products	High temperatures may produce heavy metal fume, vapor and/or dust.
Conditions Contributing to Hazardous Polymerization	Not Applicable

VII SPILL OR LEAK PROCEDURES

Steps To Be Taken If Material Is Released or Spilled
Dust material should be vacuumed, or wet swept where vacuuming is not feasible. Particulate matter should be stored in dry containers for later disposal. Do not use compressed air or dry sweeping as a means of cleaning.

| Neutralizing Chemicals | Not Applicable |

Waste Disposal Method
Dispose of toxic substances and hazardous wastes in accordance with local, state and federal regulations.

VIII SPECIAL PROTECTION INFORMATION

Ventilation Requirements
Ventilation, as described in the Industrial Ventilation Manual produced by the American Conference of Governmental Industrial Hygienists, shall be provided in areas where exposures are above the permissible exposure limits or threshold limit values specified by OSHA or other local, state and federal regulations.

SPECIFIC PERSONAL PROTECTION EQUIPMENT

Respiratory	As specified by 29CFR1910.1025 Subpart if, of the Federal Occupational Safety and Health Administration Standard for Occupational Exposure to Lead. Other local and state regulations may also apply.
Eye	Face shield or vented goggles should be used around molten metal.
Glove	Gloves should be worn when handling the product is necessary.

Other Clothing and Equipment
Coveralls, or other full body clothing, shall be worn during product use and properly laundered after use, with the wash water disposed of in accordance with local, state and federal regulations. Hard hat, safety boots and other safety equipment should be worn as appropriate for the industrial environment. Personal clothing and shoes should be protected from contamination with this product.

IX SPECIAL PRECAUTIONS

PRECAUTIONARY STATEMENTS

There are two major means of heavy metal absorption; namely, inhalation and ingestion. Most inhalation problems can be prevented with adequate use of aforementioned ventilation and respirator information. Always exercise normal, good personal hygiene prior to smoking or eating. Smoking and eating should be confined to non-contaminated areas.

Work clothes and equipment should remain in designated lead contaminated areas, and never taken home or laundered with personal clothing. Launder contaminated clothing before reuse.

Wash hands, face, neck and arms thoroughly before eating or smoking.

The product is intended for industrial, commercial and domestic use, and should be isolated from children and their environment. Caution must be exercised not to expose anyone to the smoke, fumes and dust generated from the use of this product.

Do not smoke while using this product.

Refining High Grade Sulfide Silver Ore

by Robert L. Desmarais

Having done much experimenting and research as to a most efficient refining method for high gradesulfide silver ores, I have found a method which is fairly easy and quick. Of course, different ores may cause different results. This recipe works on a galena silver sulfide ore that I have a small vein of in a mining area near Bishop in Northern California.

First of all, high grade the ore to remove as much rock and waste as possible. Then crush the ore to a fine powder, the finer grind of ore the better, since when doing a leach procedure it will quicken the action of the acid on the ore.

ROASTING

Now it may be to your advantage to do a roast on the ore. This consists of putting the crushed ore in a cast iron pan and putting it on a gas burner with a medium heat for about 45 minutes. I'm using a volume of 3 lbs. of ore in a 10" cast iron skillet. This ore, due to its density, only takes up a cup measurement. Do stir the mixture every 10 minutes or so to even the roasting action. This roasting can remove some undesirable elements that can cause a problem during leaching. Do be careful of the fumes from the roasting. These, too, can be harmful to your health.

After roasting, some of the ore might be lumped up, so may have to sift and re-grind the ore. A mortar and pestle might be useful here.

LEACHING

Next make a mixture of nitric acid and distilled water -50/50. I used about a pint total. ALWAYS POUR THE ACID INTO THE WATER or you may cause a violent reaction such as a boil over of extreme heat. Also use a heat resistant container of the glass or ceramic type, not metal since the acid will instantly attack most metals. Distilled or de-ionized water is necessary since some tap waters have other minerals or chemicals in it that can cause the silver to not come back out of the liquid.

Now you are ready to pour the liquid onto the crushed ore. Be sure you have good ventilation and the fumes are going away from you. These fumes are toxic as well as corrosive so beware. You only need to cover the ore with about 1" of liquid. When you pour the liquid into the ore container, pour slowly. Don't touch the liquid with bare hands. It will cause burns and yellow skin conditions. Nitric acid is one of the most powerful and corrosive acids.

Again, use a heat resistant container, since there will be an instant foaming and fuming action. And make sure it is quite a bit larger than your ore volume. Stir the mix using a glass rod. Stir again occasionally until all the action of the liquid has stopped. The warmer the liquid, the quicker the action but do not boil. The action is the silver being dissolved into the liquid (leaching).

THE PREGNANT SOLUTION

Now after all the action is complete and the silt has settled, 1/2 a cay to a days time, pour off the liquid into a clean container. Pour it through a paper filter, (a coffee filter will do), set in a plastic funnel over a glass or plastic container. A ribbed type funnel is quicker acting to Keep the ore from packing. Pour slowly as to not break the paper filter. They do break easily. Rinse the ore and container and let run through the filter.

If the filter did its job, the result should be a clear liquid. If the liquid is more than a little discolored just pour it through another clean filter. This liquid is still acid so be careful with it. Now it contains the silver which is dissolved in the liquid.

GETTING THE SILVER OUT OF SOLUTION

To get the silver out of the liquid, I use a method called replacement process. I simply suspend Number 1 copper into the container and let it sit until no more copper is dissolved. A day or so will usually do. Number 1 copper is a pure form with no mixes of other metals. Again a little heat will step up the action.

The copper will dissolve into the solution and the silver will fall out (precipitate). You should soon see silver flakes falling to the bottom of the container. After no more silver seems to fail, you can test the liquid to be sure all the action is complete.

Heat a small part of the liquid in a test tube to over 100 degrees F. and then put a few drops of hydrochloric acid into the liquid. If a sudden white cloud of material does not appear, all the action is complete. If the cloud does appear let the copper stay in the liquid for a longer time until no white cloud appears when tested.

Then pour off the liquid through a filter paper like before and wash the silver dust several times with distilled water. You can slowly dry the filter paper in an oven and then brush off the silver dust.

RECLAIMING THE COPPER
If you want to reclaim the copper out of the solution, put it in a tin can (tin only) and the tin will dissolve into the liquid and the copper will fall out.

To dispose of the liquid, add lye or sodium hydroxide to neutralize the acid and make it safe to dump. We must think of the environment since we have to live in it.

SMELTING THE SILVER DUST
When you have around a pound or so of silver dust, it's time to smelt the dust into solid metal or you can sell it as is. A good flux to help reduce the dust will help in keeping the loss of silver to a minimum, since part of the weight is lost in the reduction of the dust to a solid metal.

This process works on some of my ores and I hope it can do some good for you. Some day I hope to get the financing to set up a large operation at our mine and produce much silver.

Well, good luck on your prospecting and refining.

Use of Depressants on Your Table

by D. D. Peterson, Forks, WA

The use of depressants on ores run over a table is of utmost importance. Depressants are usually some form of detergent, or detergent-based products. These should be fed to the pulp feed box as well as the water distribution box (See earlier articles) on a continuous basis.

Gold, etc. when broken from its matrix, particularly if one uses some form of a hammer mill for breaking your ore down with dry classification, is often hard to wet. It will simply float on the surface of the water on your table and float on out to your tailings pond. This gold that floats is often overlooked because it does not remain with the tails, having floated off. And for this reason when one assays the tails gold will not show up in the assays.

Depressants should be fed into your system at the earliest point that it is feasible to do so. If you are using a ball mill, it is advisable to feed depressants there. Any de-watering process should be watched very closely.

GETTING THE DEPRESSANT INTO YOUR CIRCUIT

There are many ways of feeding depressants into your circuit. The best way is with metering pumps such as the ones used to feed chlorine in water systems. These pumps however are quite expensive. You could purchase one of the small submersible mild acid pumps sold by W. W. Grainger.

A large plastic garbage can is used to mix the depressant you are using in the strength you have determined is best. The submersible pump is placed in the can and a plastic line is run to your table. This should have 2 control valves. One valve is for the pulp feed box, and the other for the water distribution box. Depending on the pump used, it is sometimes advisable to make a loop of the plastic line above the valves that feed your boxes and continue this line on back to your garbage can.

This is done for 2 reasons. First, some small pumps lose their prime when they are restricted too much by your valves. The second is that the depressant is constantly being mixed by the return line so your depressant is remaining uniform in strength hour after hour.

Care should be taken to not let the pump run out of solution as it will be ruined.

There are several excellent depressants on the market. Sierra Slick sold by Sierra Chemical is one of many good ones. These depressants are in concentrate form and are mixed with water in the strength that is best for your ore.

USING DISHWASHING SOAP

The cheapest dishwashing soap that you can buy will usually work for short test runs. In this case the cheapest is the best. The reason for this is that many of the more expensive brands have various additives such as coconut oil to help soften hands. Of course, the oil is actually a form of flotation agent, so one is defeating one's purpose by using this kind of soap. The disadvantage of using cheap dishwashing soaps is the bubbles that are generated, particularly in the water distribution box.

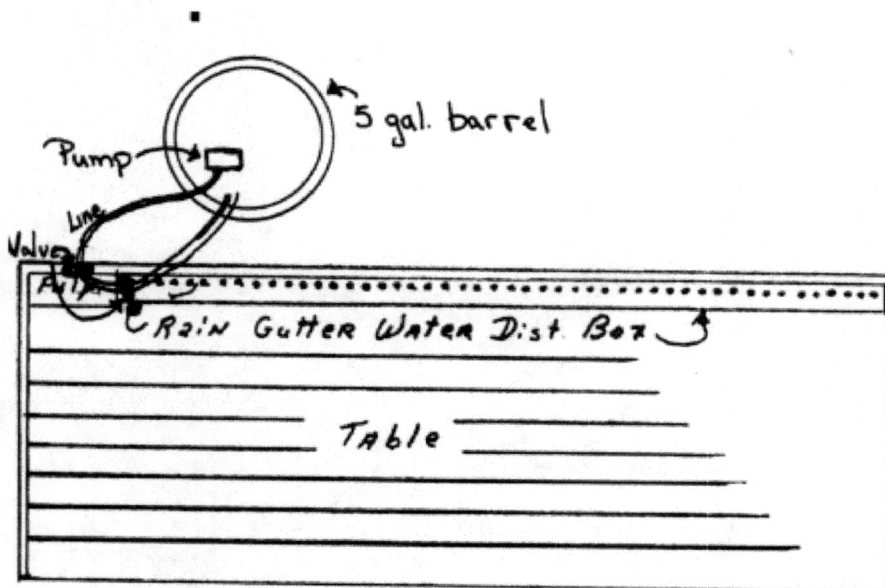

The Doodle Bug

by Gary Christopher, The Prospectors Cache Englewood, Co.

In Colorado west of Denver is a stream called Clear Creek noted for its placer gold in the stream, on the banks, and in old dredge piles. It is along this stream that machines called "Doodle Bugs" can be found in different sizes and shapes made out of wood, aluminum, sheet steel, and plastic. Basically the Doodle Bug is a concentrator that is used for bank run material and is quite a bit superior to the manufactured hydraulic concentrators that are on the market today.

Years ago when I first started in placer mining I read a book called Gold In Placer by Jack Douglas copyrighted in 1944 and 1948. Jack was one of those depression era prospectors who really knew his placer mining. In his book he described the Doodle Bug. At the time I thought he invented it but years later I saw an old movie showing a Doodle Bug being used with a flume well before the gasoline pump and engine was invented. I figure the movie must have dated from around 1906.

Basically the Doodle Bug is a 3 section machine consisting of a sluice on legs, a classifier on top of the sluice, and a hopper with spray bar on top of the classifier. Nowadays water is pumped to the Doodle Bug's spray bar through a hose from a pump and engine assembly down on the creek bank.

The Classifing Sluice Box otherwise known as the Doodlebug

All three parts of Doodlebug

TOP SECTION   MIDDLE   BOTTOM

As gold bearing gravel is dumped in the hopper, it is worked and slides down the hopper and falls flown on the grizzley bars in the classifies The oversized rocks slide down the grizzley and hit a splitter which dumps out the oversized on either side of the machine. The smaller sized material fall through the grizzley into the sluice where the gold is trapped. The drawings shown here do not show the riffles. Most Doodle Buggers use astro-turf or high-low blue carpet with flat expanded metal lathe of from 1/2" to 1" which does a fine job of catching gold. Since drawing these plans I have found that the piping and valve can be out of pvc plastic and will hold up. Also the impact board is not needed in the sluice.

My friends and I have built Doodle Bugs out of wood, aluminum, sheet metal, and plastic. Wood requires a lot of sanding, paint,

Sluice Box set up

screws, and a lining of tin and is somewhat heavy. Sheet aluminum is easy to bend and pop rivet, lightweight and stands up to acid water. However, you can beat it up if you don't watch the size of the rocks you throw at it. Sheet steel is about the same except acid water eats it up in a hurry. ABS plastic sheet 1/4" thick is the best material I have worked with. It is fairly lightweight, acid resistant, and tough. It doesn't need paint or screws. It can be filed and shaped easily and is extremely tough. One can throw 4" rocks at this machine from 10 feet away and they just bounce when hitting the hopper. The only disadvantage is the cost of the plastic and the making of the heat tape boards to bend the plastic. The heat board will probably cost you over $100 alone.

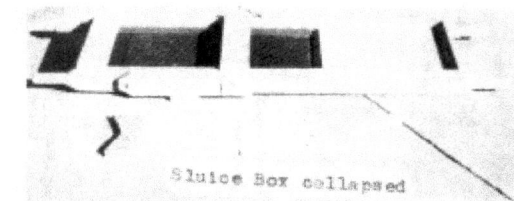

Side view of Classifier & Sluice with legs mounted.

Some improvements that have seen done on Doodle Bugs are a hinged hopper so it can be set at different angles and the splitter built out of fiat expanded metal with the long length of the diamond mesh running vertically.

If you work the bank gravels you will like this machine.

Sluice Box & Classifier with legs assembled

Sluice Box collapsed

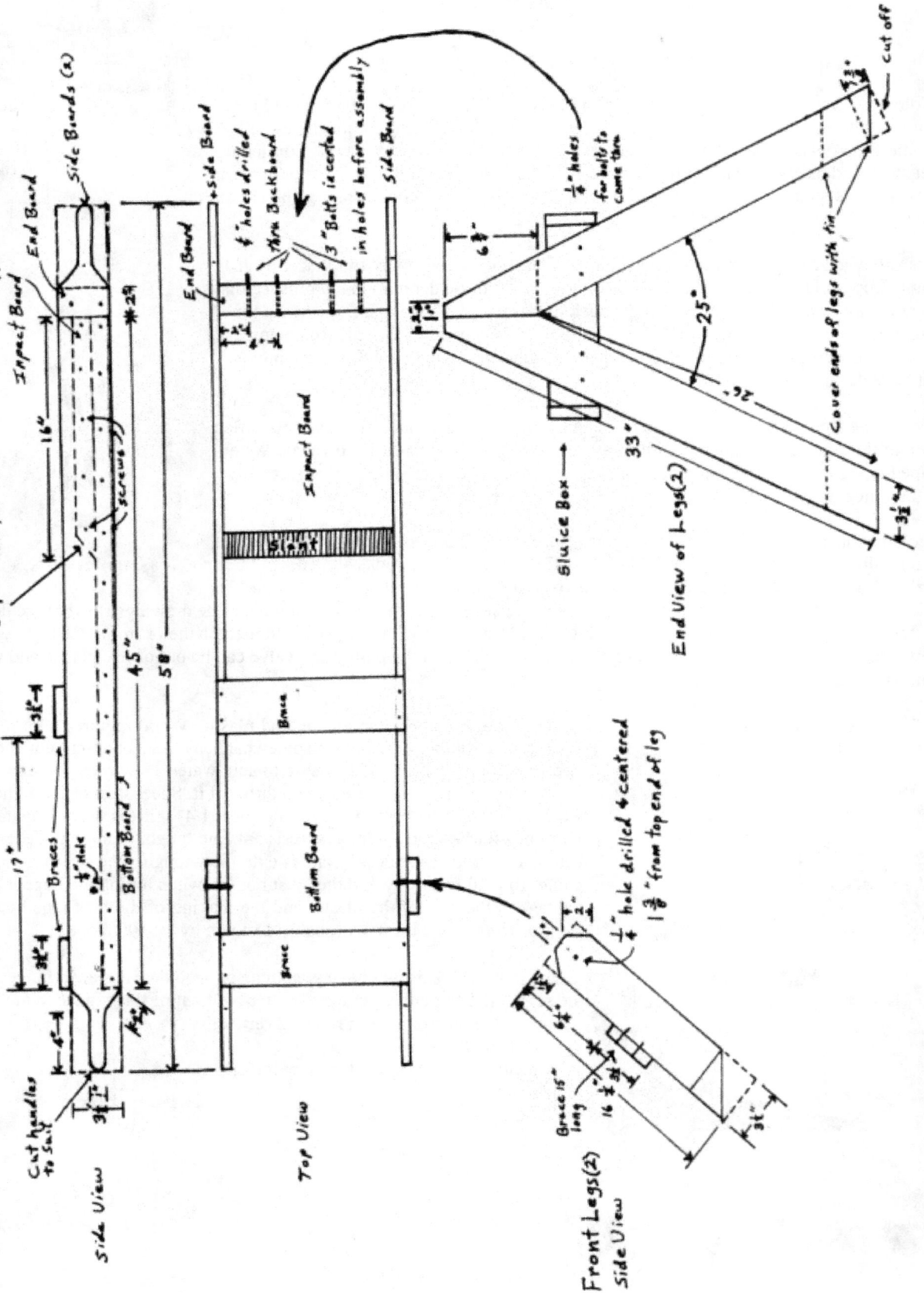

Bottom Section of Portable Classifying Sluice Box (Doodlebug)

The Sluice Box

Side Boards (x)

End Board

Impact Board

1½" zinc coated screws approx. 4" apart

slope to suit yourself

screws

16"

45"

50"

17"

Braces

¼" hole

Bottom Board

Cut handles to suit

Side View

Side Board

End Board

¼" holes drilled thru Backboard

3" Bolts in center in holes before assembly

Side Board

Impact Board

Impact Board

Brace

Bottom Board

Brace

Top View

¼" holes for bolts to come thru

6¼"

25°

26"

33"

Sluice Box

Cover ends of legs with tin

cut off

3½"

End View of Legs (2)

Front Legs (2)
Side View

¼" hole drilled & centered 1¾" from top end of leg

Brace 15" long

16

The Classifier

Middle Section of Portable Classifying Sluice Box (Doodlebug)

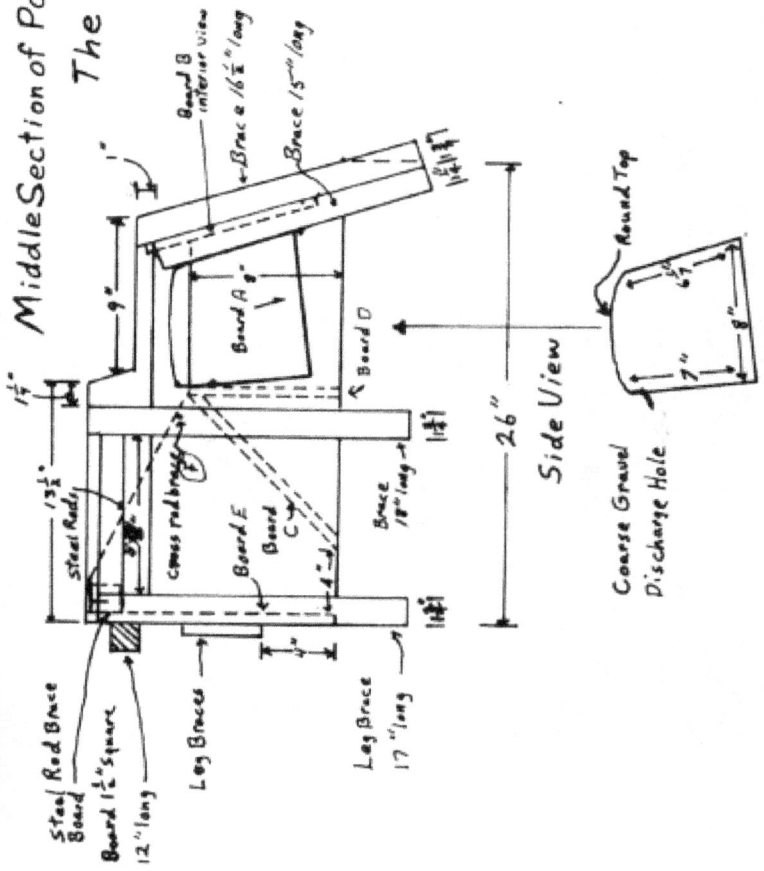

Top View looking Down

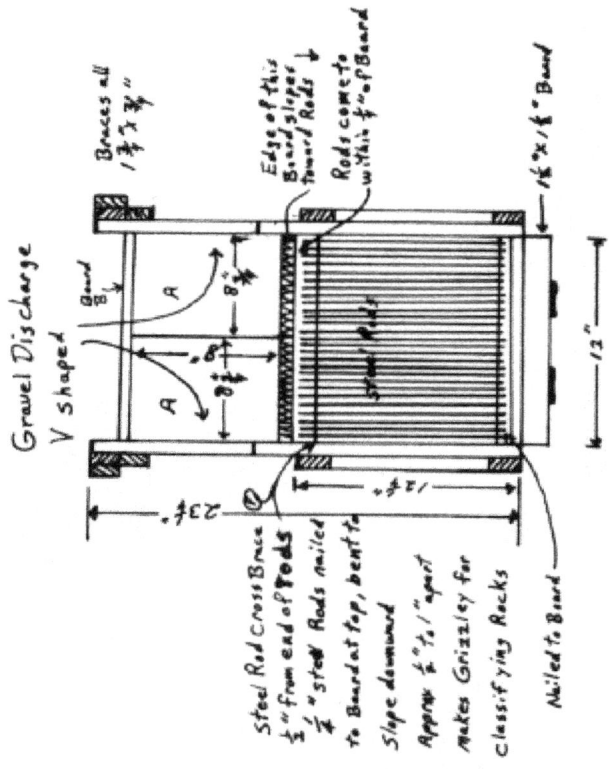

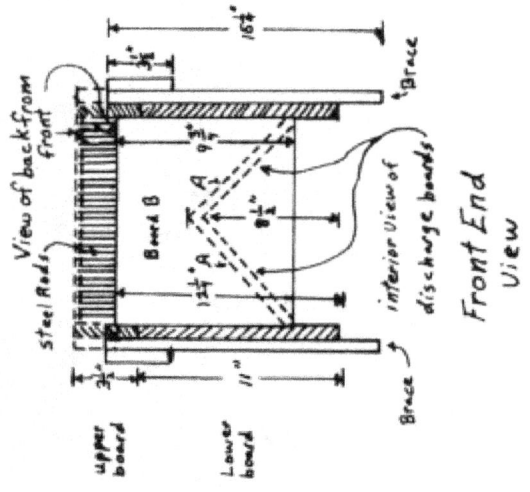

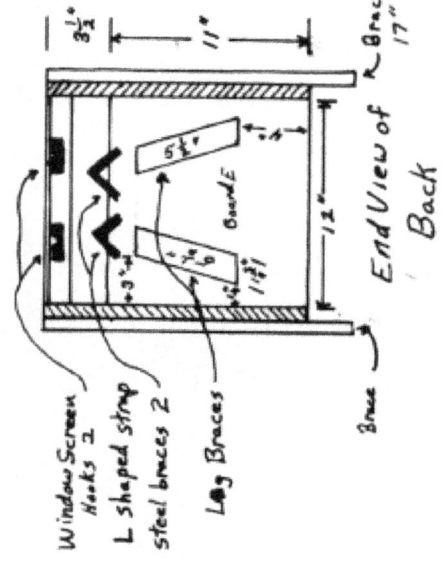

Braces all 1¾ x ¾"

Gravel Discharge V shaped

Edge of this Board slopes toward Rods

Rods come to within ¾" of Board

1¾"x ⅝" Board

Board B

A

8½"

A

Steel Rods

Steel Rod Cross Brace ½" from end of Rods

¼" steel Rods nailed to Board at Top, bent to Slope downward

Approx ⅜"±⅛" apart makes Grizzley for classifying Rocks

Nailed to Board

12"

23¾"

12¾"

End View of Back

3½"

11"

Brace Length 17"

5¼"

Board E

3¾"

1¾"

1¼"

12"

Window Screen Hooks 2

L shaped strap steel braces 2

Leg Braces

Brace

Side View

1"

9"

Board B interior View

Brace 16½" long

Brace 15" long

Board A

Board D

1¼"

13½"

Steel Rods

Cross rail brace

Board E

Board C

26"

Brace 18" long

Leg Brace 17" long

Steel Rod Brace Board

Board 1½" square 12" long

Leg Braces

Round Top

Coarse Gravel Discharge Hole

7"

8"

Front End View

View of back from front

3½"

9½"

Board B

A

A

8½"

13¾"

16¾"

11"

3½"

steel Rods

interior View of discharge boards

upper board

Lower board

Brace

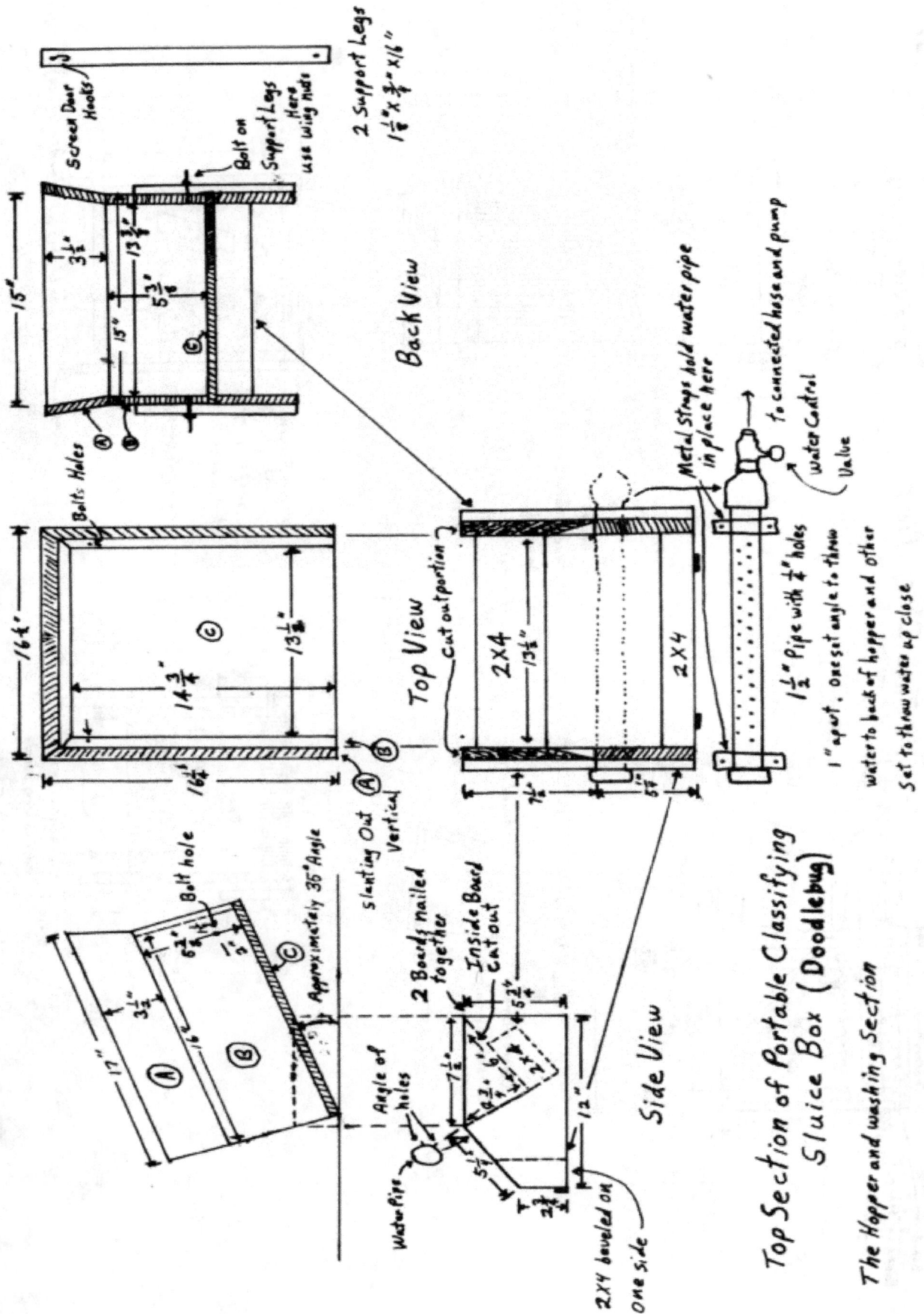

Screen Door Hooks

Bolt on Support Legs Here use wing nuts

2 Support Legs 1⅛" x ¾" x 16"

3¼"

13½"

5¾"

15"

15"

Back View

Bolts Holes

Ⓐ Ⓑ

16¼"

14¾"

13⅛"

Ⓒ

16¼"

Top View — Cut out portion

2x4   13¼"

2x4

7½"

5"

Metal Straps hold water pipe in place here

to connected hose and pump

water Control Valve

1½" Pipe with ¼" holes 1" apart. One set angle to throw water to back of hopper and other Set to throw water up close

Ⓐ Slanting Out Vertical

Ⓑ

Angle of holes

17"

3½"

6¾"

½"

Ⓐ   Ⓑ

Ⓒ

Approximately 35° Angle

Bolt hole

2 Boards nailed together

Inside Board Cut out

7½"

6"

5¼"

2x4

8½"

Water Pipe

5¼"

2¾"

12"

2x4 beveled on one side

Side View

Top Section of Portable Classifying Sluice Box (Doodlebug)

The Hopper and washing Section

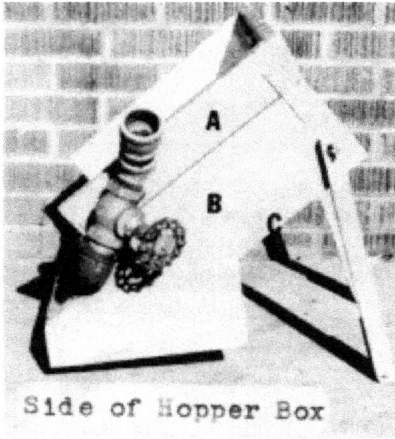

Side of Hopper Box

Side of Classifier showing coarse gravel discharge

Water spraying into hopper

Hopper Box lined with tin & showing holes in pipe water feed.

Bottom of Hopper Box

Sideview of Doodlebug showing rock splitter

Bottom of Classifier showing position of boards

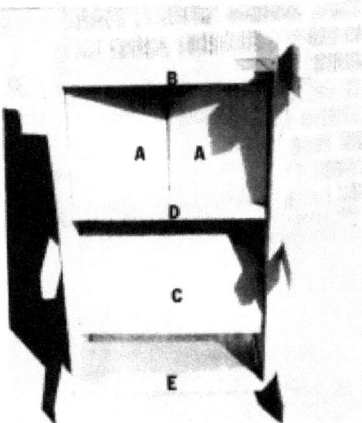

Back of Classifier

Front of Classifier showing grizzley

Classifier, with steel rod grizzley & coarse gravel discharge. All areas subject to wear covered with tin.

View of grizzley bars in classifier

Sluicebox without Riffles

by Matt Bard

During the summers of 82, '83, 84 I worked the beach placers or Nome, Alaska. The gold on these beaches is so fine it takes almost a million colors to get an ounce. This seems to be an almost impossible endeavor, but here is how it is done.

To effect correct riffle boil, the water flow must be fast enough and in sufficient volume to create what we all Know for standard sluicing operation. This will blow out most of your fines.

But I learned this technique from others and it goes like this:

A sluice is constructed of wood eight feet long and two feet wide. It is set up like any other high bank rig except the classifier screen is 1/8" or 1/16" stainless steel screen. The screen is very tough and is flat so wear and tear is no problem, and small pebbles do not lodge in the screen - they roll off.

The secret is to line the box with astro-turf or 3M Blue spaghetti carpet. This being done, all that is needed is to adjust water flow so that it just covers the top of the carpet about 1/4" should suffice.

Run the box at 12:1 drop.

In Alaska, I cleaned up every 3 days and only once failed to top an ounce. You will still lose some gold, but the majority of it will hold in the box. I'm talking #4 colors -micron size fly-shit gold at the end of the physically visible scale!!

If you spend $20,000 on a centrifuge, generators, pumps, etc., this simple device will make you feel foolish with its 95% recovery. Well, you got the idea?' Good luck.

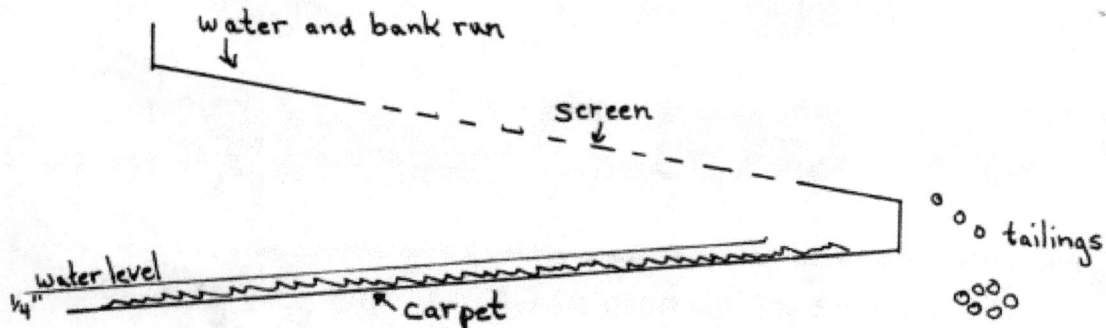

Conversion Chart – Grams, Grains, Pennyweights

TROY WEIGHTS

3.086	grains = 1 carat	= 200	milligrams
24	grains = 1 pennyweight =	1.5552	grams
20 pennyweights or 480	grains = 1 ounce	= 31.1035	grams
12 ounces or 5,760	grains = 1 pound	= 373.24	grams

GRAMS	GRAINS		PENNYWEIGHT	GRAM	GRAINS
1	15.43		1	1.55	24
2	30.9		2	3.1	48
3	46.3		3	4.7	72
4	61.7		4	6.2	96
5	77.2		5	7.8	120
6	92.6		6	9.3	144
7	108.0		7	10.9	168
8	123.5		8	12.4	192
9	138.9		9	14.0	216
10	154.3		10	15.6	240
11	169.8		11	17.1	264
12	185.2		12	18.7	288
13	200.6		13	20.2	312
14	216.0		14	21.8	336
15	231.5		15	23.3	360
16	246.9		16	24.9	384
17	262.4		17	26.4	408
18	277.8		18	28.0	432
19	293.2		19	29.5	456
20	308.6		20	31.1	480
21	324.0				
22	339.5				
23	354.9				
24	370.4				
25	385.8				
26	401.2				
27	416.7				
28	432.1				
29	447.5				
30	463.0				
31.1035	480.0	= 1 troy oz.			

Platinum, the Miracle Metal

by Jim Floyd

This is the first in a series of articles written about platinum and platinum group metals. This first article is intended as an overview to inform the reader about platinum, the miracle metal. Articles will be written with your questions in mind. Some of the upcoming articles will include fire assay techniques for detecting platinum; spot tests for platinum group in solution; other ideas are welcomed and special consultation is available for parties interested in commercial application.

The platinum metals consist of a group of six rare, closely related metals: Rhodium (Rh), Ruthenium (Ru), Palladium (Pl)» Osmium (Os), Iridium (Ir). and Platinum (Pt). Platinum is the most important member of the group. These metals are associated with gold and are similar in chemical properties.

Platinum and palladium are the most abundant of the 6 metals. Platinum was the earliest known, having been discovered by the Spaniards in the South American silver mines about 1550. The metal was called "platina del pinto" (silver-like metal from the Pinto River) but was recognized as being different from true silver because it could not be melted. Silver has a melting point of 960 C.; platinum 1773.5 C. It also caused difficulty for the Spaniards, as its density was close to that of gold (19.34) and when the gold plated, the platinum was indistinguishable from the solid metal. This lead to the production of counterfeiting gold. This was stopped by destroying all stocks of platinum.

The properties of platinum were seriously studied in the middle 18th century by Watson, Schoffe and Morggarf. In 1803, W. H. Wollaston showed that all platinum previously examined was in fact an alloy with similar metals. He managed to isolate two; one he named palladium after Pallas, a newly discovered minor planet, and the other rhodium from the Greek word rhodon, rose, because some of its salts are rose-colored.

One year later Tennant discovered two more of the metals, naming them osmium from the Greek osme, meaning "smell", as its oxides have an unpleasant odor; and iridium from iris, the Greek word for rainbow.

In 1845 Claus discovered the last metal, ruthenium, which was named after a province of his native Russia.

Platinum metals are found in their native state in alluvial deposits in the Ural Mountains and Colombia and in gold bearing rock at Witwatersrand in South Africa. Platinum occurs as its sulfide, PtS (cooperite) and a mixed sulfide containing palladium, platinum and nickel at Rusterbung in South Africa, and is found as platinum diarsenide, $PtAs_2$ (sperrylite) in the huge copper and nickel sulfide deposits in Ontario.

Platinum is a useful catalyst for a large range of chemical reactions, both organic and inorganic. For example, it changes ammonia into nitric acid. Both platinum and palladium catalysts are used extensively in the pharmaceutical industry to make drugs and vitamins. Palladium is used to create margarine.

An early use of platinum was to produce a lighter, before the invention of friction matches. It is also used in automotive catalytic conversion to control pollution caused by cars.

It is used to make a wide range of laboratory ware for nigh temperature reaction. The glass and fiberglass industries both use platinum parts to work with molten glass. Palladium and platinum are used in making jewelry and for setting diamonds. Half of all platinum imported by Japan is used in the manufacture of platinum jewelry.

Dennis, the River Rat

Sluice Box Riffle System

by C. Dennis Boring Sunnyvale, Ca.

This is the present sluice box riffle system I'm using and others too, that prefer it to the standard or stock design of various manufacturers. The picture is pretty much self-explanatory.

The majority of the gold is caught in the first 1/3rd of the system anyway and the balance is mostly real fine stuff that only needs calm water to settle out and something to catch it on, carpet and/or expanded metal sheet covering. This design is the product of many weeks of actual field testing.

Mounting Adapter for Compressor

This is a drawing of an adapter to mount a Brown compressor to a Keene bracket. It works by the plate being fastened with counter-sunk screws and then becoming part of the compressor so the bracket can be adjusted just like the compressor base with the four 3/8" 16 bolts. They're offset enough to allow use in either of two orientations, to give sufficient clearance for the motor, etc.

I use the W. Brown type compressor on my combo, dredge and on the bigger requirements the compressor that fits the bill from ITT which are quite reasonable for the new models (ie. 3.5 CFM @ 50 PSI goes for about $300) which feed air to two divers working less than 33 ft. and one diver to 50 ft. with no problem.

Popular Mining Magazine   September October 1985

REPLACE RIFFLES REMOVED WITH FLAT BAR (SUPPORT) 1/8" × LENGTH 3 PLCS AS SHOWN × 1" WIDE

THIS AREA IS OPEN BETWEEN TOP SHEET OF EXPANDED-METAL AND CARPET PRESSING EXPANDED-METAL.

ADVANCED RIFFLE DESIGN

ADAPTABLE TO ALL BOXES FROM 1½" TO 8" EITHER WITH SUCTION NOZZLE or POWER JET.

DRAWN/DESIGNED by: C. DENNIS BORING MAY 1984

RUBBER FLAP DAMPENER FULL WIDTH OF BOX INSIDE.

THIS IS OPEN RIFFLE NO EXPANDED METAL HERE EXCEPT UNDER RIFFLES.

NUGGET TRAP 1/4" Ø PERFORATED

SLUICE BOX RIFFLES (5)

ALUMINUM BOTTOM OF BOX

(1st RIFFLE) FULL-LENGTH CARPET PIECE 3/4 " ◇ EXPANDED METAL

3/4 " ◇ EXPANDED METAL COVERING BALANCE OF BOX

RIFFLE SIDE-BAR

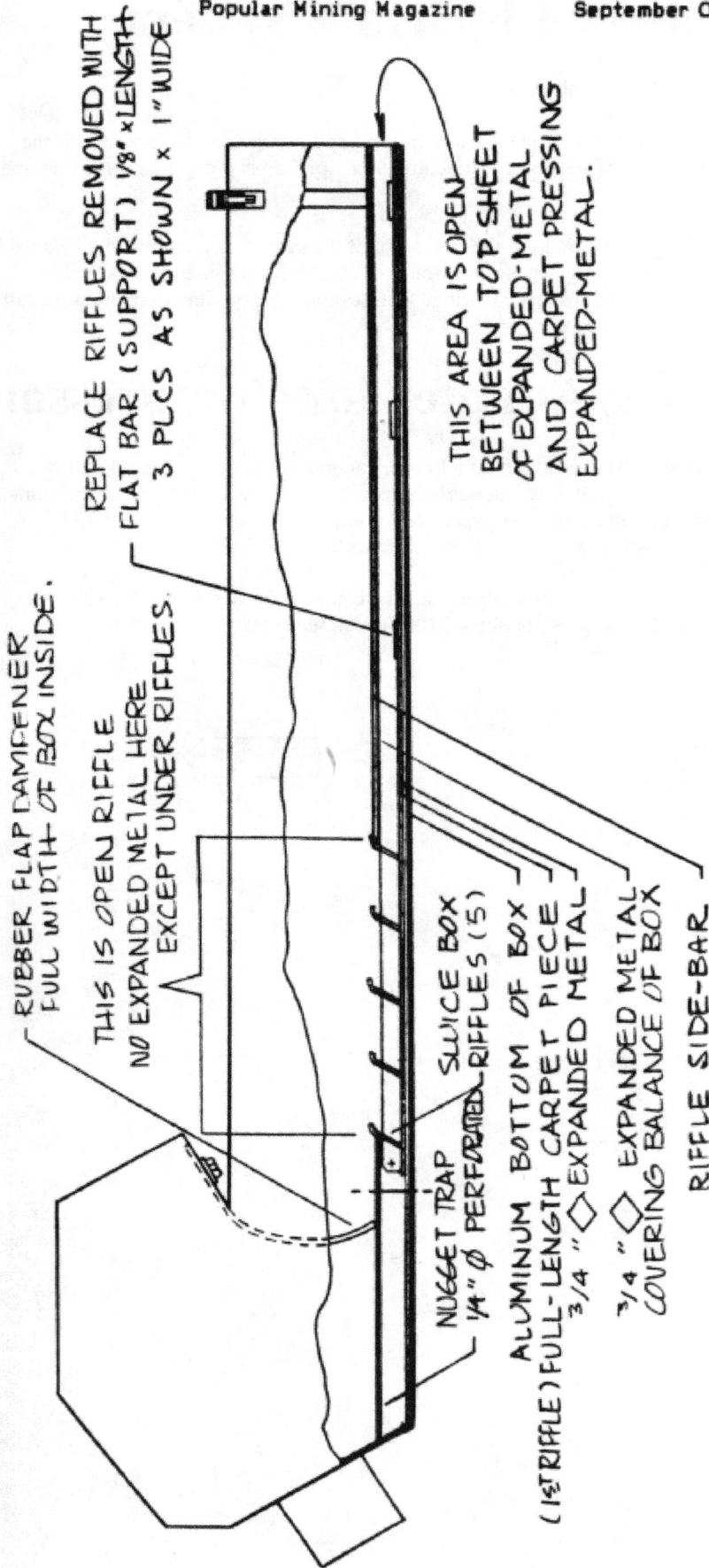

TYPICAL SLUICE BOX SIDE VIEW SHOWN WITH 'BREAK-OUT' TO VISUALIZE ITEMS. NO SCALE

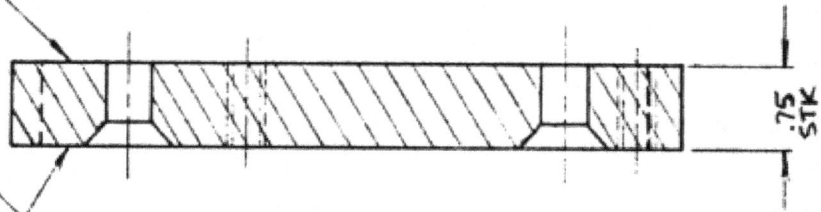

Adapter is used to mount Air Compressor to off-Brand Bracket.

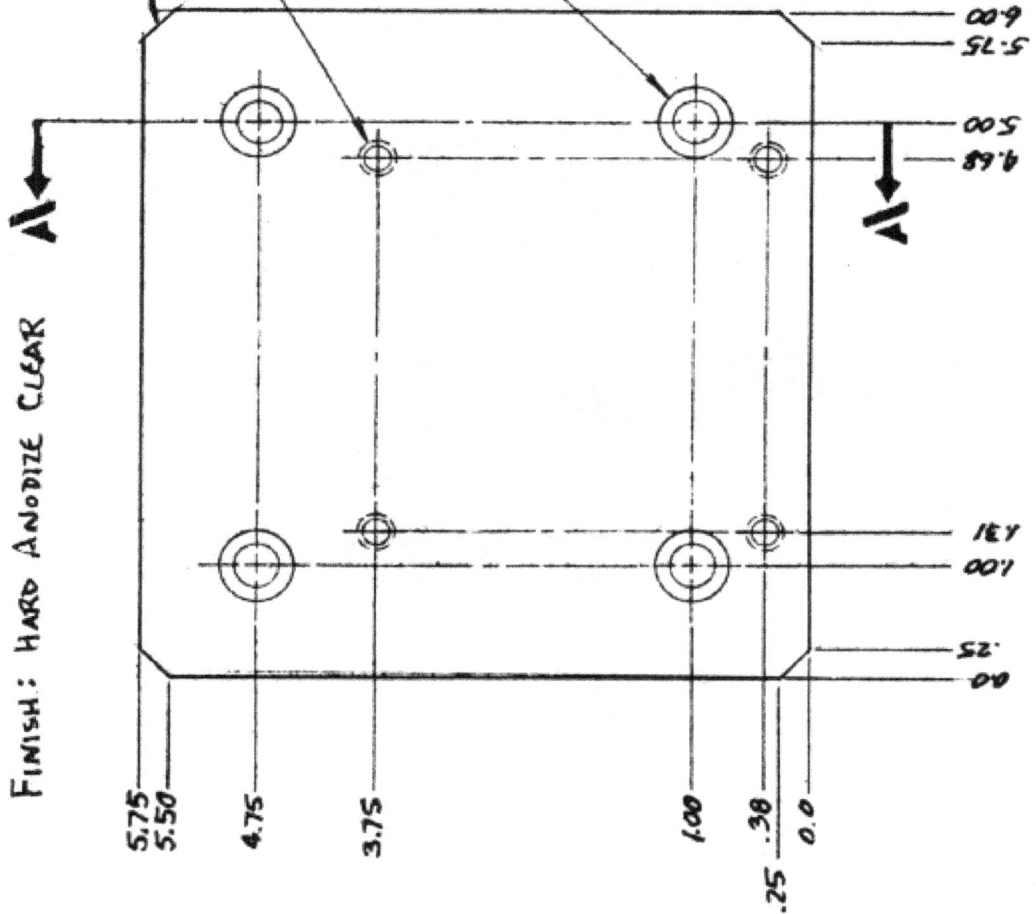

Tool Bit Finish Acceptable.
Smooth Equiv. to 100 Grit Sandpaper Wet.

Material: .75 thk stk 5052-H32 Al Aly Plate or Equiv.

Finish: Hard Anodize Clear

CHAMFER 45° ± 1° (4 PLCS)

TAP 5/16-18 UNC-2B THRU (4 PLCS)

DRILL .375 DIA THRU CISNK 82° x .20 DEEP (4 PLCS)

.75 STK

SECTION A-A

APPROVAL	DATE
DRAWN	
CHECKED	
APPROVED	

UNLESS OTHERWISE SPECIFIED
DIMENSIONS ARE IN INCHES
REMOVE BURRS AND SHARP EDGES
BREAK ALL SHARP CORNERS

6.00
5.75
5.00
4.68
4.31
1.00
.25
.00

5.75
5.50
4.75
3.75
1.00
.25 .38
0.0

SUBJECT: Micron Gold

by James Crowder, Dallas Texas

To make an attempt to recover micron size particles of gold first requires identification of the gold. A lot of speculation exists as to the form of the gold. Realizing that one system or process will not work on all ores, a quick identification is needed. Also, there are a number of waters that are carrying recoverable amounts of gold and/or silver as well as chemicals and other metals. Since we are primarily interested in gold, we will limit our discussion to gold.

GOLD IN WATER

To determine if the gold in water is metallic or in the ion form is fairly easy. As an example, salt water recovered from two oil wells. One of these wells was in Texas and the other several hundred miles away in Oklahoma. The Texas oil well salt water contained the gold metal in micron form. The Oklahoma oil well water had the gold in solution.

TESTING THE WATER

For the test, 2 beakers, one liter capacity, were placed side by side and filled with the salt water from the 2 oil wells. Sodium hydroxide was added until a pH of 8 was attained. Next, sodium borohydride was added slowly while stirring, then the solutions were allowed to settle for a minimum of 12 hours while the reducing agent, sodium borohydride, precipitated the metals and allowed the precipitates to settle.

Decanting the clear liquid and filtering the precipitates and fire assaying the filter paper that contained the precipitates resulted in obtaining a dore bead from one solution. The Texas water showed to be blank. But wait a minute, was this true? No, the Texas water had free metal and the borohydride only works on ions.

Another sample was prepared and the pH was again adjusted on both the water samples and this time cyanide was added to the solutions and stirred vigorously for 20 minutes. The beakers were set on a hot plate and heated until the vapor stage was reached, not boiling, to drive off the dissolved oxygen in the water. Again, stirring the solution gently, zinc powder was added until a mouse gray coloration was seen. Within 10 to 20 minutes the zinc was filtered out of the solutions and again the filter paper together with the zinc was fire assayed.

Both solutions were carrying gold. Deduction, it required the cyanide to reduce the metallic gold to an ion form thus allowing metallic replacement to occur on the zinc.

The above tests may or may not be a practical method for a final recovery process, but are simple enough to assay water. One milligram per liter is equal to one part per million, divided by 34.5 equals ounces per ton. Reason for the one liter test.

Salt water from oil wells is fine but who has oil wells in their backyard. Okay, substitute any kind of water or better yet let's make up some solutions from "ore".

TESTING AN ORE

Obtain a 55 gallon steel drum or any other container approximately this size. With a 5 gallon plastic pail add 30 gallons of water to the drum. Now add 1/2 can of lye or 1/2 lb. of sodium hydroxide to the water. Weigh up 50 lbs. of sand suspected of carrying gold values and add this to the drum containing the 30 gallons of water. Use a small mortar mixer, or paint mixer powered by an electric motor to thoroughly agitate and scrub the sand in the water. Even a small outboard motor propeller driven by a shaft and motor would suffice. Just simply scrub the sand for a few minutes.

When you have stopped the scrubbing action, allow the sand to settle for a few minutes. If gold (micron gold -the kind that is not visible to the eye and sometimes called flour gold), is present we should have the gold in suspension in the water.

The test is the same with oil well water and with any other water.

Now we take a pint of quart of the solution from the top of the drum and add a pinch of cyanide to the sample and stir gently for 20 to 30 minutes. Now the solution can be vacuumed or the oxygen driven off with heat. Use zinc powder to recover the gold from solution and fire assay. The one pint or one quart of liquid that was assayed will relate back to the 50 Ibs. of ore and 30 gallons of water.

Reason for this size test is to get a more representative sample and to eliminate the nugget effect.

This same test can be repeated on the same solution for several days and the results will decline at a linear rate providing there was micron gold present. According to Stokes law on settling velocities of particles in solution with a "k" factor applied to platelet or flat particles, one micron particles settle at the rate of one micron per minute. Smaller gold particles settle slower.

To provide a test for yourself make openings in the drum part way down before reaching the sand or slurry settlings and take samples of the liquid at these points at various intervals. Compare the results with your top samples to obtain some idea of settling times for the particular ore.

Just remember that when working with micron gold in water that the gold goes with the water. This applies whether it is being poured from a container or in the clay particles going downstream.

More practical methods of recovering the micron gold are being researched because of the vast amounts of micron gold in sediments. Obviously, most micron gold deposits are not high value. In sand and clay deposits high values are not required because there is no underground digging, crushing, grinding, etc. Therefore an ore containing .03 troy oz/ton ought to be a viable ore and economically feasible to have a go at it.

GOLD/SILVER WEIGHT CONVERSION TABLE
Copyright 1978 by G. M. Miller

	GRAIN	GRAM	DWT TROY	OZ TROY	LB TROY	OZ AV	LB AV	CARAT
GRAIN	1.0	0.06479	0.041667	0.0020833	0.00017	0.0022887	0.0001428	0.3240
GRAM	15.4324	1.0	0.6479	0.03215	0.0027	0.03527	0.002205	5.0
DWT TROY	24.0	1.5517	1.0	0.05	0.00416	0.0548571	0.0034285	7.7755
OUNCE TROY	480.0	31.10346	20.0	1.0	0.0833	1.09714	0.06857	155.51
POUND TROY	5780.0	373.241	240.0	12.0	1.0	13.1657	0.82286	1866.12
OUNCE AV	437.0	28.3495	18.2297	0.9114883	0.07595	1.0	0.0625	141.75
POUND AV	7000.0	453.592	291.667	14.5833	1.21107	16.0	1.0	2267.96
CARAT	3.168	0.20	0.03215	0.0064304	0.000536	0.007055	0.000441	1.0

Abbreviations: AV = Avoirdupois, DWT = Pennyweight, LB = Pound, OZ = Ounce

DIRECTIONS: To find conversion values, find basic weight in left-hand column and follow this line until it intersects with the desired weight. For example, if you desire to determine how many 'ounces Troy' there are in a 'pound Troy', locate POUND TROY in the left-hand column and follow it across until you intersect with OUNCE TROY and you will find '12.0, which is the correct number of ounces Troy in one pound Troy. Then, if you want to know how many ounces there are in 2½ pounds Troy, multiply 12.0 by 2½ and arrive at a correct figure of 30 ounces Troy. Another valuable way to use this table is for comparative purposes: for comparing avoirdupois against Troy, for example. If you want to compare the relative actual weight of a pound Troy against an avoirdupois pound, follow the POUND AV line across to the POUND TROY column and you can determine that a pound avoirdupois is equal to 1.21107 pounds Troy. Using the same method, on the other hand, you will find that an ounce avoirdupois is equal to 0.9114883 ounce Troy. Anyone who buys or sells gold or silver will find hundreds of valuable applications for this table which are not immediately apparent to the casual user.

Compliments of Exanimo Press

Reducing Gold Loss from Concentrates

by Dave Parkhurst

Most placer gold is recovered by processing equipment which utilizes gravity separation techniques and produces an initial concentrate of the combined heavy mineral content of the gravels. This "black sand" concentrate should be processed carefully to insure the highest possible recovery of the gold particles contained in the mixture. Gold losses from "cons" can be surprisingly high, especially if the person performing the final separation is unfamiliar with several basic processing techniques.

> **If a large amount of concentrate is panned at one time, most of the gold would not reach the pan bottom**

Experienced placer miners have found that maximum gold recovery from placer gravels can normally be achieved by setting the processing equipment so that all of the heavy minerals are extracted during the initial concentrating phase. This procedure reduces the loss of fine or flaky gold particles to a minimum, but it also produces a large volume of heavy concentrates. The following information is concerned with the handling and separation techniques required to insure minimal losses while extracting the gold values from the initial placer concentrate.

In order to illustrate the importance of potential gold losses from placer cons, particularly for those who are unknowingly sustaining the highest losses, I will cite a few examples from personal experience.

Over a number of years I have visited several hundred placer gold mines, ranging in size from small, pick and shovel prospects through dredging sites to operations processing well over 1,000 cubic yards per day. Wherever the final separation of gold from concentrates was performed by hand-panning (and in many where the cons were tabled or run through reverse spiral "wheels" or separators), I was always able to recover at least some gold from the discarded material (tailings) by panning. These losses ranged from about .5% to around 65% of the gold values contained in the intact gravels. These were "free" gold values, and the results were confirmed by both weight comparison and assay. Where amalgamation was used, both gold and mercury were present in the tailings - and in amounts which were sometimes greater than those being recovered.

In most cases the operators were either unaware of the losses or did not think they could be significant. This was true even where the losses exceeded 50%. In nearly every instance, the miners involved were surprised by the actual amount of gold being thrown away - particularly where large gold particles were being lost. At a couple of sites nuggets weighing between one and two pennyweights were found in the tailings and, at one site, the tailings assayed 4.8 ounces of gold per ton in three separate tests (fine gold, with amalgamation).

There are many factors responsible for these losses, but the major cause lies in the fact that gold particles do not settle rapidly in heavy and compacted materials. Consequently, procedures which work well for the primary separation of gold from bulk gravels can prove to be a disaster when applied to placer cons. There is an obvious reason for this.

For example, the specific gravity of gold in air is about 19.3 (pure) and that of quartz averages 2.65. This provides a weight ratio of $19.3/2.65 = 7.28/1$. When submerged in water, this ratio changes to $(19.3 - 1)/(2.65 - 1) =$ about 11.1 to 1. As a result, it is much easier to effect a separation of two minerals of differing specific gravities when the operation is carried out in water, since there is a much greater separating effect between the heavier and lighter components involved.

Now, let's look at the same factors as applied to heavy concentrates. Placer concentrates are a mixture of minerals having higher specific gravities but, as a general rule, iron oxides predominate even though other heavier minerals are present. So, in this example, let's use the specific gravity for magnetite and hematite (roughly 5.2) in the preceding formula. In air, this produces a ratio of: $19.3/5.2 = 3.7$ to 1; in water, it becomes: $(19.3 - 1)/(5.2 - 1) = 4.35$ to 1. The weight-ratio difference of gold to quartz as compared with gold to concentrates (in water), therefore, has been reduced to a little more than half of the ratio of gold to quartz in air. And this does not take into account the fact that the average specific gravity of most concentrates is likely to be significantly higher than 5.2

Now, imagine using the same panning techniques utilized for bank-run gravels While panning placer cons. It becomes obvious that some of the gold could not possibly settle to the bottom of the pan and, if a large amount of concentrate is panned at one time, most of the gold would not reach the pan bottom. This it particularly true for rough or flat gold particles (even larger pieces) and it also applies to heavy cons which are processed fin spiral concentrating and separating equipment.

Then there is the placer miner's Nemesis - the magnet. lf used to remove magnetic minerals from heavy, compacted cons for easier handling, the magnet alone can produce losses of over 50%. This is because magnetic lines of force attract material located beneath the gold particles and lift them out with the unwanted minerals. Besides, most platinum group minerals are slightly magnetic and, with gold, are commonly alloyed with minor amounts of iron.

GENERAL RULES

The preceding information is intended to illustrate Just how important it can be to handle and process concentrates carefully. The following general rules are cited as means of reducing gold losses from placer cons:

1) Only pro cess smaller quantities of cons at one time. The lesser mass involved provides for easier handling and a much greater efficiency in gold recovery. Many continuous-feed placer operations process cons as fast as they are produced, thereby reducing volume, compacting and difficulty in separation. Also, a greater amount of lighter concentrate can be taken in the initial separation, reducing equipment losses.

2) Subject the cons to longer periods of shaking and vibration. This gives the gold particles a much better opportunity to settle out of the mass.

In nearly every instance, the miners involved were surprised by the actual amount of gold being thrown away.

3) Run concentrate tailings through a secondary recovery stage or process them more than once. This allows an opportunity to recover any values lost in the initial recovery stage, and it also provides a check on efficiency.

4) Run periodic tests on the head materials for comparison with the actual gold recovered by the processing equipment. This will give a good estimate of potential losses. Losses can be further identified by testing at each stage of the recovery process, including the testing of waste materials discharged and samples of the "cleaned" concentrate.

5) Take periodic samples from the cleaned concentrate for lab assay and spectrographic analysis. The assays should pick up most fine gold losses and the spectro can indicate the presence of other potentially valuable minerals.

PROPER PANNING

Since a large number of small miners and prospectors use the gold pan to separate placer gold from cons, a description of the proper panning techniques for this type of processing follows:

For heavy cons, place an amount of material no more than 1" deep, or less, in a large gold pan. Larger amounts of heavier materials tend to "pack" the mass, inhibiting the downward movement of gold particles. Then place the pan flat on the ground or the top of a 5 gallon bucket, and fill with water. In this position, sharply rotate the pan about 6" clockwise, immediately reversing direction for the same distance, and continue this motion as rapidly as possible without any upward or downward movement. Occasionally jerk the pan sideways one or two inches from right to left, in the same rapid manner. If the material does not slide smoothly or is packed in some areas, break it up occasionally with the fingers. This process gives a motion similar to a shaker table, and it will slowly settle the gold particles to the bottom of the pan. It also requires less physical energy since the pan is resting on a flat surface as opposed to being held by the panner. The procedure also works well in the initial stages for panning large amounts of bulk gravel.

After doing the above for 3 or 4 minutes, start tipping the pan outward while shifting the motion to rapid side-wise movements, until the cons reach the lip of the pan. Gently dip the pan into water until it reaches a point about 1" from the edge of the material closest to the bottom of the pan. Raise the pan quickly, allowing some of the surface material to be washed out. Until familiar with this technique, it is best to wash the tailings into another container so that it can be panned again to check for losses. Do not allow much material to be washed out each time. Start the sideways rapid motion again while tilting the pan slowly backward (with the material underwater), then gradually tip the pan outward as above until the material reaches the lip of the pan. Repeat this process carefully until only a small amount of cons remain in the pan. Do not allow the water to wash over the point where the edge of the black sands touches the exposed pan bottom, and do not move the pan violently or allow any "slopping" of the material in the pan.

This procedure will concentrate the gold in a narrow crescent at the edge of the pan bottom. Next, allow about 1" of water to flow into the pan, being careful that it does not disturb the remaining concentrate. With a slow, sideways shaking movement, slowly tip the pan back until it is level. Grasp one side of the pan, holding it level with one hand, and start tapping the side of the pan holding the cons, with the heel of the other hand. Allow the pan to dip backwards at intervals so the water slowly washes the surface black sands away from the gold. Avoid any violent movement of the pan.

As the gold becomes visible, tap the pan more rapidly on the side where the gold is located. Due to the higher specific gravity of the gold, the particles will "walk" away from the black sands toward the side being tapped, while the water moves the lighter materials to the opposite side.

With practice, this method can provide a clean separation of the gold from placer concentrates. The procedure is quite tedious if a large volume of cons are involved, but it practically eliminates gold losses and provides a clean and saleable product.

It should be noted here that if much fine gold is present, air should not be allowed to reach the gold particles as they will "float" on the water surface. This is due to the fact that gold is not "wettable" and will therefore be supported by the surface tension of water (the same reason heavy ships float on water). In this case, a small amount of dish soap (or other product) can be added to the water before the final separation is performed, in order to reduce the surface tension of the water.

Another important item to consider is avoiding contamination during the handling and separation procedures, particularly oils (either processed or natural). Oil will adhere to the surface of gold particles, reduce their specific gravity, and increase their tendency to float on water. Oils will also adhere to most other surfaces, and can cause gold to "stick" to the processing equipment.

In summary, if sufficient care is taken in the processing and handling of placer concentrates and the proper separation techniques are utilized, most gold losses can be practically eliminated.

GOLD On Mercury Copper Plates

by Jim Humble

At least once a week some enthusiastic miner calls us telling about the gold that they are getting on mercury copper plates. Either they are getting the gold out of water or out of ore in some way. A guy in Texas was even selling a small kit where one just dipped a mercury plate in some water that was used to leach a small amount of ore and gold would show up on the plates.

This experiment is easy to do. One merely takes a copper plate or zinc plate and coats it with mercury. He then dips the plate in some water or some ore and water. In many cases the mercury will coat over with what appears to be a beautiful layer of gold. A slightly more sophisticated version, one connects a battery or a small power supply to the copper plate using any metal as the opposite electrode.

We first ran into this phenomena about 10 years ago. I was interested in recovery of gold from water at the time. The beautiful gold covering on the mercury fooled us and it has fooled the best of them. We didn't understand all of the tests at that time to determine gold in mercury and thus it fooled us for some time.

Several years ago a couple of miners were absolutely convinced that this gold covering was gold. I assured them that it wasn't but they persisted calling me several times. Finally, I agreed to have them come here. We then went up into the local mountains and found a stream. There they put up 50 feet of copper plates which they coated with mercury. Sure enough) soon the plates were coated with the golden hue. We scraped the plates and had several pounds of the yellow coating and mercury. But there was no way to recovery any gold from the mercury. I had learned many new tricks about mercury and I tried them all. There just plain was no gold in that mercury.

We have spent a lot of time playing with mercury. There have been thousands of tests. We have been able to take gold out of mercury after it has been triple distilled. We have found that under some conditions the gold in the mercury dissolves with nitric acid and goes into the nitric and we have found ways of recovering that. We have gone over all of the patents in the area of recovery of gold from mercury. But eventually after repeated tries and many costly assays one has got to face the fact that he is not going to get any gold out of that mercury that has a yellow coating.

The fact is that the beautiful gold covering the mercury is not gold but mercury oxide. Mercury oxide is yellow and when coated on any white metal (like mercury) makes that metal look like gold. Look it up in your chemical dictionary.

The thing that fools many people is that mercury often has gold in it that will not come out until the surface of the mercury is subjected to an electric current. The gold sometimes then goes together in large enough particles to come out in a test. The surface of mercury does sometimes collect gold out of the water and that can be very confusing. But it is not the yellow coating. When the positive side of the battery is connected to the mercury the mercury will de-plate into the water. A mercury scum will form on the surface of the mercury and this scum will collect gold.

Don't get excited about that though. It is not a commercial process. In the first place the de-plating action causes tremendous amounts of highly poisonous mercury chemicals that can kill to be released into the water. Secondly, trying to collect and process huge amounts of the mercury scum would be almost impossible. And thirdly, there are several much simpler processes.

So, the next time you see a gold covering on the outside of mercury be very cautious. Gold collects on the inside of mercury. It sometimes catches on the surface, but we have never seen it completely coat mercury. Anything is possible, but be prepared for the worst.

Gold Farms of Peru

by Paolo Greer, Fairbanks, Alaska

I've been 3 years in South America, searching for old placers in a remote section of Upper Amazon in Southeastern Peru, the provinces of the Carabaya and Sandia, cover about 125x75 miles as the condor flies.

The following article by C. Woods was published in the 1930s in the Engineering and Mining Journal. It gives the gist of how they do it in Peru.....

GOLD FARMS OF THE INAMBARI

The "chacras de oro", or "gold farms", of the Inambari River district in Peru present one of the strangest and most ancient methods of gold recovery, and a description of the procedure is well worth a place in the history of mining, writes Clarence Woods, manager Inca Mining & Development Company, Tirapata, Peru.

Owing to special geological conditions, the summit and eastern slope of the high Andes north of Lake Titicaca have been covered with gold-bearing gravel, either moraines or ancient river gravel, with occasional quartz veins containing gold. The Inambari flows parallel with the main range of the Andes, and the summit separating it from the Tambopata River, on the east, is covered with a gold-bearing gravel. The formation is cut by a series of gold-bearing quartz veins, the most important of which is the Santo Domingo vein.

Inambari River and its tributaries rise in or near the gold-bearing deposits or in a slate formation containing high-grade quartz veins. These tributaries have their source at elevations ranging from 15,000 to 18,000 ft. and they flow into the Inambari at 2,000 to 3,000 ft. elevation, within a distance of 40 miles. They are swollen in the rainy season, and the gold is flattened out and carried through these box canyons, which act as sluice boxes with poor riffles. Most of .the gold is therefore deposited along the bed of the main river or in the bars and banks during the flood season.

Along the banks of the Inambari, between the high and low-water mark, are hundreds of acres of "gold farms". These consist of riffles formed by placing flat stones on edge and normal to the flow of the river. The riffles are held securely between rows of large stones* placed about 6 feet apart and wedged securely on a bed of fine sand, as shown in the Accompanying illustration. During the rainy season the "farms" are covered with water and fine particles of gold are caught in the riffles. When the river subsides, the riffles are taken out and the fine sand is panned in wooden bateas.

Many Indians are thus occupied for 2 or 3 months a year. Actual yield of gold per square yard has not been determined, but the fact that the work continues year after year shows that it must be profitable. A team of 8 men* working on a bar near the power plant of Inca Mining &Development during the dry season of 1930, averaged an ounce of gold per day. END OF ARTICLE

Most of the chacras are inherited. To prospect a new site a minero will search the river beaches panning with his wooden batea. If he finds colors and knows his stuff he will further define the area of gold concentration and then build his "gold farm" on the spot.

They are not so symmetrical as the sketch in the old article but they are constructed in roughly outlined 2 meter squares. Cobbles 1 to 2 feet in length and half as thick compose the rectangles and smaller, flatter slate rocks are overlayed on a bed of sand in the void. See sketch from article. The last act as riffles and are lain perpendicular to the stream flow and though almost vertical, are given a slight angle as one would the riffles of most sluices.

A wall of small boulders is interposed (dug in) between the chacra and the stream to prevent erosion in flood. Likewise, the upper end of the works is tapered to eliminate the purchase of friction that the current might gain.

When the farm is cleaned it is better to take out only as many squares as folks might wash the sand beneath. If the river rises meantime, that empty spot is a weak link and the rest of the workings still in place could go with it. On a clear day, above one's head, a 4 foot wall of dirty water bearing the remains of large trees may descend the creek without a hint (it did while I was there). With 300

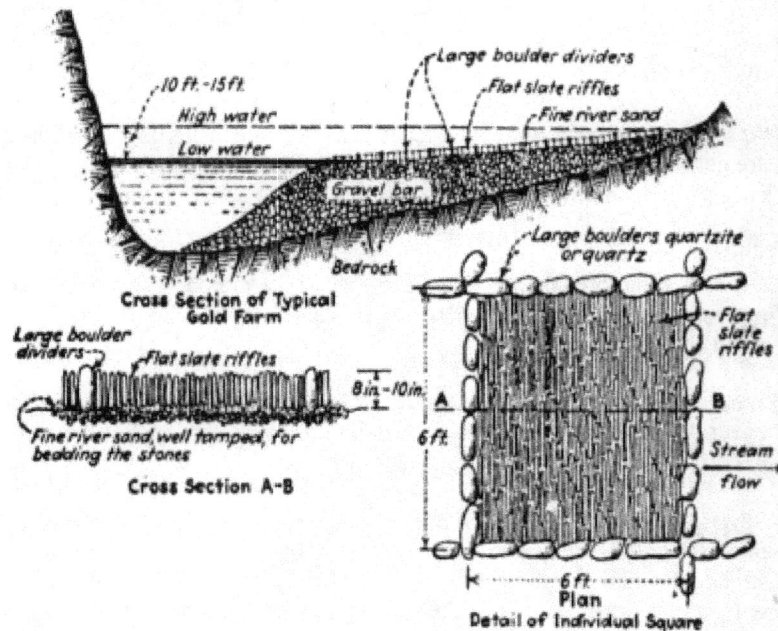

inches plus of rain a year and soft bedrock the topography is precipitous. This is high jungle and steep, short box canyons. Creeks become rivers in a moment increasing their volumes dozens of times and they are slow to subside.

One must also be cautious where to construct a chacra so he would find it in tact when the water subsides.

The chacras or "empedrados" are cleaned between floods in the relative dry season. Often, a small channel is cut in the back and enough water diverted from the river to work a small "ingenio" or sluice. Sometimes the ingenio is only a fleece (also a euphemism for the quest of Jason and his argonauts, who like the Mackenzie/Mellon/Pres. McKinley conspiracy at the turn of the century in Nome, intended expressly to rob the miners). Usually the sluice is a coarse wooden box lined with plastic, overlain with moss and sticks tied tightly in a row. It works about as well as a piano might serve for a raft. The gold is coarser than you might expect and the better empedrados give around .3 grams per square meter.

The Incas took most of their gold with patience, organization and hard work from the chacras. There is still an Inca road down to the remote area that I go. The road even continues beyond what the few locals are aware or history recounts (I've spent $30,000 of my own hard-earned in this research and though I have tried, I can find no one else who knows the mining history of the region nearly so well. Then too, for me it is a Quixotic search for the Holy Grail without material compensation and other folks are only more pragmatic). The old Inca road extension is only visible on aerial photo blow-ups (in which I've made a fair investment) and because of the great rainstorms, steep soft slopes and frequent mudslides that bring down the jungle, the Incas put their trails on the Knife ridges to last longer.

This is not the triple canopy of the lower jungle that one may walk beneath either. It is the "ceja de la selva" (eyebrow of the jungle). Every step gained is fought with a machete more slowly against the steep incline. In the mid 1700s the Indian slaves of the Spaniards uprose and especially in the outlying areas, the foreigners were massacred. Even by the last century the great placers of the Spanish and Inca empire before them were lost legends in the timeless jungle.

There is enough to explore, many of the folks there are good therapy to associate with and it's an easy way to lost 50 lbs. (I've done it 4 times) whether or not one starts out heavy.

Perhaps, uniquely, the gold in the region is specifically on bedrock of slate or within a couple feet of the surface and seldom concentrated in the least in between.

You Don't Need an Expensive Hot Plate

Jack Zillman, SJC, Calif.

Enclosed is an idea which is almost too ridiculously simple to send. However, it may be of use to some of your readers. I've enjoyed Popular Mining, having read every issue since its inception.

For a number of years, I used expensive laboratory hot plates, the current price of the least expensive units being about $85. Perhaps many have had the same thought which occurred to me about year ago when spillovers and acid fumes destroyed my hot plate.

The money saving thought 1 had resulted in the acquisition of a Hamilton Beach "fifth burner" hotplate selling for around $12 at the local discount appliance store, and of a 1/4" thick aluminum disc, purchased for $1.50 at a surplus metals yard. The disc is placed on the heating element (no big deal). A 3/16" steel disc works well as does a small cast iron frying pan. The disc thickness is somewhat important, as if too thin it will warp, depending on temperature, of course.

The temperature control functions just as well as on the expensive units. I have found no real difference in practicality between the expensive hot plate and the money saver, except for the consideration of maintaining centering of the disc on the element. This is no big problem, however. If desired, a screw which fastens the chrome-plated heat reflector dish can be removed and replaced with a countersink head screw of appropriate length and 2 nuts which enable the disc and reflector to be centered and anchored in place. It works quite well.

How to Check for Mercury in Your Ore

by Ernie Wells, Elgin, Or

In this article, I will explain how I determine it there is mercury in an ore that I am working with, to check <u>before assaying</u> how much caution I would need to use pertaining to the mercury and its fumes. However, I would still stress to avoid breathing the fumes when the door of a furnace is opened. I would certainly use a hood with an exhaust fan unless my furnace had a stack.

While there are many of us who perhaps use cautions that may not be entirely necessary, there are many who treat mercury like water. I had a friend pass on a couple of years ago and I will always believe it was the way he handled mercury. 1 tried to tell him he would kill himself but as he was older than me and stubborn as heck - he merely said, "I've been doing it for years this way and it has not hurt me yet." Well, he suddenly and unexpectedly went, but if I could have referred him to an article like the one in PM #1 and other issues he might still be around.

I have a hardrock (large low grade) mining property with very high grade pockets of mercury and some of it is native quick. I have been using preventative measures internally for <u>several years</u>.
As I am familiar with the handling of mercury (although much is still to be learned), I have owned a portable willamite screen for the last 11 years and use it regularly.

While I usually use my willamite screen here in my lab* I also carry it into the field as it is small and compact and I often like to check deposits for mercury right on the ground. It is very simple and anyone can do it in a few minutes. (It is very sensitive and will detect such small amounts of mercury that they are often missed in assays).

HOW TO MAKE IT
All you do is mortar up a small amount of the material in question. You don't even have to be careful about the screen size. Put the ground material in a flameproof container which will not burn - place the willamite screen behind the sample and heat the sample. You hold a short-wave ultraviolet light in front of the sample being heated and slightly above the bottom of it. If there is any mercury in the sample* you will see what appears to be smoke rising from the sample - showing on the screen.

IN THE LAB
In the lab I use a homemade holder which I made from an old microscope base with a clothespin bolted on to it for a stainless steel tablespoon. I clamp the handle of the spoon in the clothespin and set the screen up behind it under my exhaust hood. I put an alcohol lamp under the sample and hold a short-wave ultraviolet light in front of the sample. I turn off the lights and check to see if the sample contains mercury.

IN THE FIELD
In the field I use an inverter in my cigarette lighter to power my ultraviolet handheld light. I use the same base with a clothespin holder and the spoon - but 1 usually use a little propane torch for heat in the field. I do it after dark and have used the side of my Jeep Wagon for a backstop for the little screen, have used the inside of the rear of the vehicle with the rear door dropped, have used porches, tents, trees and even rocks. The main thing is <u>no wind</u>, having it dark and using caution in the event you have mercury so you do not breathe the fumes.

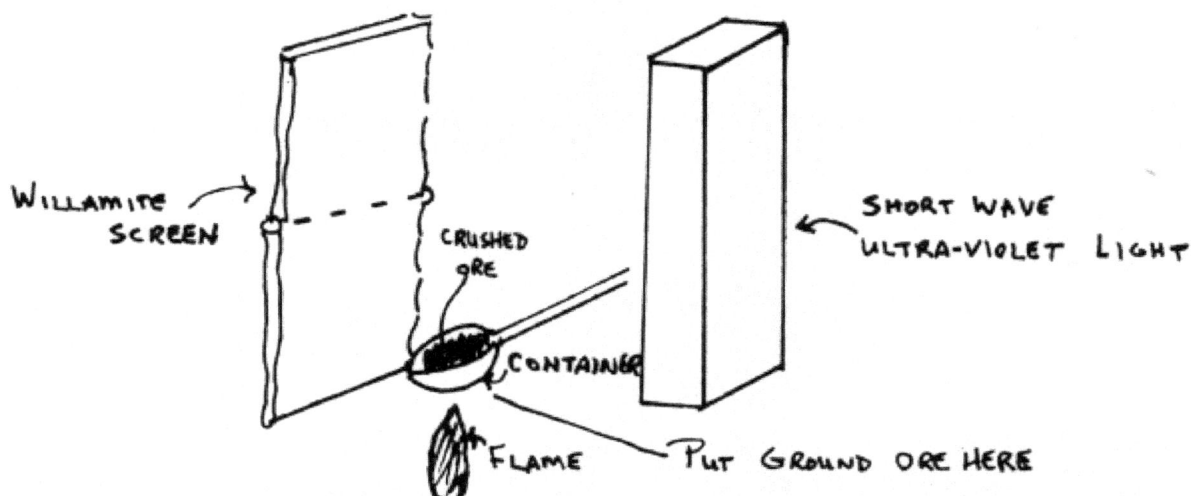

A Simple Scale You Can Build

by Oscar Bailey, Seward, Alaska

Here's a real fast, little balance that Oscar made and sent to me. You might have fun making it and playing around with it. Of course, it's not real accurate but it could help out on rough estimates and if you have a set of weights it'll be easier to adjust. Your kids would have fun with it, too.

As you can see from the picture, the scale is sitting on top of a wooden box. You store the parts in the box which makes it easy to carry around. When you are using the scale, you just set it up on the bottom side of the box.

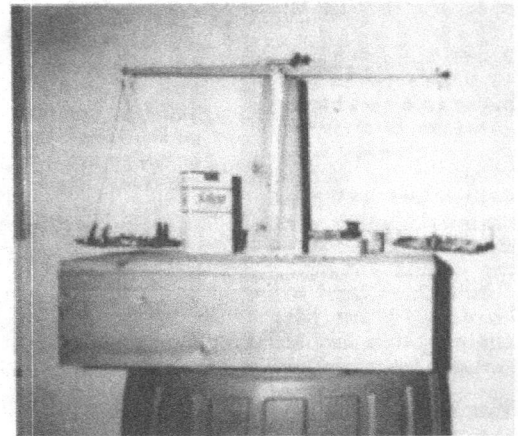

The pedestal is about 8" high and 1" wide, with a groove in the top to hold the pointer and tray bar. On the bottom of the pedestal you mark a center line and a scale like this:

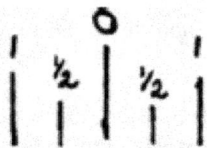

The pointer bar is made out of a piece of wood. On each end you stick a heavy-duty straight pin (one that has a large head) through the wood, glue in place and turn the pin into a hook. Get the pins in equal distances from the ends.

The pointer is cut from a piece of aluminum, and glued at the top of the bar in the center. Next to the pointer, at the top of the bar, you insert a 90° heavy wire (you can make it from a straightened out paperclip) so that you can hang and slide a weight along it to zero out your scale. (Just like you'd use in a triple beam.) The weight for zeroing is make out of aluminum - it's just a small piece folded into a V-shape.

The trays are made out of aluminum. Be sure to cut them the same size. Make them whatever size you want. Attach 3 pieces of string to each tray and tie them together at the top to another heavy-duty straight pin made into a hook. Be sure to have the strings on each tray the same length.

Now the trays can be hung on the hooks on the pointer bar and all you have to do is balance it out.
Pretty crude but fun to see how good you can get at weighing with it.

Characteristics of Particles
CHARACTERISTICS OF PARTICLES AND PARTICLE DISPERSOIDS

Particle Diameter, microns (μ)

	0.0001	(1mμ) 0.001	0.01	0.1	1	10	100	(1mm) 1,000	(1cm) 10,000

Equivalent Sizes

Ångström Units, Å: 1 — 10 — 100 — 1,000

Theoretical Mesh (Used very infrequently): 5,000 1,250 / 10,000 2,500 625

Tyler Screen Mesh / U.S. Screen Mesh: 65 35 20 10 6 3 ... 325 ... 100 48 28 14 8 ...

Electromagnetic Waves

X-Rays — Ultraviolet — Visible — Solar Radiation — Near Infrared — Far Infrared — Microwaves (Radar, etc.)

Technical Definitions

Gas Dispersoids — Solid: Fume — Dust
Liquid: Mist — Spray
Soil — Atterberg or International Std. Classification System adopted by Internat. Soc. Soil Sci. Since 1934 — Clay — Silt — Fine Sand — Coarse Sand — Gravel

Common Atmospheric Dispersoids

Smog — Clouds and Fog — Mist — Drizzle — Rain

Typical Particles and Gas Dispersoids

Rosin Smoke — Fertilizer, Ground Limestone — Oil Smokes — Fly Ash — Tobacco Smoke — Coal Dust — Metallurgical Dusts and Fumes — Ammonium Chloride Fume — Cement Dust — Sulfuric Concentrator Mist — Beach Sand — Carbon Black — Contact Sulfuric Mist — Pulverized Coal — Paint Pigments — Flotation Ores — Zinc Oxide Fume — Insecticide Dusts — Colloidal Silica — Ground Talc — Spray Dried Milk — Plant Spores — Aitken Nuclei — Alkali Fume — Pollens — Milled Flour — Atmospheric Dust — Sea Salt Nuclei — Nebulizer Drops — Hydraulic Nozzle Drops — Combustion Nuclei — Lung Damaging Dust — Pneumatic Nozzle Drops — Red Blood Cell Diameter (Adults): 7.5μ ±0.3μ — Viruses — Bacteria — Human Hair

Gas Molecules: O$_2$, CO$_2$, C$_6$H$_6$, H$_2$, F$_2$, Cl$_2$, N$_2$, CH$_4$, SO$_2$, CO, H$_2$O, HCl, C$_4$H$_{10}$

Molecular diameters calculated from viscosity data at 0°C

Methods for Particle Size Analysis

Impingers — Electroformed Sieves — Sieving — Ultramicroscope — Microscope — Electron Microscope — Centrifuge — Elutriation — Ultracentrifuge — Sedimentation — Turbidimetry — X-Ray Diffraction — Permeability — Visible to Eye — Adsorption — Scanners — Light Scattering — Machine Tools (Micrometers, Calipers, etc.) — Nuclei Counter — Electrical Conductivity

* Furnishes average particle diameter but no size distribution
** Size distribution may be obtained by special calibration

Types of Gas Cleaning Equipment

Ultrasonics (very limited industrial application) — Settling Chambers — Centrifugal Separators — Liquid Scrubbers — Cloth Collectors — Packed Beds — Common Air Filters — High Efficiency Air Filters — Impingement Separators — Thermal Precipitation (used only for sampling) — Mechanical Separators — Electrical Precipitators

Terminal Gravitational Settling* [for spheres, sp. gr. 2.0]

In Air at 25°C, 1 atm.
Reynolds Number: 10^{-12} 10^{-11} 10^{-10} ... 10^{-4} 10^{-3} ... 10^2 ... 10^3 10^4
Settling Velocity, cm/sec: 10^{-5} ... 10^{-4} ... 10^{-3} 10^{-2} 10^{-1} 10^0 10^1 10^2 10^3

In Water at 25°C
Reynolds Number: 10^{-15} 10^{-14} 10^{-13} 10^{-12} ... 10^0 10^1 10^2 10^3 10^4
Settling Velocity, cm/sec: 10^{-10} 10^{-9} 10^{-8} 10^{-7} 10^{-6} 10^{-5} 10^{-4} 10^{-3} 10^{-2} 10^{-1} 10^0

Particle Diffusion Coefficient,* cm²/sec.

In Air at 25°C, 1 atm.: 1 10^{-1} 10^{-2} 10^{-3} 10^{-4} 10^{-5} 10^{-6} 10^{-7} 10^{-8} 10^{-9} 10^{-10} 10^{-11}
In Water at 25°C: 10^{-5} 10^{-6} 10^{-7} 10^{-8} 10^{-9} 10^{-10} 10^{-11} 10^{-12}

*Stokes-Cunningham factor included in values given for air but not included for water

| 0.0001 | 0.001 (1mμ) | 0.01 | 0.1 | 1 | 10 | 100 | 1,000 (1mm) | 10,000 (1cm) |

Particle Diameter, microns (μ)

PREPARED BY C. E. APPLE

Reprinted from Stanford Research Institute Journal, Third Quarter 1961

TABLE 1

Prospectors' Maps

by Michael McKinley

Maps are a valuable source of information for the prospector. The proper use of maps can lead the prospector to areas that are very likely to be gold-bearing. These inexpensive reference tools can save the user large sums of money and valuable time. What one can learn about an area from good maps in a few minutes at home, might take days or even weeks to learn in the field. This makes it virtually necessary for every prospector to know how to use a map, and know which maps they can benefit from the most.

Geologic maps are maps which depict the various geological formations and rock-types in a given geological area. This is accomplished by coding each formation or rock type with a color and line pattern. The legend of the map will explain what the meaning of each code on the map is by outlining what geological formation or rock type each code represents. Included in the legend is the age of each rock formation. Geologic maps are usually superimposed over a topographical map which enables the reader to accurately locate any area of interest.

In addition to describing the various rock types and geological formations, geologic maps show the location of faults, folds, contacts, sheer zones, etc. Geologic maps are usually accompanied by geological reports which describe the area covered by the map. This report should be studied first, as it will tell which formations and rock types are associated with the occurrence of any gold in any specific area on the map. For example, the report may state that "several north-northwest striking systems of gold-quartz veins occur principally in slate". You would examine the legend to find the code for slate, then on. the map, locate the area of slate by matching the code in the legend with the code on the map. If the report goes on to say that the gold-quartz veins are found along fault zones, locate the fault zones within the slate area. You have now defined an area where the possibility of a gold occurrence is fairly high.

PLACER MINING

For placer miners, areas of alluvium, terrace gravel, and pediment are coded and displayed on the map. Dredgers can locate areas of heavy gold mineralization along rivers, and areas where the river bottom is soft bedrock that is likely to develop many cracks and crevices which trap and hold gold on its way downstream. Geologic maps and their accompanying reports can help lead the smart prospector to gold by identifying the areas where it is most likely to occur, and at the same time steer him away from areas where the geological formations and rock types are not associated with gold deposits. Geologic maps and reports are available from Federal and State Geological Agencies and copies of these can be found in some libraries. University libraries are likely to have many of these maps.

TOPOGRAPHICAL MAPS

Topographic maps are published by the U.S. Geological Survey. A topographic map utilizes a standard set of symbols to depict geographic (natural) and cultural (man-made) features on the surface of the earth. The special feature of topographic maps which set them apart from the others is the contour lines. These contour lines portray the varying shape and elevations of the earth's surface giving the map a 3-dimensional quality. Topographic maps are printed in many different scales, but the scale most useful to the prospector is the 7.5 minute quad. This map has a scale of 1:24,000, in which one mile measures 2 5/8", and the space between contour lines represents a 40 foot interval. These maps are ideal for the prospector as the scale makes these the most detailed and accurate maps available. The accuracy of these maps is fantastic. In order to meet the standards for horizontal accuracy not more than 10% of the tested map points can be more than 1/50th of an inch out of their correct position. For the 1:24,000 scale map this tolerance translates to a maximum error of about 40 feet on the ground. In order to meet the standards for vertical accuracy not more than 10% of the elevation test points can be more than 1/2 of a contour interval out of their correct elevation. This translates to a maximum error of 20 feet in elevation.

Many features are shown on topographic maps, but the ones of primary interest to the prospector are the following...land survey systems are shown as red lines which outline townships and sections. These lines help the prospector locate, plot, stake, and describe prospect areas and mining claims. Primary highways are shown as red lines. Secondary highways are alternately red and white lines. Light-duty roads, unimproved roads, and trails are black colored lines. These represent the access paths to any area you want to explore. The roads on a topographic map will show you how easily accessible a prospect area or claim site is. Mine symbols are in black. An area with a large number of mines is likely to be a highly mineralized area. Also some of these mines may be abandoned and not yet played out. They can be worth checking out. Contour lines, surface features and tailings are represented in brown. These tell you whether the terrain is smooth and flat, or rugged and steep. Tailings dumps often contain material which can be processed for a profit. Rivers and streams are blue. These will tell you whether water is readily available to a prospect or claim. And vegetation is represented as green patches. These green areas will give you information as to the extent and sometimes the type of vegetation growing in a specific area.

PLOT THE LOCATION OF GOLD-BEARING AREAS

Being able to properly use a topographic map will allow you to plot the exact location of good gold-bearing areas you find on a geologic map, and supply a lot of additional information about the prospect. Topographic maps and index maps are available from the U.S. Geological Survey.

Once a prospector has located a likely area to check out, he must determine whether the land is open to mineral entry. The BLM prints 2 types of maps to help the prospector make this determination. One is the Surface-Minerals Management Status map. The other is the Master Title Plats.

SURFACE-MINERALS MANAGEMENT STATUS MAP

The Surface-Minerals Management Status map portrays an area of one minute of longitude and 30 degrees of latitude. The map scale is 1:100,000 or 1.6 miles to the inch. It depicts areas managed by the BLM, Forest Service, National Park Service and others. Also shown are state and privately-owned lands. Restrictions in the management of federal lands are also shown. These maps include township and section lines, streams, towns, roads, trails, and other cultural and physiographic features. Some of them have contour lines but many do not.

The prospector uses these maps as an indication as to whether specific tracts of land are open to mineral location. Areas are color coded to indicate privately owned lands and public domain, and if public domain, it shows which governmental agency manages the land. A white area indicates land which is privately owned. Yellow areas indicate lands which are public domain administered by the Bureau of Land Management. Green areas indicate lands which are public domain administered by the U.S. Forest Service. Of primary interest to the prospector is the overprint on this map which shows the extent of federally owned mineral rights, or land that is open to claim. Minerals owned by the federal government are indicated by black vertical lines over the area. Where these lines appear over a yellow or green area, the land is open to location and administered by either the BLM or Forest Service. Where the vertical lines appear over a white area, the land is privately owned, but the govt. has retained the mineral right.

There are other symbols which may represent restrictions on mining activity in areas open to mineral location. There are many areas administered by the BLM and F.S. with federally owned mineral rights, that are power withdrawals. This is represented by red vertical wavy lines running over the black vertical lines. In this case the prospector would have to notify the managing agency if he stakes a claim in this area. He would then be notified by the agency as to whether he could mine the land, and if there are restrictions on the method of mining employed.

The Surface-Minerals Management Status overprints are published as general planning and management tools. Because of the possible lack of information regarding private land which has been re-acquired by the government, some land which is open to mineral location may not be shown on the map. Some tracts of land of 40 acres or less that are open to mineral entry may also not be shown because of the map scale. The maps are also not kept up-to-date. They should be used only to give a general idea as to the status of the parcel of land in question.

MASTER TITLE PLATS

For the final determination of land status Master Title Plats (MTP's) should be examined. MTP's are available from the BLM. They should be consulted before any prospecting or claim staking to determine whether the land is open to such activity, and to avoid trespassing on privately owned lands or public lands closed to mineral entry.

The information displayed by the MTP's is virtually the same as the Surface-Minerals Management maps, except for some very important differences. The maps are kept up-to-date. There are no omissions due to map scale as each MTP depicts one township of 36 square miles. And the mineral rights re-acquired by the government are also shown.

The MTP illustrates title information by the use of various weights and types of lines to show the area affected. Each weight or kind of line represents a specific type of status of the land that it outlines. Each status classification is also shown by an abbreviated notation. A legend of abbreviations and symbols show the meaning of the different lines and explain the abbreviations.

ON MICROFILM

Microfilm copies of the MTP's may be inspected at the regional BLM offices and copies may be purchased over-the-counter or by mail. If you order by mail be sure to request a copy of the legend of
symbols and abbreviations. You'll need it. These maps can be pretty complicated without one.

Through the use of maps you can get to know an area fairly well even before entering it. This should make your prospecting efforts more efficient, enjoyable and successful.

A Simple Concentrating Table

by Ron Lukawitski, Victoria, B.C., Canada

Enclosed you will find a few pictures of a small concentrating table that I built that uses corduroy to recover the gold.

I have a 5" Treasure Emporium triple sluice dredge and I end up with over one hundred pounds of concentrates every time I go out. My table can clean it all up in a matter of 2 or 3 hours.

This table has a corduroy top

I think that the picture of the table is enough for someone to build one from. The holes in the PVC are 1/16" and the corduroy is attached to the sides with door edge guard for cars.

I wonder if any of your readers know where I can buy some Miners Corduroy? I have lots of magnetic and non-magnetic black sand. Are there any simple tests that will tell me if they contain any gold? - Ron.

Editor's Note: All the corduroy we have ever seen can be bought at any fabric shop. Just get a heavy corduroy. It takes an assay to determine if black sand contains gold if it cannot be panned.

The table with the water running

How to Wash Leach Residue

by   David Milligan, Tucson, Az.

Efficient recovery of gold from a leach slurry is a very important step in chemical treatment of ore. Only a clear solution can be directly treated to yield a high grade concentrate which can be smelted into dore'. A clear solution can be separated from the waste solids by many methods. These include filtration or clarification. When the slurry is high in solids content, settling of the solution may yield only a small quantity of liquid. This will leave large quantities of dissolved gold with the wet solids. Efficient washing of the solids can give nearly all of the dissolved gold in a clear solution for direct treatment to dore'.

There are several methods of washing solids to recover the dissolved gold. One method is to settle the leach slurry to form a clear solution and a solid material which contains solids and solution. The clear solution is removed and treated to recover the gold. The solid material is then mixed with wash water and settled again. The clear solution is decanted for use in the grinding or leach area. Additional water may be added to the solids and the process repeated with the wash water advanced in the wash circuit. In large processing plants, a series of thickeners treat solids with a wash solution which has been advanced. This is known as a counter current decantation (CCD) circuit. Adding more water improves washing efficiency; however, if too much wash water is added, the excess wash water must be treated and discarded. This creates inefficiencies in the processing of the ore and may cause a pollution problem.

If dry ore is being processed, if wash water is used in grinding and leaching, and if no excess wash water is to be treated, the proper amount of water to be added to the last wash state is the amount of water lost with the tails. This creates a balance between the water added and the water lost -a zero discharge condition.

An important calculation is the efficiency of the washing process. This calculation involves the ratio of clear solution to the solution associated with the solids during the settling process. This is often called the wash ratio (R). It also involves the number of stages of thickening (N). The soluble losses (L) represent the percentage of the dissolved gold left with the solids compared with gold in the feed.

Several industrial flowsheets have been studied. One grinds the ore in fresh water. A second uses dry grinding or wash water to grind or leach the ore. Other flowsheets are possible. These include a large circulation of solution between the leach and first thickener or a large circulation of solution in a washing circuit within the CCD circuit. Each of these flowsheets has a particular washing efficiency.

The first leach circuit to be studied uses a grinding circuit with fresh water or treated pregnant solution (figure 1). The solids from the grinding circuit are thickened prior to leaching. This reduces the volume of the leach tanks to a minimum. The soluble losses can then be found by the following equation:

$$\text{Soluble Loss (\%)} = \frac{100}{R^n + R^{n-1} + R^{n-2} + \ldots + R^{n-n}}$$

If the wash ratio is 1 (the balanced condition), the soluble losses can be easily calculated. The soluble loss values are show in Table 1.

Table 1
Soluble Losses

Number of Stages	Losses (%)
1	50
2	33
3	25
4	20
5	17
6	14
7	12
8	11
9	10

The difficulty of efficient washing with a CCD circuit becomes apparent from Table 1. A similar relationship also exists when filtration is used instead of thickening if similar conditions exist.

The second leach circuit to be studied advances wash water into the grinding circuit or uses dry grinding of the ore (figure 2). The ground slurry is thickened and leaching continued in the leach tanks. This again produces the minimum size leach tanks. The soluble losses can be found using the following equation:

$$\text{Soluble Loss (\%)} = \frac{100}{R^n}$$

In this flowsheet, the wash ratio must exceed one to recover any solution from the first thickener. If the wash ratio is 2 and if the wash solution into the last thickener contains equal quantities of fresh water and recycled treated pregnant solution, the circuit is balanced and a zero discharge condition will occur. The soluble losses can be easily calculated. The soluble losses values are shown in Table 2.

A noted improvement in the washing efficiency is shown by comparing Table 1 and 2. The improvement is a result of the improved wash flow-rates through the thickeners.

Table 2
Soluble Losses

Number of Stages	Losses (%)
1	50
2	25
3	12
4	8
5	3
6	1

Another method used to improve washing uses a series of thickeners and recycled wash water to recover the gold while a second series of thickeners uses clean wash to recover the leach chemicals.

Still another method leaches the ore in a dilute solution. This requires an enlargement of the leaching circuit.

Each of these creates a variable R value within the wash circuit and is more difficult to calculate the washing efficiencies.

As the complexity of these calculations increase, computers can become important. Many small miners have access to small personal computers. These are ideal for the calculation of the wash efficiencies. The next article will include a computer listing for the calculation of washing efficiencies.

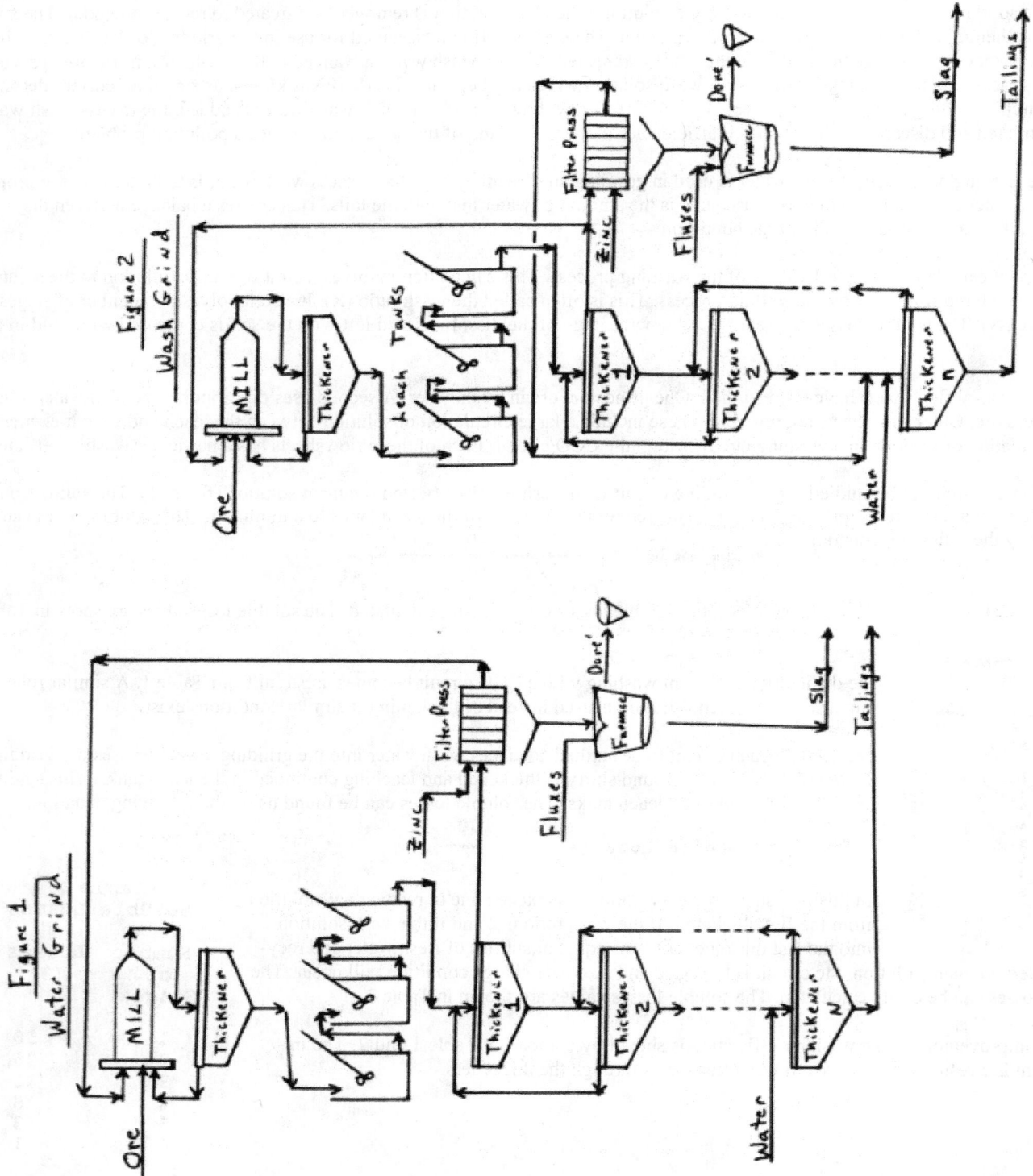

Figure 2 — Wash Grind

Figure 1 — Water Grind

Electric Gold Panner

by Chuck Galloway, Catawba, N.C.

Here is an electric panner made out of an old ice cream maker 1 bought at a flea market for $2.00, an old 5 gallon plastic bucket, an old hot water tank, and a frame I built.

The pickle bucket has grooves near the rim which provide a track for the rollers (use the rollers off of a chair or roller skate). They hold the bucket in place. Several small 1/2" slats were put inside the bucket to give it a churning effect like a cement mixer.

Simply bolt the top cap off the ice cream maker to the bottom of the bucket and set it at an angle. I made a frame from old 2"x4"s.

I used an old hot water tank to catch rain water, but you could hook your 1/4" water line up to any water source. Turn on the water and plug in the motor. Fill the bucket, with some material and let it run until the water runs clear and all of the lighter sands have washed out.

Old electric ice cream motors are plentiful and are strong, low rpm units that use about the same power as a 100 watt light bulb.

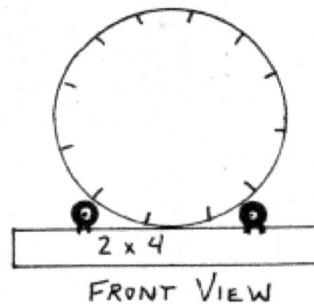

FRONT VIEW

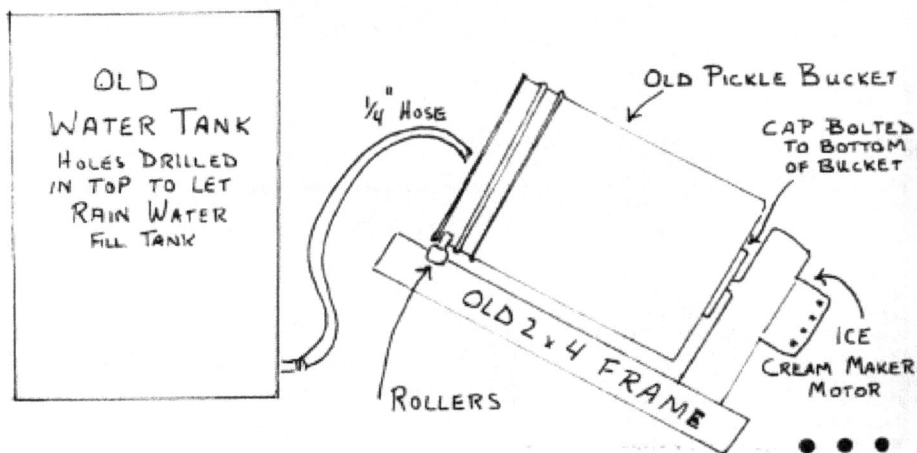

OLD WATER TANK
HOLES DRILLED IN TOP TO LET RAIN WATER FILL TANK

1/4" HOSE

OLD PICKLE BUCKET

CAP BOLTED TO BOTTOM OF BUCKET

OLD 2 x 4 FRAME

ROLLERS

ICE CREAM MAKER MOTOR

Railroad Switches You Can Make

by M. D. Isely, Trona, Ca.

A switch is used to divert a car from one track to another. An example is a track from a mine tunnel which branches; one track going to a tailings dump and the other to an ore bin. Switches can be made up on the site out of stock rail. In this way it is unnecessary to carry a supply of special switches to meet all the special requirements which are a part of mine railroading.

The part of the switch which is most likely to give trouble is the point of the tongue. This can be bent or broken off if it is not firmly against the rail when a car passes. Small fragments of rock, which are always present in mines, can hold the point away from the rail and cause premature switch point failure and sometimes derailments. It is good practice to carry a stiff fiber broom on each car to sweep out any switch which needs it.

Factory-made switches have points which are hardened on the running surface. They are sometimes made of a special railroad alloy containing more than 10% manganese. It is common practice for the entire switch assembly to be mounted on a large surface grinder where all running surfaces are ground true. Such switches serve for many years on high speed rapid transit lines where they may be thrown one way or the other every 90 seconds or so during the rush hour. Such a switch might cost more than an entire mine railroad. You don't need such a switch in a typical small mine. The speeds are much lower and so are the loads and the trips over the switch are much less frequent. The switches I will describe would be utterly unsafe on a high speed railroad but will work well in a typical mine.

When the point of a switch is not against the rail, one axle may go one direction and another the other. This is known as splitting the switch. If the car is going slowly it will stop suddenly and can be backed until it clears the switch so that the switch tongue can be properly aligned. If it is going too fast it will jump the track and may turn over and spill its load. Under such circumstances miners have been known to address the car in language not usually associated with persons of refinement. This does no good so far as the car is concerned although it may make the miner feel better. The best way to deal with this problem is to prevent it by having the switch point firmly against the rail in the first place. SWITCHES WHICH ARE PROPERLY ALIGNED SELDOM GIVE TROUBLE.

TYPES OF SWITCHES

There are several types of switches. Each has certain advantages and disadvantages. The singletongue, double throw switch is the easiest to make but the tongue pivot at the heel is difficult to reinforce. Also the point is very thin at the tip and tends to wear more rapidly and break more easily for that reason. The tongue travels a considerable distance and tends to push loose rock or dirt ahead of it which can interfere with its closing completely. If you don't have much time and traffic is light this may be an appropriate switch. The tongue can be pushed across by hand and spiked in position if it is thrown only infrequently. This is switch "A" in the drawing.

Switch "B" has the advantage of having only a single movable tongue which travels a short distance. If the travel is not great, the heel of the tongue may be attached to the stock rail by means of standard angle plates. These can flex, back and forth, considerably without fatigue. This switch involves considerable in the way of guard rails; and filler blocks, to keep them spaced from the running rails, are advisable. Switches of this type are used on main line railroads and they were quite popular as street car switches 50 years ago.

Switch C
using wrench
to throw switch

and then immersing the

8# Guard Rail

8# running rail

Weld all rail joints

Inverted 8# Rail

2" Channel

Switch C

A

B

Switch "C" is an involved switch with 2 tongues and a 30 degree divergence in the 2 tracks within the switch. I deliberately set out to build as involved a switch as one is likely to find in a small mine and I made some use of machine shop facilities todo so. This would not be absolutely necessary but it helped speed the work.

The 2 tongues are curved as are the outboard rails with which they mate. Each tongue has a 1.5" cylindrical rod 1.75" long welded to the gauge side of the web of the tongue at the heel. (The gauge side is the inboard side of the track. The outboard side is known as the field side.) This is welded just above the base of the tongue. A similar piece is welded to the web of the guard rail just below the head. A 3/8 clevis pin connects the two through a hole in the center of each cylinder which was reamed after alignment had been positively established, This pin rests on the steel plate on which the heel of the tongue rests and on which it moves as the switch is thrown. The pin may be removed with vice grips or pliers. It is held in place by gravity and the tongue can be lifted out for repair or replacement when the pin is removed.

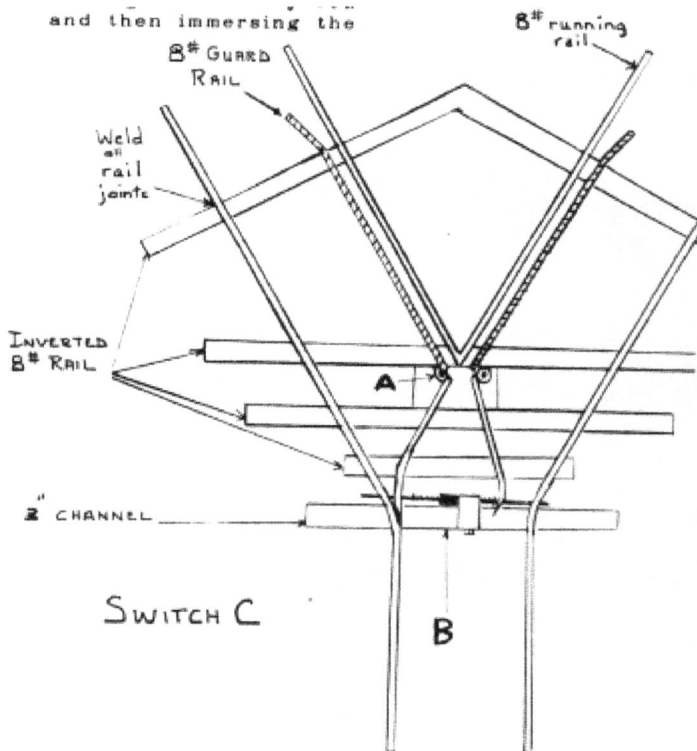

The tongue is supported by two inverted rails, the heel plate and a 3" channel, all welded to the bottom of the running rails. I hardened the point and heel of the tongue by heating to a cherry red and then immersing the part in a bucket of old crank-case oil. This isn't the best way but is better than having premature wear at those points. Another method is to set the tongue in a pan of water which almost covers the base and to then heat the head of the entire tongue by passing a flame along the top of the head. This takes more heat and equipment than I could provide. The tongue and, in fact, any part of the running surface of the rail which shows wear can be built up by welding and then ground smooth.

A 1.25" tube is welded to the underside of each tongue on the heel side of the shifting mechanism. I designed this mechanism to avoid the need for a switch stand beside the track in a mine tunnel. A sharp turn at a switch is most likely to be needed in a point where tunnels intersect. It has yet to be tested in a mine but it has worked well when a loaded car was pushed over it. It has the disadvantage of a sliding rod and two eccentric moving springs. It must be kept reasonably clean. It is quite compact and can be thrown, either way, by a single movement of a wrench which can be carried, instead of a switch hook, on a car. The wrench acts as a switch lock too. It prevents unauthorized throwing of the switch.

Notice in the section "B" detail the side view, (which is above the top view), that the top of the eccentric contacts the 1/2" plate which projects beyond the 3" channel. This keeps the eccentric from going further when the switch is thrown while the spring contacting the lower part of the eccentric is deflected against the 5/8" rod, thereby holding the eccentric from moving back. The spring may be any suitable spring. I used two old valve springs, from a Chrysler 318 engine, on each side. All dimensions shown involve 8# rail.

CONCLUSION
Look over these drawings and also examine switches installed in mines. You may find a design especially; suited to your needs. You may soon be designing! switches which will serve better than anything you have yet seen. Designing a little railroad where you don't have the Interstate Commerce Commission looking over your shoulder or the Association of American Railroads breathing down your neck can be a creative experience.

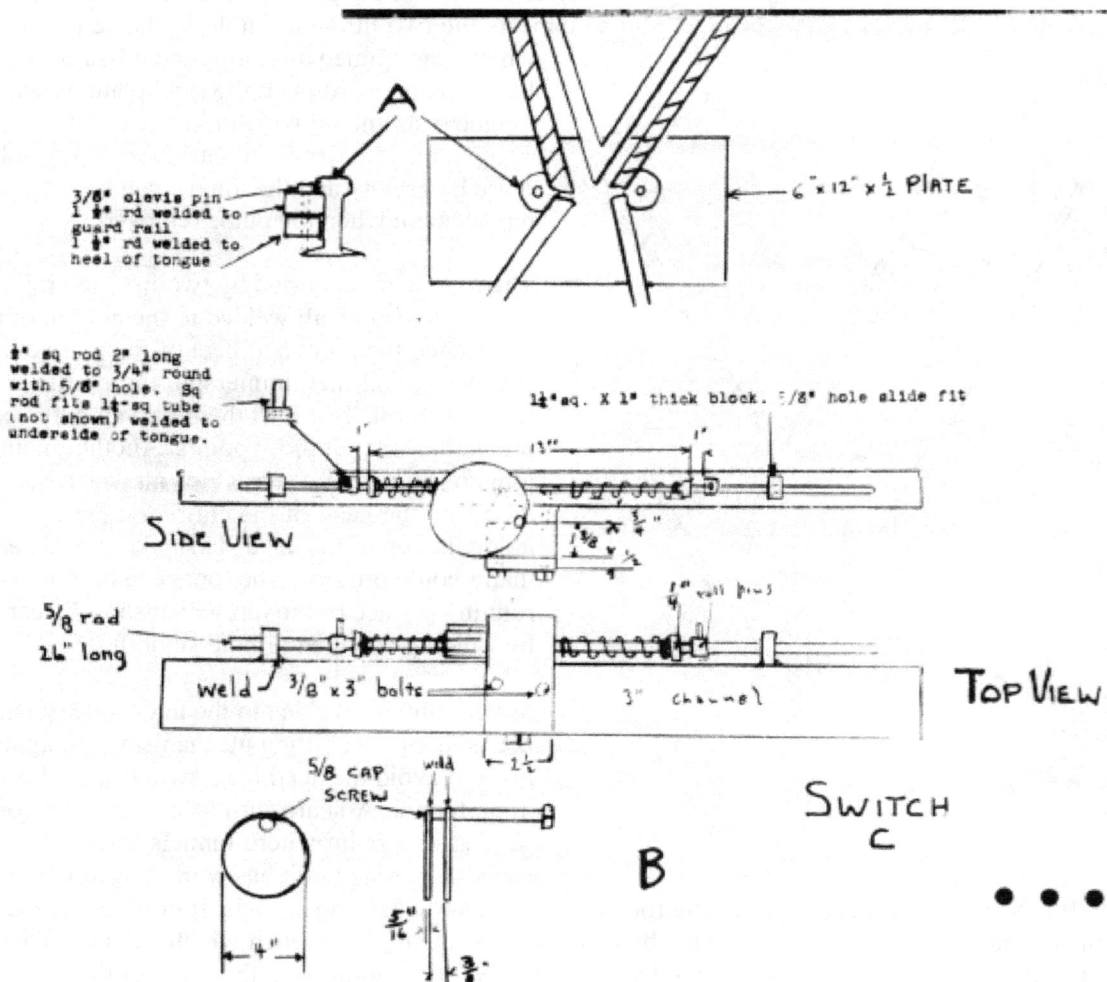

3/8" clevis pin
1 ¼" rd welded to
guard rail
1 ¼" rd welded to
heel of tongue

A

6"x12"x ½ Plate

½" sq rod 2" long
welded to 3/4" round
with 5/8" hole. Sq
rod fits 1¼ sq tube
(not shown) welded to
underside of tongue.

1¼"sq. X 1" thick block. 5/8" hole slide fit

Side View

5/8 rod
26" long

Weld 3/8" x 3" bolts

3" channel

¼" roll pins

Top View

5/8 cap
screw

weld

1½

Switch
C

B

• • •

4"

5/16"

3/4"

Extracting Gold from Water

by Karl von Mueller (Exanimo), Segundo, Co.

With respect to extracting gold from water, I have been involved in this since around 1957. IT IS VERY EASY, but there are problems. There are several ways it can be done, easily and quietly. In the beginning, we used ordinary automobile batteries, but the current drain was surprisingly high.

When alternators came into use on automobiles, we developed a "perpetual motion machine" that used a stream-powered alternator to drive a common General Motors alternator (a Ford or other automotive alternator will work just as well). The output is fed to a battery of electrodes that literally electroplate the gold, or silver, on every other electrode. More recently, we have been experimenting with 0.050 and 0.10 mil plastic sheets that are impregnated on both sides with carbon and other conductive materials' I think I would have the system "perfected " if I had the time to devote to full-time "cut-&-fit" testing.

I can tell you this: one man in Utah has had several of these installations in operation for several years. He averaged an ounce per week in the beginning using 2 units in the outflow from one mine. His initial problem was that he spaced the electrodes too far apart - 4 times what I had recommended. The good part of this system is that anybody can set up one or more units in a pregnant stream at very low cost by spacing them 50 feet apart (the units, not the electrodes) and the chances are that the last unit downstream will recover just as much as the one at the mine entrance. Three, six, or ten units in a pregnant stream is possible and profitable. If you build the units yourself, you can easily make them for $100 or less per unit. You run them just like a trapline.

One of many advantages to this method is that it is silent, clean, and inconspicuous. I have never yet heard of a Ranger or an "enforcer" finding and outlawing one of these units. There can be no squawk because they take "impurities" from the water. The only noise from a poorly designed unit is the "clop-clop-clop" of the paddle-wheel in the water as it drives the alternator. Relatively silent paddle wheel, or water-wheels, can easily be salvaged from junked squirrel-cage ventilation blowers. It is possible to drive several alternators with one large paddle wheel but the line loss of running current to the furthest cells is usually self-defeating.

IN REGARDS TO OTHER THINGS........FURNACE FIRING

With respect to firing a furnace, there are numerous ways to do it. Furnaces are easy and cheap to build in any of various types. One easy, and sometimes dangerous, way to fire a furnace is to use an ordinary garden spray to force a mist of gasoline against a baffle in the furnace. You need a relatively large flue with a barometric damper that costs around $15.00. Propane is the easiest means of firing a furnace. A valve to control the flow of propane and a long steel cone, with a built-in damper, to regulate the air will do the trick very well in experienced hands.

VIBRATION ON PLACER

Vibration is still the most promising method of extracting gold from placer material. It is more silent than most methods if done properly. It is cleaner, and the recovery rate is practically incredible. I am unusually certain that my "Vibrating Gold Concentrators" book has made hundreds (maybe thousands) of full-time placer operators out of weekend prospectors.

THE AINLEY BOWL

I don't recall ever seeing the Ainley Bowl mentioned in Popular Mining but I do want to make an observation about this device. When I was a kid at Farnam, Nebraska, Watt Ainley (a local farmer and inventor) and Chase Gish (the Ford dealer) developed this concentrator in the shop at the Ford garage and, believe it or not, tested it in the Platte River near Gothenburg, Neb. This is a long ways down the Platte from the Rockies, but they took out enough gold to encourage them. Later, Gish was the largest gold operator in Wyoming until he was shut down during WW II. Some of their units are still in use by small placer operators.

I enjoy Popular Mining very much and wish it came oftener but I understand the amount of work that goes into it. I bought the National Prospector's Gazette in 1954 and it was a lot of work and a lot of pleasure. I'm getting too old to get out into the field but I still enjoy visiting and writing about small mining. If I ever get caught up, which I won't, I'd like to submit some articles about small mining techniques that are today passe but still efficient. I plan to reprint two of Jack Douglas's books, who is mentioned in the #9 issue of PM. He was a prospector to ride the river with! The ranks are thinning of the real old-timers. Hardrock Hume, who was known and respected throughout the west, died a few years ago. Hardrock Hammond ws my pardner in the 1950s and 1960s on a number of mining projects. Cecil Banks is still with us but in very poor health at Lone Pine, Ca. Ainley and Gish have been "gone" for 20 or 25 years. Most of the old assayers are gone but in some cases their sons or aides have taken over.

I forecast the 1980 Ag and Au prices back in 1955 at a WMC meeting in Burbank, Ca. and they almost hooted me off of the floor. Now, I say that most of the younger people (under 60) will live to see $2.50 copper, $100 silver, and $1,000 gold. I could tell you why this will happen, but won't as I am often accused of being a "doomsdayer". Kindest regards, best wishes, and....Exanimo! Karl von Mueller, Exanimo Hacienda, Box 18, Segundo, Co. 81070

Silvering, Silver-plating, Desilvering

sent in by Charles Millikin, Johannesburg, Ca. The following is taken from an old book from 1907 of recipes for everything. It includes old-fashioned "trade secrets" on a great number of subjects. I thought it might be of interest. (EDITOR NOTE: we do not know the title of this book.)

BRITANNIA SILVER-PLATING
1) The article should first be cleaned and then rubbed by means of a wet cloth with a pinch of powder obtained by mixing together: nitrate of silver, 1 part; cyanide of potassium, 2 parts; chalk, 5 parts. Then wipe with a dry cloth, and polish well with rouge to give brilliancy.
2) By the electric method the metal is simply plunged into a hot saturated solution of crude potassium carbonate, and the plating is then done directly, using a strong electrical current. The potassium carbonate solution dissolves the surface of the britannia metal and thus enables the silver to take a strong hold on the article.

TO SILVER BRASS, COPPER, BRONZE, ETC.
1) In order to silver copper, brass, bronze, or coppered metallic articles, dissolve 10 parts of lunarcaustic in 500 parts of distilled water, and 35 parts of potassium cyanide (98%) in 500 parts of distilled water; mix both solutions with stirring, heat to 176° to 194° F. in an enameled vessel, and enter the articles, well cleansed of fat and impurities, until a uniform coating has formed.
2) Zinc, brass, and copper are silvered by applying a paste of the following composition: 10 parts of silver nitrate dissolved in 50 parts of distilled water, and 25 parts of potassium cyanide dissolved in distilled water; mix, stir, and filter. Moisten 100 parts of whiting and 400 parts of powdered tartar with enough of the above solution to make a paste-like mass, which is applied by means of a brush on the well-cleaned objects. After the drying of this coating, rinse off, and dry in sawdust.
3) To silver brass and copper by friction, rub on the articles, previously cleaned of grease, a paste of silver chloride, 10 parts; cooking salt, 20 parts; powdered tartar, 20 parts; and the necessary water, using a rag.

DESILVERING
1) It often happens in plating that, notwithstanding all precautions, some pieces have failed and it is necessary to commence the work again. For removing the silver that has been applied, a rapid method is to take sulfuric acid, 100 parts, and nitrate of potash, 10 parts. Put the sulfuric acid and the nitrate of potash (saltpeter) in a vessel of stoneware or porcelain, heated on the water bath. When the silver has been removed from the copper, rinse the object several times and recommence the silvering. This bath may be used repeatedly, taking care each time to put it in a stoppered bottle. When it has been saturated with silver and has no more strength, decant the deposit, boil the liquor to dryness, add the residue to the deposit, and melt in a crucible to regenerate the metal.
2) To dissolve the silver covering of a metallic object, a bath is made use of, composed of 66% sulfuric acid, 3 parts; and 40% nitric acid, 1 part. This mixture is heated to about 176° F. and the objects to be desilvered are suspended in it by means of a copper wire. The operation is accomplished in a few seconds. The objects are washed and then dried in sawdust.

PASTES FOR SILVERING
1) Carbonate of lime, 65 parts; sea salt, 60 parts; cream of tartar, 35 parts; nitrate of silver, 20 parts. Bray all in a mortar, not adding the carbonate of lime until the other substances are reduced to a fine powder. Next, add a little water to form a homogeneous paste, which is preserved in blue bottles away from the light. For use, put a little of this paste on a small pad and rub the article with it.
2) Articles of zinc, brass, or copper may also be silver-plated by applying to them a pasty mass of the following composition: first dissolve 10 parts, by weight, of nitrate of silver in 50 parts, by weight, of distilled water; also 25 parts, by weight, of potassium cyanide in sufficient distilled water to dissolve it. Pour the two together, stir well, and filter. Now 100 parts, by weight, of whiting or levigated chalk and 400 parts, by weight, of potassium bitartrate, finely powdered, are moistened with the above solution sufficiently to form a soft paste, which may be applied to the objects, previously well cleansed, with a brush. After this coating has dried well, rinse it off, and dry the object in clean sawdust.

RESILVERING
1) Take 100 parts, by weight, of distilled water and divide it into 2 equal portions. In the one dissolve 10 parts of silver nitrate and in the other 25 parts of potassium cyanide. The two solutions are reunited in a single vessel as soon as completed. Next prepare a mixture of 100 parts of Spanish white, passed through a fine sieve, 10 parts of cream of tartar, pulverized, and 1 part of mercury. This powder is stirred in a portion of the above liquid so as to form a. rather thick paste. The composition is applied by means of the finger, covered with a rag, on the object to be silvered. The application must be as even as possible. Let the object dry and wash in pure water. The excess of powder is removed with a brush.
2) The following is a process used when the jeweler has to repair certain pieces from which silvering has come off in places, and which he would like to repair without having recourse to the battery, and specially without having to take out the stones or pearls: Take nitrate of silver, 25 parts, by weight; cyanide of potassium, 50 parts; cream of tartar, 20 parts; Paris white, 200 parts; distilled water, 200 parts; mercury, 2 parts. Dissolve the nitrate of silver in half of the distilled water and the cyanide in the other half; mix the two liquids; next bray well in a mortar the mercury, Paris white, and cream of tartar. Preserve the products of these two operations separately, and when you wish to use them make a rather soft paste of the two, which apply with a little cotton or a brush on the portion to be silvered. Let dry and subsequently rub with a soft brush.

Increasing Mill Production

Here are some suggestions that might help a small operator. The suggestions have been not only tried but used commercially and they work and work well. Mill tests were constantly exceeded. This does not mean that in every case something and perhaps many things might work better.

One property we successfully mined and milled using gravity - by official reports it could not be done. The gangue was harder than the ore values we wished to save.

We installed a large high speed hammer mill and "impact shattered" the ore to 1/4 minus. Hammer mills have more maintenance than other crushers and they must be constantly maintained. (We had spare hammers with cat springs and hard facing welded on the striking surface always ready and changed the hammers after a set number of hours).

We took a 5 ft. conical ball mill (self-classifying harding) ended up turning it 30.6 rpm - built tube-type mechanical screen-type classifiers and attached this unit right to the mill discharge and had each size go to a separate set of Wilfley tables (one primary and one secondary in series).

Now for the very important step. We sorted out ball charge going from 3" balls down to 1" (the 1" won't stay that size and as you operate you will only need to add 3"ers), lowered the ball charge by approximately 15% which is under the recommended amount, and increased the water flow into the ball mill by approximately 4 times. The circuit was closed and the over-grind was constantly returned to the feed of the ball mill. It would take some experimenting to get the right amount of water for each ore body.

In this way we turned a 15-17 ton mill into a 40 ton mill! Your mill RPM must be adjusted to the proper ball cascade. You can use a very simple belt feed with an adjustable gate. Also the mill feed must be constant and the mill noise will tell you the exact load.

Hanging over each table was an adjustable flood lamp (adjustable for height and for any spot from halfway on the table to the discharge end) and they would swing to within 3" of the low side of the table. You find the exact location for each lamp on each table. Then standing back 15 to 20 feet from the tables on a platform slightly higher than the table tops, you can actually see if you are sliming any quantities of ore. The lights must be within a foot to a very few inches from the table tops.

Something to always remember. Do not run mud or very wet ore through a hammer mill unless you are prepared for very high maintenance.

The mill that used this process was located at John Day, Oregon.

Diver Control Switch

by C. Dennis Boring

I think everyone running a 6" or bigger suction dredge will be able to benefit from this. Maybe others too. The diver can control the run/idle capability or use it as a signal device too. Possibilities are endless! The drawing is quite self-explanatory.

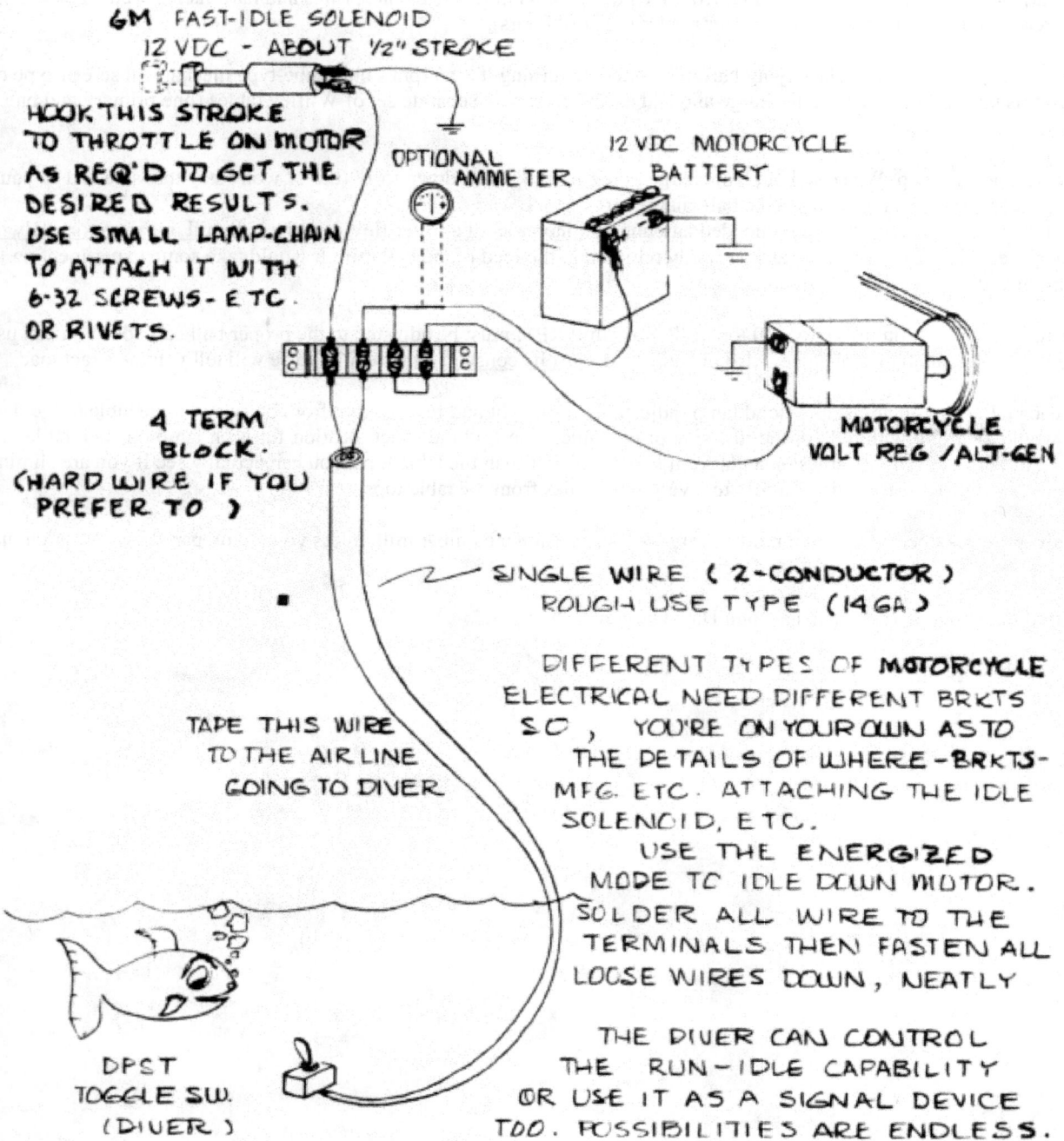

GM FAST-IDLE SOLENOID
12 VDC - ABOUT ½" STROKE

HOOK THIS STROKE
TO THROTTLE ON MOTOR
AS REQ'D TO GET THE
DESIRED RESULTS.
USE SMALL LAMP-CHAIN
TO ATTACH IT WITH
6-32 SCREWS - ETC.
OR RIVETS.

OPTIONAL
AMMETER

12 VDC MOTORCYCLE
BATTERY

MOTORCYCLE
VOLT REG / ALT-GEN

4 TERM
BLOCK.
(HARD WIRE IF YOU
PREFER TO)

SINGLE WIRE (2-CONDUCTOR)
ROUGH USE TYPE (14 GA)

DIFFERENT TYPES OF **MOTORCYCLE**
ELECTRICAL NEED DIFFERENT BRKTS
SO , YOU'RE ON YOUR OWN AS TO
THE DETAILS OF WHERE - BRKTS -
MFG. ETC. ATTACHING THE IDLE
SOLENOID, ETC.
    USE THE ENERGIZED
MODE TO IDLE DOWN MOTOR.
SOLDER ALL WIRE TO THE
TERMINALS THEN FASTEN ALL
LOOSE WIRES DOWN , NEATLY

TAPE THIS WIRE
TO THE AIR LINE
GOING TO DIVER

THE DIVER CAN CONTROL
THE RUN-IDLE CAPABILITY
OR USE IT AS A SIGNAL DEVICE
TOO. POSSIBILITIES ARE ENDLESS.

DPST
TOGGLE SW.
(DIVER)

C. DENNIS BORING
5-1-84

The Mac-Mac Carbon/Resin Incinerator

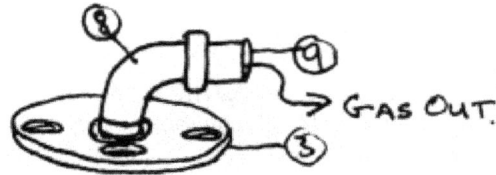

→ GAS OUT.

PARTS.

① BODY BARREL
①A FLANGES DRILLED FOR BOLTS.
② ASBESTOS SEAL.
③ TOP COVER - WITH GAS HOLE.
④ BASE PLATE - WITH AIR INLET.
⑤ STEEL MESH DISC.
⑥ SCREEN SUPPORTS. (3)
⑦ LEGS. (3)
⑧ INLET/OUTLET ELBOWS.
⑨ NIPPLE- HOSE CONNECTORS.
⑩ AIR CONTROL VALVE.

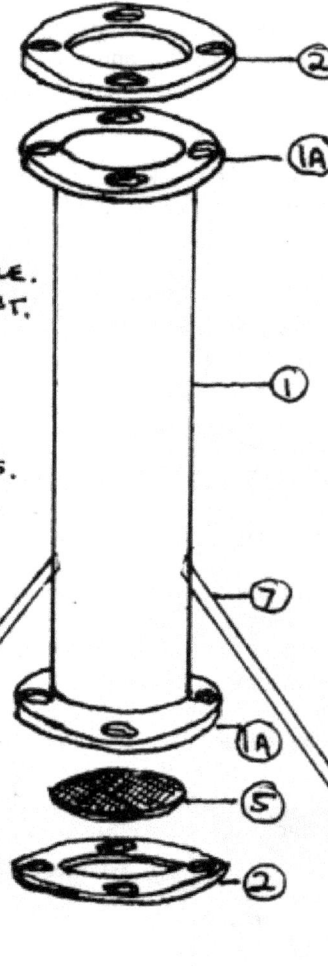

OPERATION.

Ⓐ BOLT BASE PLATE SEAL AND MESH SCREEN IN POSITION.

Ⓑ CONNECT INLET TO AIR SUPPLY (HAIR-DRIER OR COMPRESSOR.)   ⑩

Ⓒ IGNITE BARBEQUE CHARCOAL LUMP WITH TORCH AND DROP ONTO MESH-SCREEN.

Ⓓ GENTLY 3/4 FILL BODY WITH CARBON OR RESIN.

Ⓔ BOLT DOWN TOP COVER + CHECK SEALS.

Ⓕ CONNECT HOSE TO GAS OUT ⑨ INSERT OPEN END IN CONTAINER OF WATER.

Ⓖ INCINERATE UNTIL COOL. RECOVER RESIDUE FROM BODY AND WATER CONTAINER.

Ⓗ FLUX AND SMELT RESIDUE.

Ⓘ RE-INCINERATE MESH (PLUS) RESIDUES WITH NEXT BATCH.

MAC-MAC CARBON/RESIN INCINERATOR.

NOT TO SCALE.

[signature] OCT 85.

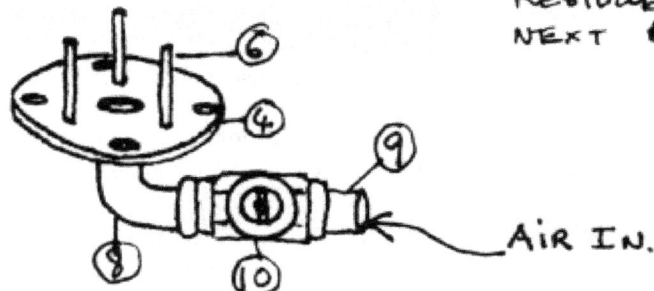

AIR IN.

• • •

Let's Talk Clean-up - Stamp Mills

by B. W. McKibbin, South Africa

For the small worker or the weekender, CLEAN UP is the most exciting and potentially lucrative form of gold recovery available in the eighties.

WHAT IS IT?

There are many forms.....let us consider a well known example. All mining folk have at least seen a picture of the famous California Stamp Mill. Used throughout the mining world to crush rock into sand in order to access the Au. Many are still operating effectively all over the world. As many as are still in use; there are dozens, derelict in bush or desert. Lonely monuments to brave miners of yesteryear.

Consider this....those tough old diggers were not recreating at a holiday camp. They were seeking gold at $20.00 or less per oz. and to make ends meet, needed 5-10 oz. per day, to warrant the cost of erecting and operating that lonely mill. They were working only rich ore by today's standards.

HOW DID IT WORK?

Simple, ore was mined by blasting out of mother earth. The broken rock was conveyed to the stamp mill which pounded it to fine sand in a cast iron mortar box. This was done by dropping heavy stamps (up to 1250 lbs. each) onto the ore in the box.

Mercury was generally placed into the box to "catch" the freed gold. The newly crushed, fine sand was then washed through a screen and over a mercury-coated copper plate to trap additional freed gold. Below the plates, one could find a mercury trap and often a corduroy strake table, also to trap very fine gold. This happy enterprise stamped away, 24 hours a day, delivering up to 50 tons per day of sand and as many ounces of gold per day. Often year in and year out.

So what? Well, tough as those old miners were, they were also mortal folks like you and me, perhaps more so. They WASTED, knowingly and unknowingly. (See D. Parkhurst, Reducing Gold Loss From Concentrates). Think....every time that mortar box cover was lifted (perhaps 2 or 3 times a day) to check for blockages on the screens, or to move a stubborn chunk of quartz, maybe (not always, but maybe) 1/4 oz. or more of gold-loaded mercury splashed out. Ever tried to recover a globule of mercury that dropped to mother earth? You are getting the message; the same sort of thing happened on the copper plates.

Picture this scene......2 AM frosty morning, the dog-watch plate minder, whose job it was to see that the mercury on the plates remained bright and fluid and to ensure that the screens kept free of blockage, correct water flow, etc. is on the job. Now, it's as cold as a frog's hind tit, so he stokes up the fire in the half 55 gallon drum and stretches out to warm the "'ole of 'im". ZAM; he's out. Later.....3:30 AM fires down to cold ashes, the chill creeps into his bones and slowly he emerges from slumber. The stamps are droning away, lovely relaxing vibration, except for the damn cold. Blow up
the fire, flames dancing merrily, toast butt and hands.....ah, better. Check the plates and screens. Oh Jeez... blocked , sludge pouring over the top, plates are sickened and lumpy. A large chunk of "silver" broken away from the plates. Quick, clear the screens, it'll take 45 minutes, then scrape and redress the plates. Hurry, hurry. Be lucky to finish before the day-shift takes over. If the boss finds out....that's tickets. Do you, dear reader, believe this did not happen once, twice, three times per month? I know it happened often.

Here's another scene......
Saturday night, week to go out to plate cleaning. Got to ride into town for a drink of rye with the boys from "The Golden Snake". Problem, all last month's "dust" was used to pay for that new haulage cable. Oh, what the hell; scrape a shovel full of amalgam off the plates. Stoke up that same drum fire, shovel on, to burn off the mercury. Cannot hang around these fumes, just go and check the ore bin and feed hopper whilst that cooks up. Should give enough for a bottle or three of red-eye. Result...fires too hot, mercury spits, shovel tips, 50% of amalgam drops into fire and on through to the ground. Damn, kick some dust over the mess. If the boss finds out...Sure it happened. Still does to you and me, right? No problem now because this is where we score on clean up.

Ultimately that mine was abandoned and that old mill was left to the elements. Muse over the many possible reasons., .war, death, bankruptcy, too many shovelsful of red-eye. Does it matter? Right now it is a golden opportunity.

PLAN YOUR ACTION AS FOLLOWS:

MORTAR BOX

Now rusted and flaky. Well, chip, scrape, steel brush and do it all again. Down to bright shiny metal, inside and out. Be meticulous, take it apart. It's hard work and uncomfortable to get into the corners, but get on with it. Save every speck of rust and dust.

STAMPS

Like the box, scrape, chip, brush. Give particular attention to the slots and slits in the stamps and stamp heads. Use a screw driver and broken hack-saw blade. Get it spotless.

POWER BLOCK
That's the foundation. Sometimes concrete. If timber (better) scrape out every crevice, wash it down and save every grain, every speck. Now do it again, but really do it thoroughly this time.

FEED END
Dig up 15 yards square, right up to the power block timbers/wall. Go deep...6 feet. Pan a handful, if the merest speck of colour shows, go deeper. That 24 hr. per day dropping of 1250 lbs. of stamp every minute vibrated the earth for 50 yards around. So, six feet is shallow. Remember, a micron of gold still has a specific gravity of 19 and this tends to go down, helped by that vibration.

DISCHARGE END
Dig up under the old plates and cord, table positions. Down to 12 feet or bedrock, whichever comes first. Go deep and wide, those mortar box splashes went out 15 feet sometimes. Watch out for mercury pools, 132 1/2 gallon is not uncommon and to see it trickling into your new excavation is most frustrating. It is remarkable how bright and silvery fluid it remains, after 50-60 years underground.

TIMBERING
If it is a really old stamp, rotted and dangerous, it may be wise to pull her down. This is a sad operation, almost like desecrating an old gravestone and I don't enjoy it. However, should you decide it safer to pull her down, do take care. You are dealing with heavy machinery and if pinned down you will definitely be badly maimed, if not killed. A melancholy note, so be warned. Saw up all the timbering over a plastic drop sheet. Man, this is really hard work, even with a chain saw. Incinerate the wood, every last bit, saw dust and all. (See Mac Mac Incinerator at beginning.) Remember, an open fire loses gold , which seems to vaporize or get carried off on the oily smoke, like float gold out of your pan.

METAL PARTS
All the metal parts will be rusted and flaky, especially bolts and stays holding the timbering together. Soak them all in a plastic or stainless steel tub. Use hydrochloric acid...pool acid works fine. Wash that lot real good, brush off all flecks down to clean iron. Use a face mask, rubber gloves, goggles, and apron. Acid is dangerous.

FINAL RECOVERY
Wood ash....smelt.

Metal scrapings, rust, flakes...... smelt.

Excavated material...heap leach with cyanide or thiourea, or concentrate and sell.

When panning or concentrating, watch out for quartzy grains of white sands which roll over the black concentrates in the pan. A magnifying glass will show this to be oxidized amalgam. It is 50% Au. which is porous and spongy and gets coated with water which seems to trap air and allow it to easily roll out of the pan. Quite big lumps, 1/4" square, can fool a greenhorn.

SAFETY FACTORS
Mercury is poisonous, deadly. Do not handle it with bare hands. Do not breathe the fumes when checking the spongy oxidized lumps by burning in an old spoon. Cyanide and mercury give off bad medicine, beware. Respect this and live happy and sane. Did you know that the "Mad Hatter" expression derives from Hatters utilizing mercury oxide to brush-up the pile on top hats? Most Hatters were scatty within 3 years of taking up the trade. Don't you get caught out. One of the first symptoms of mercury poisoning is drooling....like an idiot. Then, teeth drop out.

Those old stamps are mighty heavy equipment. Pull the king-posts down with a Turfor and cable and the mortar box and stamps simply totter at a crazy angle. You then march in through the dust and grime and the Number 3 stamp shaft, 2" of solid steel, 12 feet long, rusted these past 50 years, snaps off....and drives you into the dirt like a 6 inch nail. Not nice for the folks back home, or for your buddies. Plan it carefully.

RESULTS
Variable. A good site will yield 25 oz. per ton of excavation. Excavation should yield 100 tons, ie. 2500 oz. At $300. oz., that's big bucks. Not Mega but getting there. A 10 or 20 stamp should do it. Incidentally, don't be put off if the site has been previously dug up. Most old clean ups only went down 6 inches. There is still good value beneath that. I have taken 3 grams amalgam from 20 grams scalings in the slots of a five stamp erected as a national monument. Just a sliver of rust picked out with my pen-knife as a check, you understand? Each slot holds 20 oz. of gunge and there are 10 slots in a five stamp. Two hours work and one has some capital for a bigger scale venture. That's the beauty of CLEAN UP, no end of possibilities and hell exciting. TRY IT!

A Lift Mechanism for Your Pot Furnace Crucible

This is a lifting mechanism for your crucibles that are too big to be lifted with regular or oversize tongs. One quarter inch steel should be used for all sizes of crucibles up to 200 pounds. Three-eighths inch steel should be used on larger crucibles. The drawings are mostly self-explanatory. Most everything is built around your crucible, whatever size it may be. One set of tongs should be able to handle at least two different sizes of crucibles, possibly three. The furnace (earlier article) is a very good furnace to use with this mechanism.

Operation of this mechanism is reasonably simple. The tongs are scissor, but both handles come off on one side. The tongs are lowered over the crucible and are then locked closed by clamping the two handles together. One then turns the crank on the WORM GEAR DRIVE CABLE HOIST until the crucible has been lifted above the furnace. (It is important to used the WORM GEAR DRIVE CABLE HOIST, because it does not require a ratchet that might fail at a crucial moment. The worm gear cannot be budged from the crucible end). When the crucible is above the furnace, it is pulled on the roller until it is clear of the furnace and over the device for pouring. The crank on the WORM GEAR DRIVE CABLE HOIST is then turned, letting the crucible down to rest in the cradle of the pouring mechanism.

This kind of a lift mechanism, if built correctly, can be used for any size crucible, but do not try to pour the crucible from this lift mechanism. It requires a simple pouring mechanism that will be given next issue. The crucible sits in a cradle that is mounted with a swivel midways on the crucible. This allows it to be tipped easily with complete control and safety.

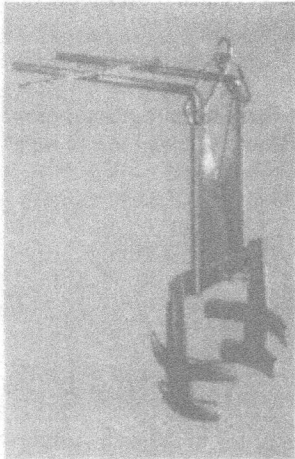

Be sure the neck of the scissor tongs is long enough to stick down into your furnace. If necessary the 5/8" rods can both be bent making a neck to stick into the furnace. Everything can be cut to fit and then taken to a welder for welding. In which case welding should not be more than $30.00.

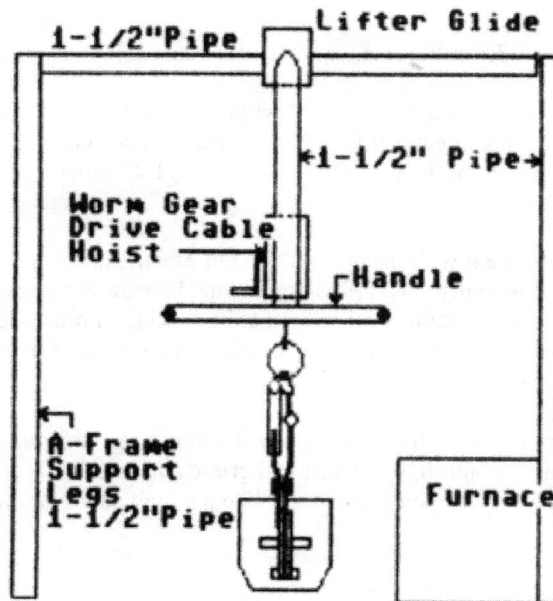

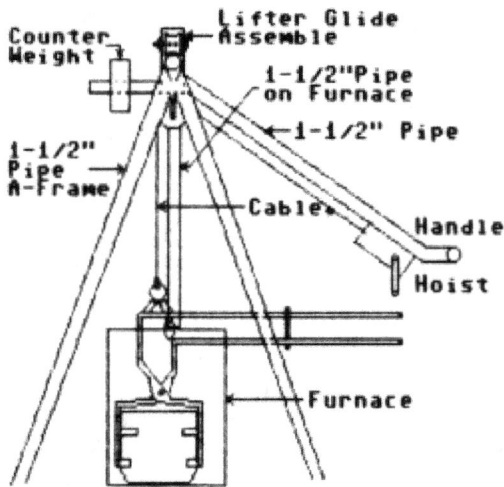

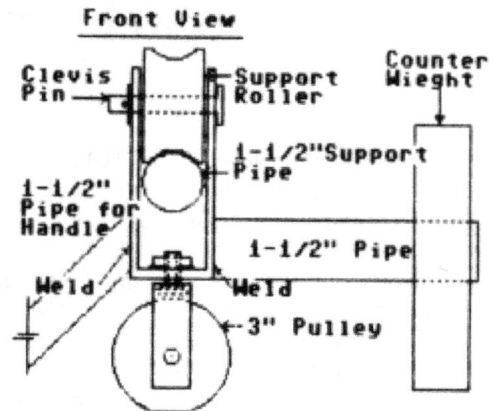

LIFTER GUIDE ASSEMBLE Ⓐ

Front View

Side View

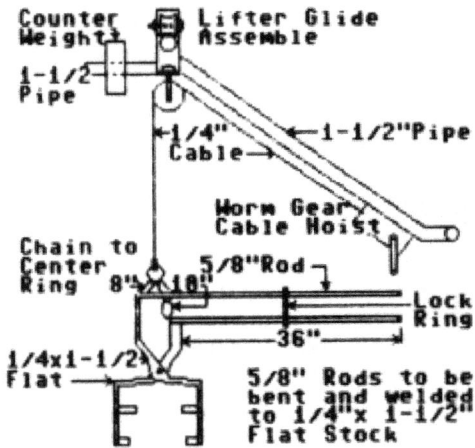

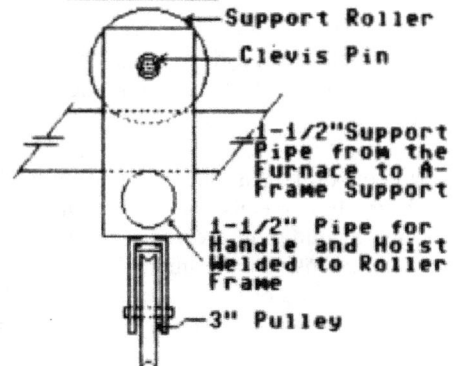

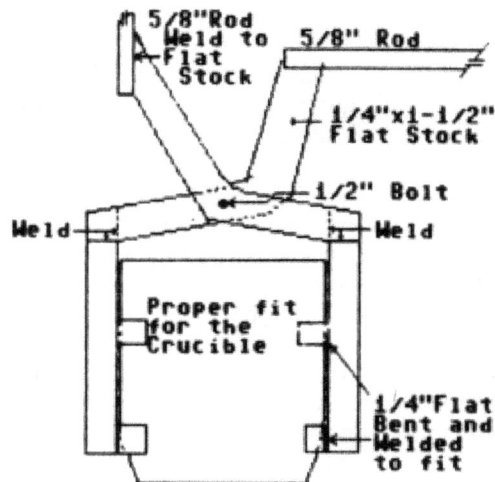

● ● ●

Determining Platinum
Group Metals
in Nickel Ores and Concentrates

by Jim Floyd, Texas

The following was published in November of 1940 by The Canadian Journal of Research. Although this seems to be a very dated article, the procedures are not extremely difficult and results will tell the practical prospector and miner whether or not the ore will contain PGM's (platinum group metals). Jim Floyd

"Most published methods for the determination of platinum and associated metals are not of general application to ores and concentrates, and their accuracy as applied to unknown material is in many cases seriously open to question. The methods here described are chiefly based upon experience with the products of the Sudbury area, but are believed to be applicable without substantial modification to all sulfide ores of nickel and to smelter and refinery products derived therefrom. Methods of concentrating the PGM's are described, an outline is given of an approximate method of determination based on sulfuric acid parting, and details of a more reliable method in which nitric acid is the parting acid are set out. The results obtained are compared with those of Beamish and Associates on similar material.

So far as the writer (F. E. Lathe) is aware, no simple and at the same time reliable methods for the determination of PGM's in ores and concentrates have yet been published. If such methods exist, those who might be expected to know most about them are apparently disinclined to share their knowledge with others. There is in the literature much that is uncertain, contradictory, or inapplicable to any particular case.

Gilchrist & Wichers of the National Bureau of Standards have published separations which appear to be almost as simple and accurate as those for the base metals, but their results were obtained in the analysis of solutions in which all the PGM's were present in the form of chlorides, a condition rarely if ever encountered in practice. Beamish and Assoc. of the University of Toronto have published the results of very extensive experimental work with the PGM's; their observations are of interest chiefly to the specialist. These authors did not find the methods of Gilchrist and Wichers entirely satisfactory. The methods described below are neither simple nor novel, but they are offered because they have given satisfactory results in the hands of the writer (and myself, Jim Floyd, I might add) and
various others to whom they have been made privately available from time to time. They are based upon experience with the Granby Consolidated Mining, Smelting and Power Co. and the British American Nickel Corp. and in particular with the last, where the writer for some years investigated methods of analysis and concentration of the PGM's. The B.A.N.C. methods were in turn based upon those used at the nickel refinery at Kristiansand, Norway, but the latter were found to require substantial modification when applied to the ores and products of the Sudbury district.

The problem is essentially threefold. First, while the assayer engaged in the determination of gold and silver in ores and metallurgical products (scrap), seldom encounters metals of the platinum group, he wishes to know whether, if they occur in any particular case, their presence will be detected in the ordinary course of analysis or if they will merely contaminate the gold and silver and be reported as such.

Second, when it is known or suspected that the PGM's are present, a partial or complete determination is required. The nature or amount of the sample available frequently does not justify an attempt to make an accurate determination of all the platinum metals; it is therefore of interest to know what information regarding the nature and quantity of platinum metals can be obtained with a minimum of effort, and in what respects the assays are likely to be in error.

Third, in the case of shipments of nickel ore, matte, concentrates, or refinery slimes, the amount of money involved is sufficient to justify the observance of elaborate precautions in the separation and determination of gold, silver, and as many metals of the platinum group as are present in appreciable quantity and will be paid for by the purchaser. In an extreme case, to be described below, the amount of platinum metals present was a deciding factor in the sorting of ore prior to smelting.

Evidence is not wanting to show that even assayers of wide experience may make serious errors in determining the metals of the platinum group. For example, a sample of the matte first produced by the B.A.N.C. (concentrate in a ratio of about 25:1 from the ore) was sent to one of the most reliable assayers in the U.S.; a certificate was received showing the approximate platinum metals content of the matte, together with an accompanying letter, in which it was explained that on such material the assayer could do little more than "express an opinion" based on the evidence available. Again, a sample was sent to a prominent refiner of platinum metals, who reported about 50% more platinum than palladium - whereas, palladium was actually present in the larger proportion. In a third case, shipments were made of B.A.N.C. refinery slimes containing 5% to 8% of the PGM's, and a maximum difference of about 1% (25% to 30% of the weight of each metal present) was found between the assays of shipper and refiner for both platinum and palladium. In this instance, it was eventually recognized that the shipper's methods were the more reliable. (Note: I, Jim Floyd, personally experienced similar situations in 1981-82; it seems little changed even after 40 years!).

PROPERTIES OF PLATINUM AND ASSOCIATED METALS
It is remarkable that wherever (so far as the writer is aware) nickel has been found in the form of sulfide, there are associated with it iron, cobalt, copper, silver, gold, and the whole group of platinum metals. The relation of these metals to one another is shown in Table I.

TABLE I
Periodic Grouping of Platinum & Associated Metals, with Atomic Numbers

Group VIII			Group I
Fe	Co	Ni	Cu
26	27	28	29
Ru	Rh	Pd	Ag
44	45	46	47
Os	Ir	Pt	Au
76	77	78	79

The chemist who assays for the platinum metals will do well to bear this grouping in mind, for it is closely tied up with the properties and separation of the metals. Ruthenium and osmium are the two platinum metals of highest melting points, but their tetra oxides (OsO_4, RuO_4) are readily volatile at slightly elevated temperatures. The order of increasing melting points of the metals - palladium, platinum, rhodium, iridium, ruthenium, and osmium - is approximately that of increasing resistance to acid attack. The last four of these metals (in the proportions in which they occur in nickel sulfide ores) are almost completely insoluble in both nitric acid and aqua regia, thereby providing a convenient method of separation from platinum and palladium, both of which dissolve fairly readily under the conditions to be described. In fact, palladium, the least noble of the whole group, is soluble in hot nitric acid, even when in dense form, such as foil. Nickel and palladium are precipitated by dimethylglyoxime (DMG), and platinum is also precipitated under certain conditions. Copper, silver, and gold are readily attacked by soluble cyanides, but are highly resistant to fused alkali (sodium hydroxide, etc).

THE DETECTION OF PLATINUM METALS IN ASSAYING FOR GOLD
The platinum metals will on rare occasion, be present in sufficient quantity to affect the appearance of the silver bead, but reliance should not be placed upon this method of detection. In small proportion they give the silver bead a slightly dull appearance. If the color of the gold bead obtained in parting with nitric acid is abnormal, the possibility of the presence of platinum metals should always be investigated.

Two other indications obtained in parting are of greater significance. If the proportion of platinum metals to gold be even a few parts per hundred, their presence will ordinarily be indicated in two ways - by "flouring" of the gold and by the reddish-brown color of the solution. If neither of these conditions be observed it may be safely concluded that little if any, platinum or palladium, the commonest metals of the platinum group, is present.

"Flouring" of the gold is the breaking up of the bead to such a state of fineness that it becomes difficult, if not impossible, to wash it safely by the usual method of decantation. The reddish-brown coloration is caused by solution of platinum and palladium in nitric acid. (Note: platinum can be dissolved by nitric when in alloyed form; this is also true when dissolving silver with platinum in it. J.F.) Palladium is particularly effective in coloring the solution (wine color. J.F.)

APPROXIMATE DETERMINATION BY SULFURIC ACID METHOD
When the presence of the platinum metals is thus detected it is usually necessary to determine at least the approximate total amount of metals of the group present. For this purpose it is desirable to run a separate assay, adding to the crucible or scorifier used at least 20 times as much silver as the total amount of gold and platinum metals believed to be present, - a greater proportion does no harm. The silver bead is then parted with concentrated sulfuric acid and the residue weighed as combined gold and platinum metals. If palladium be present with the platinum, as is usually the case, it will largely go into solution and will make its presence evident by the reddish-brow color of the solution. It may be determined by precipitation with DMG, as described below. An outline of this method is given in Figure 1.

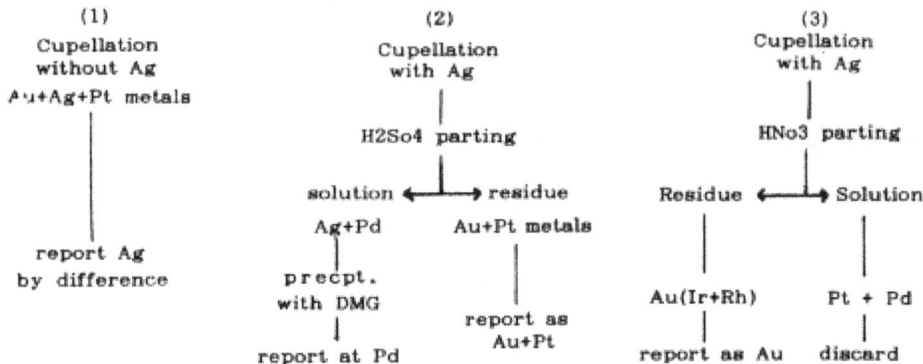

```
       (1)                        (2)                          (3)
   Cupellation               Cupellation                  Cupellation
   without Ag                  with Ag                      with Ag

  Au+Ag+Pt metals                |                             |
                           H2So4 parting                 HNo3 parting
                                 |                             |
                      solution ←——→ residue        Residue ←——→ Solution
                                 |                             |
                        Ag+Pd       Au+Pt metals   
      |                   |                           Au(Ir+Rh)      Pt + Pd
   report Ag           precpt.            |              |            |
   by difference       with DMG       report as      report as Au  discard
                          |           Au+Pt
                     report at Pd
```

FIGURE 1

If the quantity of metals of the platinum groups proves to be sufficient to justify further work, a partial separation may be made by treatment of the residue with aqua regia, which dissolves gold, platinum, and palladium, but for a more accurate determination reference is made to the methods to be described below.

PRELIMINARY CONCENTRATION OP THE PLATINUM METALS

The methods of concentration to be adopted will depend chiefly on the grade of material being assayed. It is desirable for an accurate separation and determination of the platinum metals to have a total quantity of 5 to 10 mg. of these metals with which to work, and this usually requires the use of a large sample. Even in the case of the ores of the Sudbury area, which now leads the world in the production of platinum metals, the proportion present is usually so small that a weight of ore in the order of 50 to 100 assay tons should preferably be used. This may be concentrated by fusing with excess litharge flux and combining the lead buttons, followed by repeated scorification to eliminate nickel and copper, or the whole quantity may be fused to a matte in a single operation, as in a large crucible in a gas-fired furnace.

In the determination of platinum metals in rock-house waste, referred to above, the writer used triplicate samples of 25 kg. each and reduced these to matte by smelting with suitable fluxes, thereby effecting a concentration of 10:1. A further concentration of 5:1 was made by treating the matte with hydrochloric acid (1 part water/1 part HCl) to remove iron and nickel sulfides, following which the residues were fused in several portions with lead oxide in crucibles and repeatedly scorified with test lead to remove copper. The charges were cooled in the crucibles and the latter were broken to release the lead buttons - a desirable precaution in order to avoid possible loss of iridium, rhodium, and osmium which do not alloy with lead. This treatment substantially eliminates nickel, which according to Seath and Beamish, prevents complete collection of platinum metals with lead. The method of concentration is shown in Figure II.

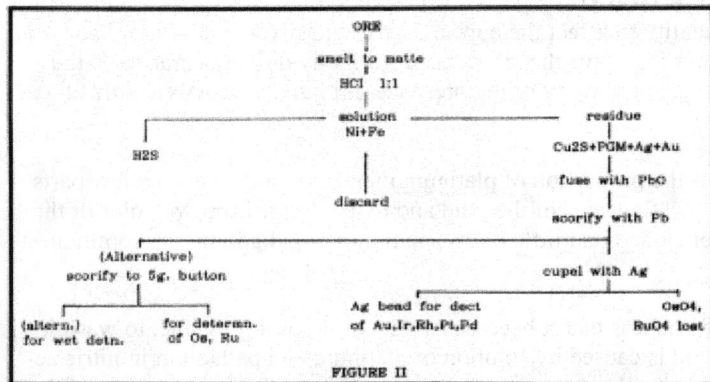

FIGURE II

If, on the other hand, a very rich material is being assayed, such as concentrated refinery slimes, a sample as small as 100 to 200 milligrams may be used. This is scorified with test lead, or if a determination of ruthenium and osmium is not required, it may be wrapped in lead foil and cupelled directly with 20 grams of test lead and the necessary for parting. A quantity of silver 20 times the total weight of gold and platinum metals present is recommended.

Determinations of ruthenium and osmium are not often required, but if the loss of these metals (as volatile oxides) must be prevented the final lead buttons resulting from any method of concentration should not be cupelled below a weight of 3 to 5 grams. The final buttons may be dissolved directly in nitric acid (1 part acid/10 parts distilled water). This procedure is inconvenient, and it is customary to eliminate the lead by cupellation, using a temperature somewhat higher than normal. Most of the osmium is likely to be lost in this operation, but if the material being assayed is a converter matte the osmium may indeed have been lost previously during the converter blow. Experimental evidence on this point is lacking.

The cupel should preferably be of a material completely acid soluble, and the silver bead should be removed from it with care in order to avoid the loss of iridium and rhodium, which may be loosely adherent. Possible loss by spitting should be avoided by covering the cupel with an empty hot cupel. The presence of platinum metals greatly increases the tendency to spit.

PROOFS OR STANDARDS

For accurate work the necessity of using proofs can scarcely be over-stressed. A separation that is entirely satisfactory for two metals present in certain proportions may become very incomplete if the proportions are substantially different or an additional metal of the platinum group is present. This fact no doubt accounts for many of the discrepancies found in the literature. The use of proofs containing in approximately the right proportions all the platinum metals present is therefore always essential if accurate determinations are required. On unknown material this will of course involve a preliminary assay. In important work the writer has sometimes used as many as three proofs, these containing silver, gold, and the platinum metals in slightly different proportions.

RUTHENIUM AND OSMIUM

The separation of ruthenium and osmium should begin with the lead button, which, as Russell Beamish and Seath have shown, cannot safely be cupelled below 3 gm in weight if as much as 5 milligrams of osmium be present. With greater quantities of osmium, the button should be left still larger. Ruthenium and osmium are separated from the other platinum metals by oxidation and distillation. In the presence of appreciable quantities of iridium and rhodium, which usually accompany ruthenium and osmium, the lead button maybe dissolved by gently warming with nitric acid (1:10), where upon these metals will be found in the residue and maybe filtered off and fused with sodium hydroxide and sodium peroxide in a silver or nickel crucible at a dull red heat. In the absence of iridium and rhodium, or in case of doubt, both the lean solution and the residue should be examined for ruthenium and osmium. Two methods of distillation are available.

According to the method recommended by J. Bishop & Co., the melt is dissolved in hot water, the osmium and ruthenium are distilled together in a current of chlorine and are collected in a train of Woulff bottles containing, successively; water, sodium bisulfate, and sodium hydroxide. The distillates are combined and acidified with hydrochloric acid and the metals are reduced with zinc and filtered off. The osmium is dissolved with hydrogen peroxide, leaving a residue of ruthenium, and the osmium is finally precipitated with hydrogen sulfide.

Gilchrist and Wichers separate osmium from ruthenium by distilling the former from a boiling nitric acid solution of moderate concentration (1 part nitric/1 part water), through which a stream of air is passed. The osmium is collected in hydrochloric acid solution (1:1) saturated with sulfur dioxide, and after the destruction of the sulfite compounds is precipitated as OsO_2, from hydrochloric acid solution by means of sodium bicarbonate. Elaborate precautions are necessary to prevent the loss of osmium. The solution from which the osmium was distilled is then evaporated with hydrochloric acid to destroy the nitric acid, sodium bromate is added, the ruthenium is distilled, and it is then collected and precipitated as was the osmium.

The writer has had insufficient experience with ruthenium and osmium to permit a definite recommendation as to the method to be adopted. The chlorine distillation is that usually followed. Gilchrist and Wichers state that the chlorine distillation may give a low result for ruthenium, since some may be precipitated with iridium hydroxide when the solution becomes nearly neutral. They, however, introduce the iridium as chloride, whereas in the procedure described above it remains in the insoluble residue.

NITRIC ACID SEPARATION
No known acid, under the conditions usually encountered, will completely dissolve any of the platinum metals from the silver bead and leave the others wholly in the residue. The writer has had the best success with nitric acid, but his experience is in accord with that of Seath and Beamish, that two or three treatments are ordinarily required to dissolve all the platinum. An outline of the separation is given in Fig III.

The silver bead is transferred without cleaning to a beaker and is warmed with 10ml of nitric acid (1:1) until action ceases; 6 ml of sulfuric acid (1:1) is added and the liquid is evaporated to fumes. Unnecessary heating is objectionable, as some rhodium may go into solution. Silver is precipitated with a slight excess of hydrochloric acid in the diluted solution still containing the residue of platinum metals. It is then filtered, dried, scorified with 25 grams of test lead, cupelled, and parted again in the same way. A second precipitation of silver and a third cupellation and parting are required for accurate work. On taking up the silver sulfate with water after the final parting, no hydrochloric is added, but the liquid is filtered and the residue treated as described below. The first two filtrates may be united, but the third is kept separate and silver is precipitated in it as chloride, filtered off, and discarded. The final filtrate may then be combined with the others, which contain the buck of the palladium and platinum. In the above parting, the expulsion of the nitric acid is desirable before precipitation of the silver in order to avoid the formation of aqua regia and the possible solution of some gold. If the precipitation of silver is made as described, before combining the filtrates, the separation of the platinum and palladium from silver is complete; otherwise, some palladium will be carried down with the silver chloride.

AQUA REGIA SEPARATION
The residue, which contains gold, iridium and rhodium (also some ruthenium and osmium, if present and not already separated), is washed with hot strong ammonium acetate solution, to dissolve any lead sulfate present, then with ammonia to remove any possible silver chloride, and finally
with hot water. The clean residue is dried, the paper is burned, and heating is continued for a short time at dull red heat to oxidize and expel osmium, and possibly ruthenium. The residue is warmed for one hour with aqua regia (1 part A.R./5 parts water) to dissolve the gold; the insoluble iridium and rhodium are then filtered off and washed.

GOLD
The gold solution is evaporated three times nearly to dryness with HCl to expel nitric acid. Gold is likely to separate as metal if it becomes quite dry. The solution is taken up with water and gold is precipitated in any convenient way, as by sulfur dioxide, oxalic acid, hydroquinone, or a ferrous salt. The writer prefers to precipitate the gold with hydrogen sulfide and to filter, scorify with lead and silver, cupel, and part with nitric acid. The gold is then coherent and easy to wash by decantation.

IRIDIUM AND RHODIUM
A quantity of two or three grams of alkali bisulfate is fused in a quartz crucible until spattering ceases, then the residue of iridium and rhodium is added and fusion is continued for 10 minutes. If, on cooling, the lead is distinctly yellow with dissolved rhodium, the fusion is repeated until the final melt is white. The melt is dissolved in hot dilute HCl (1:10) and the residue is filtered off and weighed as iridium. If much ruthenium is known to have been present originally, some may still remain. This can be removed by fusion with potassium nitrate in a silver dish, but it will not in any case represent the total amount of ruthenium originally present and may therefore

be discarded. The filtrate is evaporated to dryness and the sulfates are dissolved in 20 ml of hot water and 1 ml of concentrated HCl. If an appreciable quantity of rhodium be present, the solution will be red, and the writer has on occasion been able to determine the amount by comparing the color with that of the proofs. Zinc dust is added and warming is continued for at least one hour. Any excess of zinc is then dissolved with hydrochloric acid and the rhodium filtered off, ignited at a low temperature, treated with a drop of formic acid to reduce any oxide formed, dried and weighed as metal.

PLATINUM AND PALLADIUM

The filtrates from the three nitric acid partings are combined and evaporated to strong fumes of sulfuric acid. After dilution, the solution is warmed well to dissolve any palladium sulfate, which goes into solution rather slowly, and is allowed to stand for the separation of lead sulfate. This is filtered off and washed with dilute sulfuric acid (1:20). A drop of HCl is added and the solution is allowed to stand for some hours for the separation of any silver that may remain. If a precipitate forms it is filtered off.

The dimethylglyoxime precipitation of palladium is very satisfactory if precautions are taken to avoid the precipitation of platinum. The reagent is not always of high purity and should be re-crystallized by cooling a hot saturated alcoholic (ethyl alcohol) solution. The solution of palladium and platinum should be brought to the proper degree of acidity, as by neutralizing with ammonia and then adding 3 to 5% of its volume of concentrated hydrochloric acid. Precipitation should be carried out at room temperature by adding 10 ml or more of a 1% alcoholic solution of dimethylglyoxime, and the lemon-yellow precipitate is filtered off after standing only 10-15 minutes. A relatively open filter paper is satisfactory.

The platinum compound, when formed, is anistropic, being blue in one direction and bronze in the other. A greenish tint to the palladium precipitate, therefore, usually indicates the presence of platinum. In such a case the precipitate should be dissolved in hot hydrochloric acid (1:1), adjusted to the proper acidity and re-precipitated. But if the above precautions are taken, the platinum will remain in solution.

The palladium precipitate, if large, may be weighed as such, or if small, may be ignited gently to metal, but since the carbon is difficult to burn off completely and the ignition of this precipitate frequently leads to the loss of a little palladium, it is safest to dissolve it in hot HCl (1:1), oxidize it with a little nitric acid, expel these by evaporating with sulfuric acid and finally precipitate in hot slightly acid solution with hydrogen sulfide. It is well to continue passing the gas until the solution reaches room temperature and then allow it to stand overnight before filtering. When the precipitate has been ignited it should be moistened with a little formic acid to reduce surface oxide, and dried at about 60° C in order to avoid re-oxidation, which takes place very readily.

To the filtrate from the dimethylglyoxime precipitate there is added 10% of its volume of concentrated HCl, since platinum sulfide is more insoluble in such a solution than in weaker acid. The liquid is then heated to boiling and hydrogen sulfide is passed in until the solution is cold. After standing overnight the precipitate is filtered off and ignited at the lowest possible temperature. The metal is then treated with a little nitric acid (1:3) to dissolve any lead or silver that may possibly be present, and is filtered, ignited, and weighed as platinum. Some may prefer to scorify the sulfide with test lead and silver, cupel and part with sulfuric acid.

SILVER

Owing to the loss of silver in cupellation when appreciable quantities of platinum metals are present, this element cannot be accurately determined in the usual way. The lead button containing the silver may be scorified to 3 to 5 grams and the latter is then dissolved in nitric acid (1:4). The insoluble residue is filtered off and the silver is precipitated in the filtrate as chloride. This is freed from palladium by solution in ammonia and re-precipitation and is finally weighed as chloride or cupelled to metal. The latter may, as a check, be parted with nitric acid, but any residue should be negligible in quantity."

End of Article

A few comments on safety are necessary as this article was primarily written for the informed assayer. Hydrogen sulfide is by far more dangerous than hydrogen cyanide. Any sulfide ore when treated with acids releases H₂S - I'm sure many of you have treated such ores and smelled the odor of rotten eggs—this is H₂S. DANGER. It has the property of deadening the smell receptors in the nose—you can no longer smell the H₂S, which rapidly reaches toxic and deadly concentrations!! With these warnings in mind, I will describe an apparatus for the production of H₂S as used in the above procedures:

A flask of 250 ml or more is fitted with a 2 holed stopper, (see drawing) and a thistle tube is placed in one side & "L" shaped tube in the other. Then 10-15 grams of PeS, iron sulfide (made by fusing iron and sulfur) is placed in the flask. Enough dilute HCl is added to cover the FeS. The gas will start generating immediately. A rubber tube with glass end can then be used to bubble through the solution in the above procedure. (This information can be obtained in any high school chemistry book.) The solution may require slight heating.

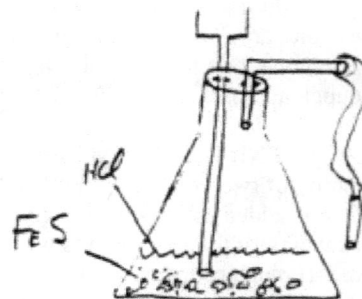

HCl

FeS

FIGURE III

```
                                    Ag bead
                                       |
                               HNO3 + H2SO4
                          Hcl (to ppt. Ag as AgCl)
        _____|_____
       |                                                                |
 Agcl, Au, Ir, Rh (pt,                                              Solution
      pd,ru, os)                                                     Pt + Pd
       |
 scorify, cupel part--_____        solution
      RuO4,OsO4                                        Pt + Pd
    and ppt. Agcl
       |                              RuOy,OsO4
 Agcl, Au, Ir, Rh (Pt,Ru,Os)         Solution
       |                             Ag + Pt
 scorify, cupel and part<_____          |_____
       |                           PPt, Agcl      solution, Pt
       |                               |
 Residue, Au, Ir, Rh (Ru,           discard                    Pd + Pt
       Os)                                                         |
       |                                                    Dimethylglyoxime
 Ignite--RuO4, OsO4                                    _____|_____
       |                                              |                         |
   Aqua regia                                      PPt. Pd                 Solution Pt
    ____|_____                     Hcl soln                    H2S
   |                          |                        |                        |
 Soln.Au                   Residue                    H2S                    Ignite
 H2S, scorify          Ir, Rh (Ru,Os)                  |                       HNO3
 cupel, part                  |                 ignite, reduce            _____|_____
    |                 Ignite--RuO4, OsO4               |                  |           |
  [AU]                  Fuse KHSO4                   [Pd]            Solution      Residue
                    _____|_____                                Ag, Pb        [Pt]
                   |                 |                                   |
             Solution, Rh     Residue Ir (Ru)                       Discard
                   |                 |
                Zn dust          Fuse KNO3
                   |          _____|_____
                 [Rh]        |               |
                          Residue      Solution, Ru
                            |                 |
                          [Ir]            Discard
```

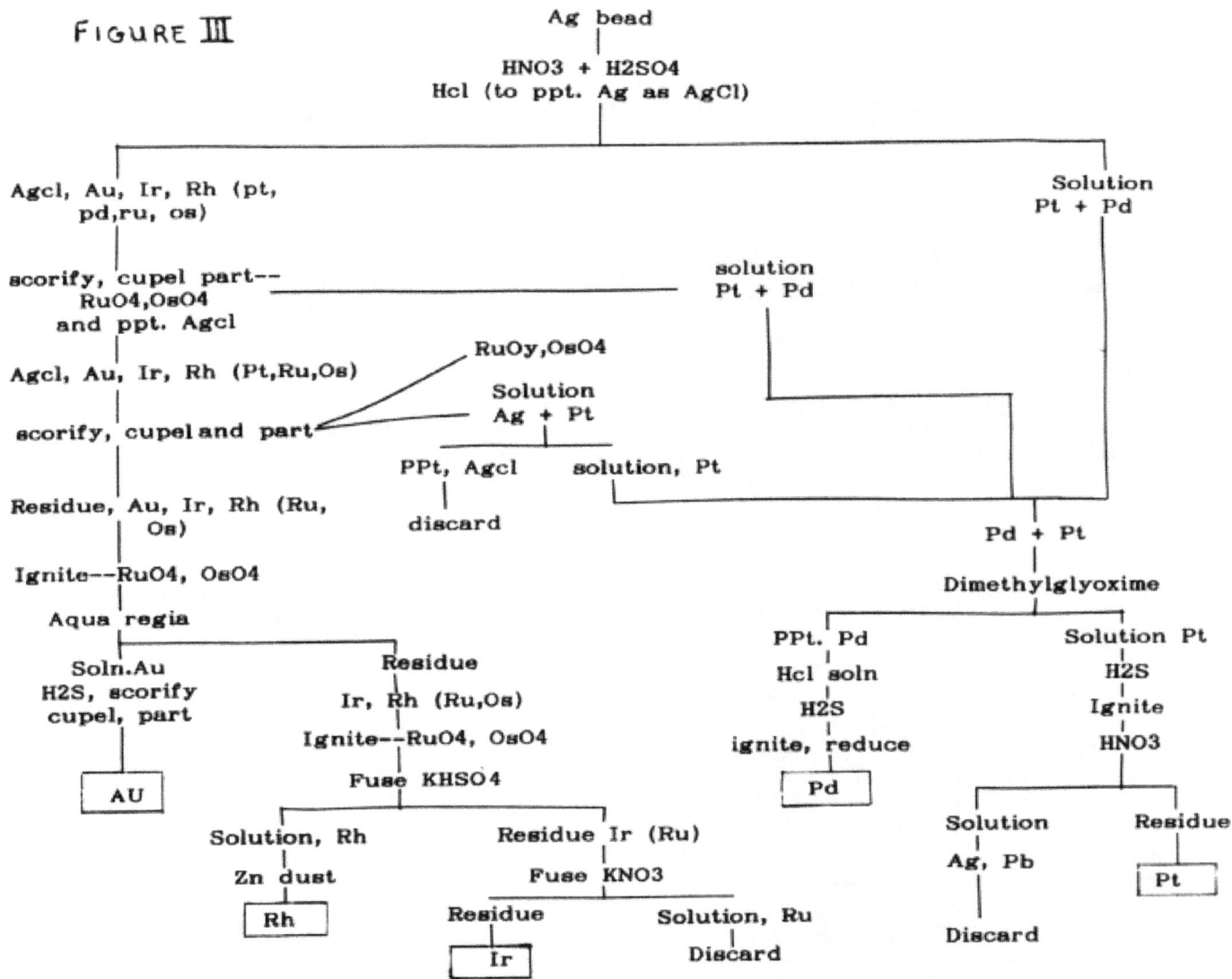

NOTE: This should only be done under a fume hood, or out-of-doors with plenty of wind!!! This stuff can kill you!! Please be very careful!!!!

I'm sure the above article will help many of you with some of the ones that have so far been mysterious. As the article states, this is not a simple procedure, but not above even the amateur. However, I'm sure it will create many questions.

'Til next article, have fun and BE SAFETY MINDED!!! Jim Floyd

WELDING ENGINEER

ENGINEERING DATA SHEET

Temperature Data*

This Useful Chart Includes Melting Points And Other Temperature Data

THE accompanying chart not only lists the melting points of various alloys and metals but also serves as a convenient means for conversion between Centigrade and Fahrenheit temperature scales. In addition, on the far right of the scale are shown the color designations that are commonly used in judging the temperatures of hot metal by color.

Melting Points

This chart should prove useful to all welding operators for it contains basic information on working with metals at elevated temperatures. Reference to the chart, for instance, shows why aluminum and aluminum alloys, because of their low melting points, give little or no indication by change in color when they approach welding heat. On the other hand, the high melting point of wrought iron explains why considerably more heat is required to weld this metal than is required for cast iron, for instance.

Temperature Color Scale

Another use for the chart is in estimating the temperature by color. For instance, instructions may require that the part be preheated to 1,100 deg. F. before welding. If you are without a thermocouple or other means for accurately measuring high temperatures, reference to the chart shows that the part, at 1,100 deg. F. would have a blood-red color. With a little experience, you can estimate this fairly closely by eye. In this connection, it should be mentioned that the color scale is for observations made in a fairly dark place and without welding goggles. As the light increases, the color groups on the scale will apply to higher temperatures.

Conversion Data

Finally, the chart is a ready means for converting Fahrenheit to Centigrade, and vice versa. Suppose you are familiar with the Fahrenheit scale, yet instructions call for the quenching of a welded part from 900 deg. C. Reference to the chart shows this to be approximately 1,650 deg. F.

_____

*Reprinted from *Oxy-Acetylene Tips*, February, 1942

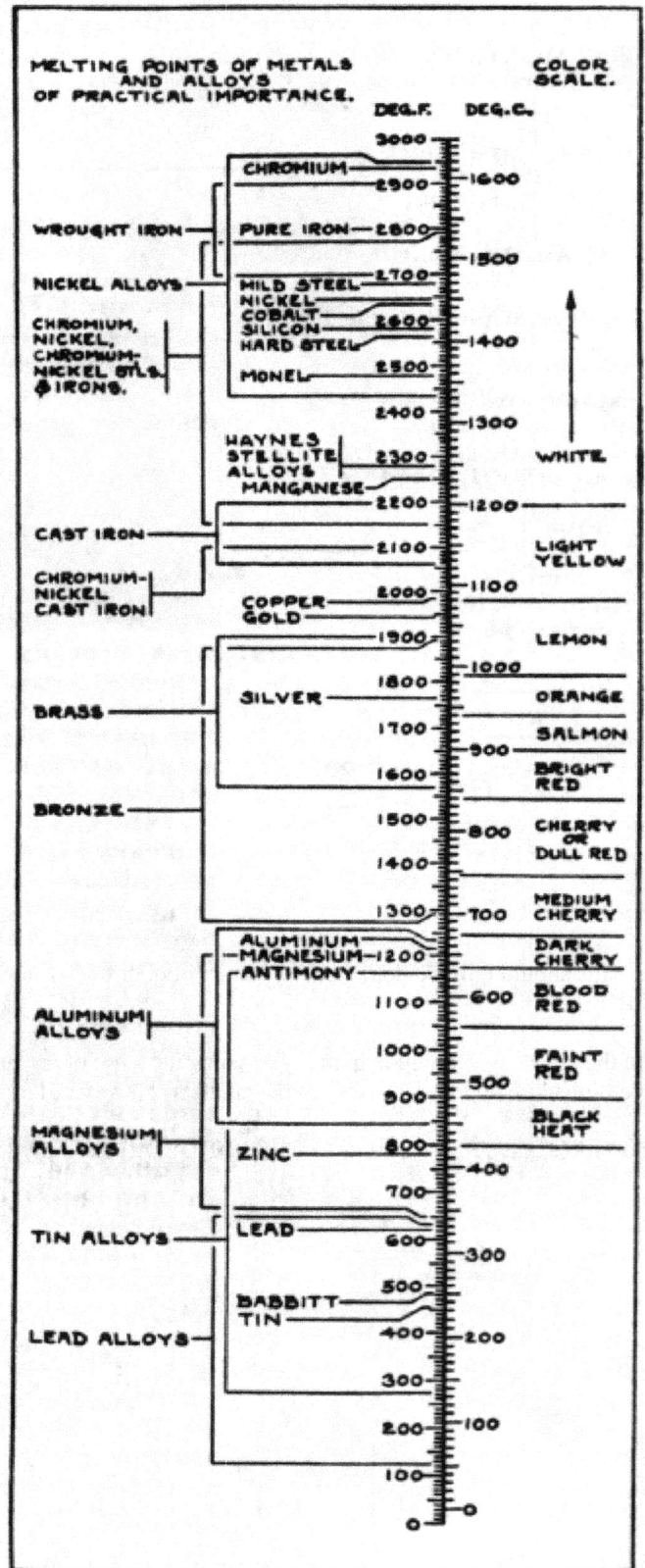

MELTING POINTS OF METALS AND ALLOYS OF PRACTICAL IMPORTANCE.

COLOR SCALE.

Index

www.ingramcontent.com/pod-product-compliance
Lightning Source LLC
Chambersburg PA
CBHW080545220326
41599CB00032B/6365